W FARWELL 367

Understanding
Media

Marshall McLuhan, 1944 (Percy Wyndham Lewis)

Understanding Media

THE EXTENSIONS OF MAN

Critical Edition

Marshall McLuhan

edited by W. Terrence Gordon

GINGKO PRESS

GINGKO PRESS Inc., September 2003
5768 Paradise Drive, Suite J,
Corte Madera, CA 94925
Phone (415) 924-9615, Fax (415) 924-9608
email: books@gingkopress.com
www.gingkopress.com

ISBN: 1-58423-073-8

© 1964, 1994 Corinne McLuhan
Editor's Introduction, Chapter introductions © 2003 W. Terrence Gordon

Marshall McLuhan Project, General Editors
W. Terrence Gordon, Eric McLuhan, Philip B. Meggs

With very special thanks to Corinne McLuhan and Matie Molinaro

Marshall McLuhan, 1944 © Wyndham Lewis and the Estate of
Mrs. G.A. Wyndham Lewis, by kind permission of the Wyndham Lewis Memorial
Trust (a registered charity)

Book design: Julie von der Ropp
Printed in Germany by Clausen & Bosse

All rights reserved. No part of this book may be reproduced or utilized in any form
or by any means, mechanical or electronic, including photocopying and recording,
or by any information storage and retrieval system, without permission in writing
from the publisher.

LIBRARY OF CONGRESS CATALOGING IN PUBLICATION DATA:

McLuhan, Marshall, 1911-1980
Understanding media : the extensions of man / by Marshall McLuhan ;
edited by W. Terrence Gordon — Critical ed.
p. cm.

Includes bibliographical references and index.
ISBN: 1-58423-073-8
1. Mass media. I. Gordon, W. Terrence, 1942- II. Title.
P90.M26 2003
302.23—dc21
2003012174

The author wishes to thank the publishers of the *Times Literary Supplement* for
granting him permission to reprint the editorial of July 19, 1963, which appears in
the chapter on The Printed Word in this book. Special acknowledgements are due to
the National Association of Educational Broadcasters and the U.S. Office of Education,
who in 1959-1960 provided liberal aid to enable the author to pursue research in the
media of communication.

Understanding
Media

Contents

Appendix

Editor's Introduction

As millions watched the first U.S. television appearance of the Beatles, a few months after the assassination of John F. Kennedy, one viewer thought to connect the two events. If the new rhythms, the lyrics, and the haircuts of the Liverpool four brought the first genuine distraction for some from the senselessness of Dallas, for Marshall McLuhan they brought confirmation that the medium is the message. McLuhan was about to publish *Understanding Media* and stake a place for himself amid the upheavals of the 1960s, explaining them as effects of electronic technology on the human senses and sensibilities.

From McLuhan's writings, *Bartlett's Familiar Quotations* records "The new electronic interdependence recreates the world in the image of a global village" and "The medium is the message." Like so many of the aphorisms that McLuhan favored, "the medium is the message," is a paradox. It invites reflection and compels us to plumb its depths, to interpret it, to put it on, to understand it by becoming its content—the very principle the saying articulates. The influence of Canadian political economist Harold Innis, though not made explicit in the main text of *Understanding Media* (cf. Appendix: Report on Project in Understanding New Media, pages 513, 527), is apparent in the notion that the medium is the message. Innis taught that a medium of communication exerts a powerful influence on the spread of knowledge over space and time, and he emphasized the necessity of studying the features of the medium to assess that influence on the social and cultural milieu where it operates. McLuhan took the germ of the idea applied by Innis to media of communication, redefined media as any and all technological extensions of body and mind, and fashioned the framework of this book.

The appearance of *Understanding Media* obliged McLuhan to set the record straight on what he meant by the phrase he was already well known for. Publicly, he pontificated on the interplay of sensorial modalities; privately, he tended to the poetic, making the analogy between the creation of new environments under technology and the appearance of the last ring in the growth of a tree trunk. When McLuhan spoke of language as mankind's first technology, he used the same tree analogy to prolong a Joycean echo: utterings-outerings-outer rings.

Innis and Joyce. They seem unlikely bedfellows as sources for a book on media, unless we remember that the metaphor for McLuhan's life and work is the sailor / fisherman in Edgar Allan Poe's *A Descent into the Maelstrom*. Beginning in 1946 (see Publications of Marshall McLuhan — "Footprints in the Sands of Crime"), McLuhan referred repeatedly to the story of the man who survived the maelstrom by learning from everything around him. In the preface to his first book, *The Mechanical Bride*, he links this strategy to his own method: "Poe's sailor saved himself by studying the action of the whirlpool and by cooperating with it. The present book likewise makes few attempts to attack the very considerable currents and pressures set up around us today by the mechanical agencies of the press, radio, movies, and advertising. It does attempt to set the reader at the center of the revolving picture created by these affairs where he may observe the action that is in progress and in which everybody is involved."

McLuhan the maelstrom observer, McLuhan the media investigator, was still an undergraduate at the University of Manitoba when he declared in his diary that he would never be an academic. He felt that he was learning in spite of his professors, but he would become a professor of English in spite of himself. After Manitoba came graduate work at Cambridge University, and there the seed was planted for McLuhan's move toward media analysis. Looking back on his years in Cambridge, he reflected that a principal aim of many of its faculty in the twentieth century could be summarized as the training of perception. There could be no more apt phrase to summarize his own aim throughout his career — whether in teaching English literature to undergraduates or explaining to corporate executives how little they understood about their corporation.

If McLuhan needed a catalyst to move him decisively toward media analysis, it came from the shock he experienced in his first teaching post. The students he faced at the University of Wisconsin, in 1936, were his juniors by no more than five to eight years, yet he felt that he was teaching them across a gap. He sensed that this had something to do with ways of learning, ways of understanding, and he felt a compulsion to investigate it . The investigation led him back to the subtlest of lessons on the training of perception from his Cambridge professors, such as I. A. Richards, and forward to discoveries from James Joyce, Harold Innis; back to antiquity and the myth of Narcissus, forward to the mythic structure of Western culture under electric technology. Along the way there would be detours through the impoverished world of Dagwood and the rich linguistic world of the Trobriand Islanders (see Appendix).

Understanding Media occupies a central place in McLuhan's work. It first appeared just over twenty years after he wrote his doctoral thesis at Cambridge on sixteenth-century dramatist, satirist, and pamphleteer Thomas Nashe, and just over twenty years before McLuhan's ideas found their final form of expression in the posthumous *Laws of Media*, written with his son, Eric McLuhan. In retrospect, there is a line of intellectual inquiry running back to the Nashe work and through to *Laws of Media*. It is illuminated by *Understanding Media*.

The book defies summary. McLuhan wanted it that way. When we are faced with information overload, he taught, the

mind must resort to pattern recognition to achieve understanding. *Understanding Media* illustrates the point by its
style. The reader must reach for the ideas it expresses each
time they whirl past. There is the notion of electric light
escaping attention as a medium of communication in chapter
one. That inattention is explained in chapter five. There is a
discussion of the notion of closure as equilibrium, as displacement of perception, as completion of image (see Glossary).
There is Narcissus, a regular in *Understanding Media* and
McLuhan's metaphor for the failure to understand media as
extensions of the human body and the failure to perceive the
message (new environments) created by media (technology).
There is McLuhan explaining that response to increased
power and speed of bodily extensions begets still more extensions. Chaplin, Joyce, Chopin, Pavlova, Eliot, and Charles
Boyer — all in the space of a single paragraph. When the
McLuhan maelstrom subsides, a raft may be fashioned
by the reader prepared to lash together the ideas that have
surfaced:

1

We think of media principally as media of communication:
press, radio, and television. McLuhan thought of a medium as
an extension of the human body or the mind: clothing extends
skin, housing extends the body's heat-regulating mechanism.
The stirrup, the bicycle, and the car are all extensions of the
human foot. A medium, or a technology, can be any extension
of the human being.

2

Media come in pairs, one "containing" the other. So, telegraph
contains the printed word, which contains writing, which contains speech. The contained medium is the message of the

containing one, but the effects of the latter are obscured for the user, who focuses on the former. Because those effects are so powerful, any message, in the ordinary sense of "content" or "information," has far less impact than the medium itself. Thus, "the medium is the message."

3

Not all media work in pairs. McLuhan finds two exceptions. In the example above, speech is the medium contained by writing, but there the chain of media ends. Thought is non-verbal and pure process. The second exception is that of electric light, which permits activities that would be impossible in the dark. While these activities might be considered as the "content" of the light, this simply reinforces McLuhan's principle that any medium exerts its most potent effect by changing the form, scale, and speed of human relations and activities.

4

Media are agents of change in our experience of the world, in our interaction with each other, in our use of our physical senses — the senses that media extend. They must be studied for their effects, because the constant and inevitable interplay among media obscures those effects and hampers our ability to use media effectively.

5

McLuhan teaches that a new medium typically does not displace or replace another as much as it complicates its operation. It is precisely such interaction that obscures media effects. Early mankind's technology — the one from which writing, print, and telegraph all derived — was speech. Transformed into writing, speech acquired a powerful visual bias, producing effects in social and cultural organization that endure to the present. But the gain in power came with a loss, for writing separated speech from the other physical senses. And when the development of radio permitted the extension of speech, a similar loss occurred, for the radio reduced speech to one sense—the auditory-aural. Radio is not speech (because we only listen), but it creates the illusion, like writing, of containing speech.

6

Many reviewers found McLuhan's observations penetrating, when he published *The Mechanical Bride*; just as many dismissed them as wrong. When *Understanding Media* first appeared, some disagreed even with its basic classification of media as hot vs. cool. The contrast hinges on special senses of the words "definition" and "information" and on physical senses more than word senses. McLuhan uses the phrase "high definition" as found in the technical language of television, meaning well-defined, sharp, solid, detailed, etc., in reference to any visual form. Letters of the alphabet, numbers, photographs, and maps, for example, are high definition. If forms and shapes and images are not so distinct, they are of low definition. In this case, our eyes scan what is visible and fill in what is missing, as in the case of sketches, cartoons, etc.

When McLuhan speaks of a medium transmitting information, he is not referring to facts or knowledge but to the response of our physical senses to the medium. The examples above are limited to the visual sense, but the principle applies to sounds as well. A high definition medium gives a lot of information and requires little of the user, whereas a low-definition medium provides little information, making the user work to fill in what is missing. This is the basis of the contrast between hot and cool media: high definition is hot; low definition is cool. Here are some of McLuhan's examples:

COOL	HOT
telephone	radio
speech	print
cartoons	photographs
television	movies
seminar	lecture

The lecture / seminar contrast indicates that hot media are low in participation, whereas cool media are high in participation. The other examples remind us that participation does not refer primarily to intellectual involvement but, like "definition" and "information," to how a medium engages our physical senses.

7

Discussing the myth of Narcissus, McLuhan points out the common error in which Narcissus is said to have fallen in love with *himself*. In fact, it was his failure to recognize his image that proved to be his undoing. He succumbed to the typical numbing effect of any technology, as it lulls the user into a trance. Technologies create new environments, the new environments create pain, and the body's nervous system shuts down to block the pain. The name Narcissus derives from the Greek word *narcosis*, meaning numbness.

The story of Narcissus encapsules mankind's obsessive fascination with new extensions of the body, but it also shows how these extensions are inseparable from what McLuhan calls "amputations." The wheel, for example, extended the human foot and relieved it of the pressure of carrying loads. At the same time, it created new pressure by separating or isolating the extended foot from the rest of the body. Pedaling a bicycle or throttling a car down a freeway involves the foot in such a specialized task as to deprive it of the ability to perform its basic function of walking. The medium gives power through extension but immobilizes and paralyzes what it extends. In this sense, technologies both extend and amputate. Amplification turns to amputation. The central nervous system reacts to the pressure and disorientation of the amputation by blocking perception. Narcissus, *narcosis*.

McLuhan finds a further lesson on the power of media in the Greek myth of king Cadmus, who sowed dragon teeth and harvested an army. It was also Cadmus who introduced the phonetic alphabet (from Phoenicia) to Greece. The dragon's teeth may, therefore, represent an older form of hieroglyphic writing from which the much more powerful alphabet grew (see Glossary).

8

Beyond Narcissus, McLuhan finds another dimension of opposing effects resulting from the transition from mechanical to electronic technologies. This transition, accompanied by a relentless speeding-up of all human activity, brought the expansionist pattern associated with the old order into conflict with the contracting forces of the new one. Explosion (whether of population or knowledge) has reversed into implosion, because electronic technology has created a global village where knowledge must be synthesized instead of being splintered into isolated specialties.

McLuhan offers examples of overheated technologies and overextended cultures, and the reversals that result. Overextended road turns cities into highways and highways into cities. In the industrial society of the nineteenth century, with its emphasis on fragmented procedures in the workplace, both the commercial and social world began to put new emphasis on unified and unifying forms of organization (corporations, monopolies, clubs, societies). As for the twentieth century, McLuhan characterizes the theme of Samuel Beckett's *Waiting for Godot* as the destructive aspects of the vast creative potential opened by the electronic age. (This type of reversal is crucial in the integrated laws of media formulated late in McLuhan's career.)

9

McLuhan's analysis focuses on media effects in all areas of society and culture, but his starting point is always the individual, because media are defined as technological extensions of the body. So, McLuhan often puts both his inquiry and his conclusions in terms of the ratio between our physical senses (the extent to which we depend on them relative to each other) and the result of modifications to that ratio. This inevitably involves a psychological dimension. For example, when the alphabet was invented and brought about the intensification of the visual sense in the communication process, sight took such priority over hearing that the effect carried over from language and communication to reshape literate society's conception of space.

McLuhan stresses sense ratios and the effects of altering them: in Africa, the introduction of radio, a hot medium distorting the sensory balance of oral culture, produced the inevitable disorienting effect and rekindled tribal warfare; in dentistry, a device called an audiac consists of headphones bombarding the patient with enough noise to block pain from the dentist's drill; in Hollywood, the addition of sound to silent pictures impoverished and gradually eliminated the role of mime, with its tactility and kinesthesis.

These examples involve the relationships among the five physical senses, which may be ranked in order of the degree of fragmentation of perceptions received through them. Sight comes first, because the eye is such a specialized organ. Then come hearing, touch, smell, and taste, progressively less specialized senses. By contrast with the enormous power of the eye and the distances from which it can receive a stimulus, the tongue is thought capable of distinguishing only sweet, sour, bitter, and salt, and only in direct contact with the substance providing the stimulus.

10

Western culture, with its phonetic literacy, when transplanted to oral, nonliterate cultures, fragments their tribal organization and produces the prime example of media hybridization and its potent transforming effects (see Chapter 5). At the same time, electricity has transformed Western culture, dislocating its visual, specialist, fragmented orientation in favor of oral and tribal patterns. McLuhan retains the metaphors of violent energy in speculating on the final outcome of these changes — the fission of the atomic bomb and the fusion of the hydrogen bomb.

The hybridization of cultures occupies McLuhan most fully, but he offers other examples, such as electric light restructuring existing patterns of social and cultural organization by liberating the activities of that organization from dependence on daylight.

McLuhan emphasizes that media as extensions of the body not only alter the ratios among our physical senses, but that when the media combine they establish new ratios among themselves. This happened when radio came along to change the way news stories were presented and the way film images were presented in the talkies. Then television came along and brought big changes to radio.

When media combine, both their form and use change. So do the scale, speed, and intensity of the human endeavors affected. And so do the environments surrounding the media and their users. The hovercraft is a hybrid of the boat and the airplane. As such it eliminates not only the need for the stabilizing devices of wings and keels but the interfacing environments of landing strips and docks.

McLuhan wrote with no knowledge of galvanic skin response technology, terminal node controllers, or the Apple

Newton. He might not have been able even to imagine what a biomouse is. But he pointed the way to understanding all of these, not in themselves, but in their relation to each other, to older technologies, and above all in relation to ourselves—our bodies, our physical senses, our psychic balance. When he published *Understanding Media* in 1964, he was disturbed about mankind shuffling toward the twenty-first century in the shackles of nineteenth century perceptions. He might be no less disturbed today. And he would continue to issue the challenge that confronts the reader at every page of this book to cast off those shackles.

— W. Terrence Gordon
HALIFAX, 2003

Part I

Introduction
to the First Edition

*In these few opening pages, McLuhan barely hints at the mate-
rial that he will develop chapter by chapter, mentioning only
extensions of skin, hand, or foot and the world of advertising.
But he gives us a clear statement of one unifying theme ("the
Western world is imploding") and the consequence of that state
of affairs ("this is the Age of Anxiety for the reason of the electric implosion"), linking them to the therapeutic purpose of his
book ("it explores the contours of our own extended beings in
our technologies, seeking the principle of intelligibility in each
of them"). We also find the first of what will be constant references to literary works ("the Theater of the Absurd dramatizes
this recent dilemma of Western man, the man of action who
appears not to be involved in the action"). Such references consolidate McLuhan's theme that the artist is always ahead of his
time in recognizing the effects of new media. Though McLuhan
is often regarded as a technological optimist or even as a harbin-
ger of a "techno-utopia," his caution is explicit ("whether the
extension of consciousness ... will be a 'good thing' is a question
that admits of a wide solution"), and his faith is clearly not in
electric technology but in realizing "the aspiration of our time
for wholeness, empathy, and depth of awareness," spawned by
that technology. The McLuhan wit is at work here, transforming
an obscure term from Latin grammar (ablative absolute) into an
illuminating pun. The same term will reappear much later in
an arresting linguistic metaphor for media effects (see Wheel,
Bicycle, and Airplane, p. 250). McLuhan is in full flight already
in this introduction, challenging us to plunge with him into
what he calls "the creative process of knowing."*

— *(Editor)*

4

James Reston wrote in *The New York Times* (July 7, 1957):

> A health director ... reported this week that a small
> mouse, which presumably had been watching tele-
> vision, attacked a little girl and her full-grown cat....
> Both mouse and cat survived, and the incident is
> recorded here as a reminder that things seem to be
> changing.

After three thousand years of explosion, by means of
fragmentary and mechanical technologies, the Western world
is imploding. During the mechanical ages we had extended
our bodies in space. Today, after more than a century of
electric technology, we have extended our central nervous
system itself in a global embrace, abolishing both space and
time as far as our planet is concerned. Rapidly, we approach
the final phase of the extensions of man—the technological
simulation of consciousness, when the creative process of
knowing will be collectively and corporately extended to the
whole of human society, much as we have already extended
our senses and our nerves by the various media. Whether
the extension of consciousness, so long sought by advertisers
for specific products, will be "a good thing" is a question
that admits of a wide solution. There is little possibility
of answering such questions about the extensions of man
without considering all of them together. Any extension,
whether of skin, hand, or foot, affects the whole psychic and
social complex.

Some of the principal extensions, together with some of
their psychic and social consequences, are studied in this
book. Just how little consideration has been given to such
matters in the past can be gathered from the consternation of
one of the editors of this book. He noted in dismay that
"seventy-five per cent of your material is new. A successful
book cannot venture to be more than ten per cent new." Such

a risk seems quite worth taking at the present time when the stakes are very high, and the need to understand the effects of the extensions of man becomes more urgent by the hour.

In the mechanical age now receding, many actions could be taken without too much concern. Slow movement insured that the reactions were delayed for considerable periods of time. Today the action and the reaction occur almost at the same time. We actually live mythically and integrally, as it were, but we continue to think in the old, fragmented space and time patterns of the pre-electric age.

Western man acquired from the technology of literacy the power to act without reacting. The advantages of fragmenting himself in this way are seen in the case of the surgeon who would be quite helpless if he were to become humanly involved in his operation. We acquired the art of carrying out the most dangerous social operations with complete detachment. But our detachment was a posture of noninvolvement. In the electric age, when our central nervous system is technologically extended to involve us in the whole of mankind and to incorporate the whole of mankind in us, we necessarily participate, in depth, in the consequences of our every action. It is no longer possible to adopt the aloof and dissociated role of the literate Westerner.

The Theater of the Absurd dramatizes this recent dilemma of Western man, the man of action who appears not to be involved in the action. Such is the origin and appeal of Samuel Beckett's clowns. After three thousand years of specialist explosion and of increasing specialism and alienation in the technological extensions of our bodies, our world has become compressional by dramatic reversal. As electrically contracted, the globe is no more than a village. Electric speed in bringing all social and political functions together in a sudden implosion has heightened human awareness of responsibility to an intense degree. It is this implosive factor that alters the position of the Negro, the teenager, and some other groups. They can no longer be *contained*, in the political sense of limited association. They are now *involved* in our lives, as we in theirs, thanks to the electric media.

This is the Age of Anxiety for the reason of the electric implosion that compels commitment and participation, quite regardless of any "point of view." The partial and specialized character of the viewpoint, however noble, will not serve at all in the electric age. At the information level the same upset has occurred with the substitution of the inclusive image for the mere viewpoint. If the nineteenth century was the age of the editorial chair, ours is the century of the psychiatrist's couch. As extension of man the chair is a specialist ablation of the posterior, a sort of ablative absolute of backside, whereas the couch extends the integral being. The psychiatrist employs the couch, since it removes the temptation to express private points of view and obviates the need to rationalize events.

The aspiration of our time for wholeness, empathy and depth of awareness is a natural adjunct of electric technology. The age of mechanical industry that preceded us found vehement assertion of private outlook the natural mode of expression. Every culture and every age has its favorite model of perception and knowledge that it is inclined to prescribe for everybody and everything. The mark of our time is its revulsion against imposed patterns. We are suddenly eager to have things and people declare their beings totally. There is a deep faith to be found in this new attitude — a faith that concerns the ultimate harmony of all being. Such is the faith in which this book has been written. It explores the contours of our own extended beings in our technologies, seeking the principle of intelligibility in each of them. In the full confidence that it is possible to win an understanding of these forms that will bring them into orderly service, I have looked at them anew, accepting very little of the conventional wisdom concerning them. One can say of media as Robert

7

Theobald[1] has said of economic depressions: "There is one additional factor that has helped control depressions, and that is a better understanding of their development." Examination of the origin and development of the individual extensions of man should be preceded by a look at some general aspects of the media, or extensions of man, beginning with the never-explained numbness that each extension brings about in the individual and society.

---

1. See Chapter 2, page 41 and Name Index.

Introduction
to the Second Edition

As several passages below indicate, this text addresses critical reaction to the first edition of the book and outright misunderstanding on various points. There is likely no other McLuhan text in which he so readily concedes to make the sort of explicit statements of his ideas that readers crave: " 'the medium is the message' means …" He no doubt made such a concession with great reluctance. The reader who will come to understand by the end of the book the multiple meanings of the medium is the message *will come to understand at the same time that McLuhan's preference for a prose style that explores instead of explaining is inseparable from the theme summarized in* the medium is the message.

— (Editor)

EXPLORE

EXPLAIN

Jack Paar mentioned that he once had said to a young friend, "Why do you kids use 'cool' to mean 'hot'?" The friend replied, "Because you folks used up the word 'hot' before we came along." It is true that "cool" is often used nowadays to mean what used to be conveyed by "hot." Formerly a "hot argument" meant one in which people were deeply involved. On the other hand, a "cool attitude" used to mean one of detached objectivity and disinterestedness. In those days the word "disinterested" meant a noble quality of fairmindedness. Suddenly it got to mean "couldn't care less." The word "hot" has fallen into similar disuse as these deep changes of outlook have developed. But the slang term "cool" conveys a good deal besides the old idea of "hot." It indicates a kind of commitment and participation in situations that involves all of one's faculties. In that sense, one can say that automation is cool, whereas the older mechanical kinds of specialist or fragmented "jobs" are "square." The "square" person and situation are not "cool" because they manifest little of the habit of depth involvement of our faculties. The young now say, "Humor is not cool." Their favorite jokes bear this out. They ask, "What is purple and hums?" Answer, "An electric grape." "Why does it hum?" Answer, "Because it doesn't know the words." Humor is presumably not "cool" because it inclines us to laugh *at* something, instead of getting us emphatically involved in something. The story line is dropped from "cool" jokes and "cool" movies alike. The Bergman and Fellini movies demand far more involvement than do narrative shows. A story line encompasses a set of events much like a melodic line in music. Melody, the *melos modos,* "the road round," is a continuous, connected, and repetitive structure that is not used in the "cool" art of the Orient. The art and poetry of Zen create involvement by means of the *interval*, not by the *connection* used in the visually organized

Western world. Spectator becomes artist in oriental art because he must supply all the connections.

The section on "media hot and cool" confused many reviewers of *Understanding Media* who were unable to recognize the very large structural changes in human outlook that are occurring today. Slang offers an immediate index to changing perception. Slang is based not on theories but on immediate experience. The student of media will not only value slang as a guide to changing perception, but he will also study media as bringing about new perceptual habits.

The section on "the medium is the message" can, perhaps, be clarified by pointing out that any technology gradually creates a totally new human environment. Environments are not passive wrappings but active processes. In his splendid work *Preface to Plato* (Harvard University Press, 1963), Eric Havelock contrasts the oral and written cultures of the Greeks. By Plato's time the written word had created a new environment that had begun to detribalize man. Previously the Greeks had grown up by benefit of the process of the *tribal encyclopedia*. They had memorized the poets. The poets provided specific operational wisdom for all the contingencies of life— Ann Landers in verse. With the advent of individual detribalized man, a new education was needed. Plato devised such a new program for literate men. It was based on the Ideas. With the phonetic alphabet, classified wisdom took over from the operational wisdom of Homer and Hesiod and the tribal encyclopedia. Education by classified data has been the Western program ever since.

Now, however, in the electronic age, data classification yields to pattern recognition, the key phrase at IBM. When data move instantly, classification is too fragmentary. In order to cope with data at electric speed in typical situations of "information overload," men resort to the study of configurations, like the sailor in Edgar Allan Poe's *Maelstrom*. The drop-out situation in our schools at present has only begun to develop. The young student today grows up in an electrically configured world. It is a world not of wheels but of circuits, not of fragments but of integral patterns. The student today *lives* mythically and in depth. At school, however, he encounters a situation organized by means of classified information. The subjects are unrelated. They are visually conceived in terms of a blueprint. The student can find no possible means of involvement for himself, nor can he discover how the educational scene relates to the "mythic" world of electronically processed data and experience that he takes for granted. As one IBM executive puts it, "My children had lived several lifetimes compared to their grandparents when they began grade one."

"The medium is the message" means, in terms of the electronic age, that a totally new environment has been created. The "content" of this new environment is the old mechanized environment of the industrial age. The new environment reprocesses the old one as radically as TV is reprocessing the film. For the "content" of TV is the movie. TV is environmental and imperceptible, like all environments. We are aware only of the "content" or the old environment. When machine production was new, it gradually created an environment whose content was the old environment of agrarian life and the arts and crafts. This older environment was elevated to an art form by the new mechanical environment. The machine turned Nature into an art form. For the first time men began to regard Nature as a source of aesthetic and spiritual values. They began to marvel that earlier ages had been so unaware of the world of Nature as Art. Each new technology creates an environment that is itself regarded as corrupt and degrading. Yet the new one turns its predecessor into an art form. When writing was new, Plato transformed the old oral dialogue into an art form. When printing was new the Middle Ages became an art form. "The Elizabethan world view" was a view of the Middle Ages.

13

And the industrial age turned the Renaissance into an art form as seen in the work of Jacob Burckhardt. Siegfried Giedion, in turn, has in the electric age taught us how to see the entire process of mechanization as an art process. (*Mechanization Takes Command*)

As our proliferating technologies have created a whole series of new environments, men have become aware of the arts as "anti-environments" or "counter-environments" that provide us with the means of perceiving the environment itself. For, as Edward T. Hall has explained in *The Silent Language*, men are never aware of the ground rules of their environmental systems or cultures. Today technologies and their consequent environments succeed each other so rapidly that one environment makes us aware of the next. Technologies begin to perform the function of art in making us aware of the psychic and social consequences of technology.

Art as anti-environment becomes more than ever a means of training perception and judgment. Art offered as a consumer commodity rather than as a means of training perception is as ludicrous and snobbish as always. Media study at once opens the doors of perception. And here it is that the young can do top-level research work. The teacher has only to invite the student to do as complete an inventory as possible. Any child can list the effects of the telephone or the radio or the motor car in shaping the life and work of his friends and his society. An inclusive list of media effects opens many unexpected avenues of awareness and investigation.

Edmund Bacon, of the Philadelphia town-planning commission, discovered that school children could be invaluable researchers and colleagues in the task of remaking the image of the city. We are entering the new age of education that is programmed for discovery rather than instruction. As the means of input increase, so does the need for insight or pattern recognition. The famous Hawthorne experiment, at the General Electric plant near Chicago, revealed a mysterious

14

effect years ago. No matter how the conditions of the workers were altered, the workers did more and better work. Whether the heat and light and leisure were arranged adversely or pleasantly, the quantity and quality of output improved. The testers gloomily concluded that testing distorted the evidence. They missed the all-important fact that when the workers are permitted to join their energies to a process of learning and discovery, the increased efficiency is phenomenal.

Earlier it was mentioned how the school drop-out situation will get very much worse because of the frustration of the student need for participation in the learning process. This situation concerns also the problem of "the culturally disadvantaged child." This child exists not only in the slums but increasingly in the suburbs of the upper-income homes. The culturally disadvantaged child is the TV child. For TV has provided a new environment of low visual orientation and high involvement that makes accommodation to our older educational establishment quite difficult. One strategy of cultural response would be to raise the visual level of the TV image to enable the young student to gain access to the old visual world of the classroom and the curriculum. This would be worth trying as a temporary expedient. But TV is only one component of the electric environment of instant circuitry that has succeeded the old world of the wheel and nuts and bolts. We would be foolish not to ease our transition from the fragmented visual world of the existing educational establishment by every possible means.

The existential philosophy, as well as the Theater of the Absurd, represents anti-environments that point to the critical pressures of the new electric environment. Jean-Paul Sartre, as much as Samuel Beckett and Arthur Miller, has declared the futility of blueprints and classified data and "jobs" as a way out. Even the words "escape" and "vicarious living" have dwindled from the new scene of electronic involvement. TV engineers have begun to explore the braille-like character of the TV image as a means of enabling the blind to see by having this image projected directly onto their skins. We need to use all media in this wise, to enable us to see our situation.

In Chapter One there are some lines from *Romeo and Juliet* whimsically modified to make an allusion to TV. Some reviewers have imagined that this was an involuntary misquotation.

The power of the arts to anticipate future social and technological developments, by a generation and more, has long been recognized. In this century Ezra Pound called the artist "the antennae of the race." Art as radar acts as "an early alarm system," as it were, enabling us to discover social and psychic targets in lots of time to prepare to cope with them. This concept of the arts as prophetic, contrasts with the popular idea of them as mere self-expression. If art is an "early warning system," to use the phrase from World War II, when radar was new, art has the utmost relevance not only to media study but to the development of media controls.

When radar was new it was found necessary to eliminate the balloon system for city protection that had preceded radar. The balloons got in the way of the electric feedback of the new radar information. Such may well prove to be the case with much of our existing school curriculum, to say nothing of the generality of the arts. We can afford to use only those portions of them that enhance the perception of our technologies, and their psychic and social consequences. Art as a radar environment takes on the function of indispensable perceptual training rather than the role of a privileged diet for the elite. While the arts as radar feedback provide a dynamic and changing corporate image, their purpose may be not to enable us to change but rather to maintain an even course toward permanent goals, even amidst the most disrupting innovations. We have already discovered the futility of changing our goals as often as we change our technologies.

1

The Medium
is the Message

The text of this chapter rescues the title from the status of cliché, to which it degenerated soon after McLuhan coined the phrase, by providing an important qualifier: "the medium is socially the message." This expansion recalls and consolidates a key observation from McLuhan's introduction: "Any extension ... affects the whole psychic and social complex." This is the entire rationale for a project in understanding media. Here, as elsewhere, the memorable formulation in the chapter title has overshadowed equally pithy and crucial statements in the text, such as "nothing follows from following except change," McLuhan's resuscitation of David Hume's observations on fallacious notions of causality. It is repeated in conjunction with the explicit mention of Hume in Chapter 9, The Written Word, and explored elsewhere in the book in relation to the interplay of technology and culture, particularly the split between inner and outer causality as characteristic of nonliterate vs. visual cultures respectively, and the meaning of causality under electric technology. Refining the notion of causality in relation to media continued to occupy McLuhan's thinking for years after the first publication of this book. His major themes of organic patterns, fragmentation, automation, myth (see Glossary) are all introduced in this chapter, and a giant in the field of communications media, RCA's David Sarnoff, is trounced, relegated to the crowded gallery of those of whom McLuhan is fond of declaring that they understand exactly nothing about the fundamental operation of media. Among those spared from such judgment is Kenneth Boulding. The first of two references to his book The Image, occurs here, with a quotation indicating that he provided McLuhan with a valuable idea to explore and integrate into his own thinking: "The meaning of a message is the change which it produces in an image."

— *(Editor)*

In a culture like ours, long accustomed to splitting and dividing all things as a means of control, it is sometimes a bit of a shock to be reminded that, in operational and practical fact, the medium is the message. This is merely to say that the personal and social consequences of any medium — that is, of any extension of ourselves — result from the new scale that is introduced into our affairs by each extension of ourselves, or by any new technology. Thus, with automation, for example, the new patterns of human association tend to eliminate jobs, it is true. That is the negative result. Positively, automation creates roles for people, which is to say depth of involvement in their work and human association that our preceding mechanical technology had destroyed. Many people would be disposed to say that it was not the machine, but what one did with the machine, that was its meaning or message. In terms of the ways in which the machine altered our relations to one another and to ourselves, it mattered not in the least whether it turned our cornflakes or Cadillacs. The restructuring of human work and association was shaped by the technique of fragmentation that is the essence of machine technology. The essence of automation technology is the opposite. It is integral and decentralist in depth, just as the machine was fragmentary, centralist, and superficial in its patterning of human relationships.

The instance of the electric light may prove illuminating in this connection. The electric light is pure information. It is a medium without a message, as it were, unless it is used to spell out some verbal ad or name. This fact, characteristic of all media, means that the "content" of any medium is always another medium. The content of writing is speech, just as the written word is the content of print, and print is the content of the telegraph. If it is asked, "What is the content of speech?" it is necessary to say, "It is an actual process of

19

thought, which is in itself nonverbal." An abstract painting represents direct manifestation of creative thought processes as they might appear in computer designs. What we are considering here, however, are the psychic and social consequences of the designs or patterns as they amplify or accelerate existing processes. For the "message" of any medium or technology is the change of scale or pace or pattern that it introduces into human affairs. The railway did not introduce movement or transportation or wheel or road into human society, but it accelerated and enlarged the scale of previous human functions, creating totally new kinds of cities and new kinds of work and leisure. This happened whether the railway functioned in a tropical or a northern environment, and is quite independent of the freight or content of the railway medium. The airplane, on the other hand, by accelerating the rate of transportation, tends to dissolve the railway form of city, politics, and association, quite independently of what the airplane is used for.

Let us return to the electric light. Whether the light is being used for brain surgery or night baseball is a matter of indifference. It could be argued that these activities are in some way the "content" of the electric light, since they could not exist without the electric light. This fact merely underlines the point that "the medium is the message" because it is the medium that shapes and controls the scale and form of human association and action. The content or uses of such media are as diverse as they are ineffectual in shaping the form of human association. Indeed, it is only too typical that the "content" of any medium blinds us to the character of the medium. It is only today that industries have become aware of the various kinds of business in which they are engaged. When IBM discovered that it was not in the business of making office equipment or business machines, but that it was in the business of processing information, then it began to navigate with clear vision. The General Electric

Company makes a considerable portion of its profits from electric light bulbs and lighting systems. It has not yet discovered that, quite as much as AT&T, it is in the business of moving information.

The electric light escapes attention as a communication medium just because it has no "content." And this makes it an invaluable instance of how people fail to study media at all. For it is not till the electric light is used to spell out some brand name that it is noticed as a medium. Then it is not the light but the "content" (or what is really another medium) that is noticed. The message of the electric light is like the message of electric power in industry, totally radical, pervasive, and decentralized. For electric light and power are separate from their uses, yet they eliminate time and space factors in human association exactly as do radio, telegraph, telephone, and TV, creating involvement in depth.

A fairly complete handbook for studying the extensions of man could be made up from selections from Shakespeare. Some might quibble about whether or not he was referring to TV in these familiar lines from *Romeo and Juliet*:

> But soft! what light through yonder window breaks?
> It speaks, and yet says nothing.

In *Othello*, which, as much as *King Lear*, is concerned with the torment of people transformed by illusions, there are these lines that bespeak Shakespeare's intuition of the transforming powers of new media:

> Is there not charms
> By which the property of youth and maidhood
> May be abus'd? Have you not read Roderigo,
> Of some such thing?

In Shakespeare's *Troilus and Cressida*, which is almost completely devoted to both a psychic and social study of communication, Shakespeare states his awareness that true social and political navigation depend upon anticipating the consequences of innovation:

> The providence that's in a watchful state
> Knows almost every grain of Plutus' gold,
> Finds bottom in the uncomprehensive deeps,
> Keeps place with thought, and almost like the gods
> Does thoughts unveil in their dumb cradles.

The increasing awareness of the action of media, quite independently of their "content" or programming, was indicated in the annoyed and anonymous stanza:

> In modern thought, (if not in fact)
> Nothing is that doesn't act,
> So that is reckoned wisdom which
> Describes the scratch but not the itch.

The same kind of total, configurational awareness that reveals why the medium is socially the message has occurred in the most recent and radical medical theories. In his *Stress of Life*, Hans Selye[1] tells of the dismay of a research colleague on hearing of Selye's theory:

> When he saw me thus launched on yet another enraptured description of what I had observed in animals treated with this or that impure, toxic material, he looked at me with desperately sad eyes and said in obvious despair: "But Selye, try to realize what you are doing before it is too late! You have now decided to spend your entire life studying the pharmacology of dirt!"
>
> (Hans Selye, *The Stress of Life*)

---

1. References within the text of this edition have been left in the form used in earlier editions but corrected and/or given in fuller form, including date of original publication, whenever possible, in the Works Cited section, p. 569.

As Selye deals with the total environmental situation in his "stress" theory of disease, so the latest approach to media study considers not only the "content" but the medium and the cultural matrix within which the particular medium operates. The older unawareness of the psychic and social effects of media can be illustrated from almost any of the conventional pronouncements.

In accepting an honorary degree from the University of Notre Dame a few years ago, General David Sarnoff made this statement: "We are too prone to make technological instruments the scapegoats for the sins of those who wield them. The products of modern science are not in themselves good or bad; it is the way they are used that determines their value." That is the voice of the current somnambulism. Suppose we were to say, "Apple pie is in itself neither good nor bad; it is the way it is used that determines its value." Or, "The smallpox virus is in itself neither good nor bad; it is the way it is used that determines its value." Again, "Firearms are in themselves neither good nor bad; it is the way they are used that determines their value." That is, if the slugs reach the right people firearms are good. If the TV tube fires the right ammunition at the right people it is good. I am not being perverse. There is simply nothing in the Sarnoff statement that will bear scrutiny, for it ignores the nature of the medium, of any and all media, in the true Narcissus style of one hypnotized by the amputation and extension of his own being in a new technical form. General Sarnoff went on to explain his attitude to the technology of print, saying that it was true that print caused much trash to circulate, but it had also disseminated the Bible and the thoughts of seers and philosophers. It has never occurred to General Sarnoff that any technology could do anything but *add* itself on to what we already are.

Such economists as Robert Theobald, W. W. Rostow, and John Kenneth Galbraith have been explaining for years how it is that "classical economics" cannot explain change or growth. And the paradox of mechanization is that although it is itself the cause of maximal growth and change, the principle of mechanization excludes the very possibility of growth or the understanding of change. For mechanization is achieved by fragmentation of any process and by putting the fragmented parts in a series. Yet, as David Hume showed in the eighteenth century, there is no principle of causality in a mere sequence. That one thing follows another accounts for nothing. Nothing follows from following, except change. So the greatest of all reversals occurred with electricity, that ended sequence by making things instant. With instant speed the causes of things began to emerge to awareness again, as they had not done with things in sequence and in concatenation accordingly. Instead of asking which came first, the chicken or the egg, it suddenly seemed that a chicken was an egg's idea for getting more eggs.

Just before an airplane breaks the sound barrier, sound waves become visible on the wings of the plane. The sudden visibility of sound just as sound ends is an apt instance of that great pattern of being that reveals new and opposite forms just as the earlier forms reach their peak performance. Mechanization was never so vividly fragmented or sequential as in the birth of the movies, the moment that translated us beyond mechanism into the world of growth and organic interrelation. The movie, by sheer speeding up the mechanical, carried us from the world of sequence and connections into the world of creative configuration and structure. The message of the movie medium is that of transition from lineal connections to configurations. It is the transition that produced the now quite correct observation: "If it works, it's obsolete." When electric speed further takes over from mechanical movie sequences, then the lines of force in structures and in media become loud and clear. We return to the inclusive form of the icon.

To a highly literate and mechanized culture the movie appeared as a world of triumphant illusions and dreams that money could buy. It was at this moment of the movie that cubism occurred, and it has been described by E. H. Gombrich (*Art and Illusion*) as "the most radical attempt to stamp out ambiguity and to enforce one reading of the picture — that of a man-made construction, a colored canvas." For cubism substitutes all facets of an object simultaneously for the "point of view" or facet of perspective illusion. Instead of the specialized illusion of the third dimension on canvas, cubism sets up an interplay of planes and contradiction or dramatic conflict of patterns, lights, textures that "drives home the message" by involvement. This is held by many to be an exercise in painting, not in illusion.

In other words, cubism, by giving the inside and outside, the top, bottom, back, and front and the rest, in two dimensions, drops the illusion of perspective in favor of instant sensory awareness of the whole. Cubism, by seizing on instant total awareness, suddenly announced that *the medium is the message*. Is it not evident that the moment that the sequence yields to the simultaneous, one is in the world of the structure and of configuration? Is that not what has happened in physics as in painting, poetry, and in communication? Specialized segments of attention have shifted to total field, and we can now say, "The medium is the message" quite naturally. Before the electric speed and total field, it was not obvious that that medium is the message. The message, it seemed, was the "content," as people used to ask what a painting was *about*. Yet they never thought to ask what a melody was about, nor what a house or a dress was about. In such matters, people retained some sense of the whole pattern, of form and function as a unity. But in the electric age this integral idea of structure and configuration has become so prevalent that educational theory has taken up the matter. Instead of working with specialized "problems" in arithmetic, the structural approach now follows the line of force in the field of number and has small children meditating about number theory and "sets."

Cardinal Newman said of Napoleon, "He understood the grammar of gunpowder." Napoleon had paid some attention to other media as well, especially the semaphore telegraph that gave him a great advantage over his enemies. He is on record for saying that "Three hostile newspapers are more to be feared than a thousand bayonets."

Alexis de Tocqueville was the first to master the grammar of print and typography. He was thus able to read off the message of coming change in France and America as if he were reading aloud from a text that had been handed to him. In fact, the nineteenth century in France and in America was just such an open book to de Tocqueville because he had learned the grammar of print. So he, also, knew when that grammar did not apply. He was asked why he did not write a book on England, since he knew and admired England. He replied:

> One would have to have an unusual degree of philo-sophical folly to believe oneself able to judge England in six months. A year always seemed to me too short a time in which to appreciate the United States properly, and it is much easier to acquire clear and precise notions about the American Union than about Great Britain. In America all laws derive in a sense from the same line of thought. The whole of society, so to speak, is founded upon a single fact; everything springs from a simple principle. One could compare America to a forest pierced by a multitude of straight roads all converging on the same point. One has only to find the center and everything is revealed at a glance. But in England the paths run criss-cross, and it is only by travelling down each one of them that one can build up a picture of the whole.

De Tocqueville, in earlier work on the French Revolution, had explained how it was the printed word that, achieving cultural saturation in the eighteenth century, had homogenized the French nation. Frenchmen were the same kind of people from north to south. The typographic principles of uniformity, continuity, and lineality had overlaid the complexities of ancient feudal and oral society. The Revolution was carried out by the new literati and lawyers.

In England, however, such was the power of the ancient oral traditions of common law, backed by the medieval institution of Parliament, that no uniformity or continuity of the new visual print culture could take complete hold. The result was that the most important event in English history has never taken place; namely, the English Revolution on the lines of the French Revolution. The American Revolution had no medieval legal institutions to discard or to root out, apart from monarchy. And many have held that the American Presidency has become very much more personal and monarchical than any European monarch ever could be.

De Tocqueville's contrast between England and America is clearly based on the fact of typography and of print culture creating uniformity and continuity. England, he says, has rejected this principle and clung to the dynamic or oral common-law tradition. Hence the discontinuity and unpredictable quality of English culture. The grammar of print cannot help to construe the message of oral and nonwritten culture and institutions. The English aristocracy was properly classified as barbarian by Matthew Arnold because its power and status had nothing to do with literacy or with the cultural forms of typography. Said the Duke of Gloucester to Edward Gibbon upon the publication of his *Decline and Fall*: "Another damned fat book, eh, Mr. Gibbon? Scribble, scribble, scribble, eh, Mr. Gibbon?" De Tocqueville was a highly literate aristocrat who was quite able to be detached from the values and assumptions of typography. That is why he alone understood the grammar of typography. And it is only on those terms, standing aside from any structure or medium, that its principles and lines of force can be discerned. For any medium has the power of imposing its own assumption on the unwary. Prediction and control consist in avoiding this subliminal state of Narcissus trance. But the greatest aid to this end is simply in knowing that the spell can occur immediately upon contact, as in the first bars of a melody.

A Passage to India by E. M. Forster is a dramatic study of the inability of oral and intuitive oriental culture to meet with the rational, visual European patterns of experience. "Rational," of course, has for the West long meant "uniform and continuous and sequential." In other words, we have confused reason with literacy, and rationalism with a single technology. Thus in the electric age man seems to the conventional West to become irrational. In Forster's novel the moment of truth and dislocation from the typographic trance of the West comes in the Marabar Caves. Adela Quested's reasoning powers cannot cope with the total inclusive field of resonance that is India. After the Caves: "Life went on as usual, but had no consequences, that is to say, sounds did not echo nor thought develop. Everything seemed cut off at its root and therefore infected with illusion."

A Passage to India (the phrase is from Whitman, who saw America headed Eastward) is a parable of Western man in the electric age, and is only incidentally related to Europe or the Orient. The ultimate conflict between sight and sound, between written and oral kinds of perception and organization of existence is upon us. Since understanding stops action, as Nietzsche observed, we can moderate the fierceness of this conflict by understanding the media that extend us and raise these wars within and without us.

Detribalization by literacy and its traumatic effects on tribal man is the theme of a book by the psychiatrist J. C. Carothers, *The African Mind in Health and Disease* (World

Health Organization, Geneva, 1953). Much of his material appeared in an article in *Psychiatry* magazine, November, 1959: "The Culture, Psychiatry, and the Written Word." Again, it is electric speed that has revealed the lines of force operating from Western technology in the remotest areas of bush, savannah, and desert. One example is the Bedouin with his battery radio on board the camel. Submerging natives with floods of concepts for which nothing has prepared them is the normal action of all our technology. But with electric media Western man himself experiences exactly the same inundation as the remote native. We are no more prepared to encounter radio and TV in our literate milieu than the native of Ghana is able to cope with the literacy that takes him out of his collective tribal world and beaches him in individual isolation. We are as numb in our new electric world as the native involved in our literate and mechanical culture.

Electric speed mingles the cultures of prehistory with the dregs of industrial marketeers, the nonliterate with the semi-literate and the postliterate. Mental breakdown of varying degrees is the very common result of uprooting and inundation with new information and endless new patterns of information. Wyndham Lewis made this a theme of his group of novels called *The Human Age*. The first of these, *The Childermass*, is concerned precisely with accelerated media change as a kind of massacre of the innocents. In our own world as we become more aware of the effects of technology on psychic formation and manifestation, we are losing all confidence in our right to assign guilt. Ancient prehistoric societies regard violent crime as pathetic. The killer is regarded as we do a cancer victim. "How terrible it must be to feel like that," they say. J. M. Synge took up this idea very effectively in his *Playboy of the Western World.*

If the criminal appears as a nonconformist who is unable to meet the demand of technology that we behave in uniform and continuous patterns, literate man is quite inclined to see others who cannot conform as somewhat pathetic. Especially the child, the cripple, the woman, and the colored person appear in a world of visual and typographic technology as victims of injustice. On the other hand, in a culture that assigns roles instead of jobs to people—the dwarf, the skew, the child create their own spaces. They are not expected to fit into some uniform and repeatable niche that is not their size anyway. Consider the phrase "It's a man's world." As a quantitative observation endlessly repeated from within a homogenized culture, this phrase refers to the men in such a culture who have to be homogenized Dagwoods in order to belong at all. It is in our I.Q. testing that we have produced the greatest flood of misbegotten standards. Unaware of our typographic cultural bias, our testers assume that uniform and continuous habits are a sign of intelligence, thus eliminating the ear man and the tactile man.

C. P. Snow, reviewing a book of A. L. Rowse (*The New York Times Book Review*, December 24, 1961) on *Appeasement* and the road to Munich, describes the top level of British brains and experience in the 1930s. "Their I.Q.'s were much higher than usual among political bosses. Why were they such a disaster?" The view of Rowse, Snow approves: "They would not listen to warnings because they did not wish to hear." Being anti-Red made it impossible for them to read the message of Hitler. But their failure was as nothing compared to our present one. The American stake in literacy as a technology or uniformity applied to every level of education, government, industry, and social life is totally threatened by the electric technology. The threat of Stalin or Hitler was external. The electric technology is within the gates, and we are numb, deaf, blind and mute about its encounter with the Gutenberg technology, on and through which the American way of life was formed. It is, however, no time to suggest strategies when the threat has not even been acknowledged to exist. I am in the position of Louis Pasteur telling doctors that their greatest enemy was quite invisible, and quite un-recognized by them. Our conventional response to all media, namely that it is how they are used that counts, is the numb stance of the technological idiot. For the "content" of a medium is like the juicy piece of meat carried by the burglar to distract the watchdog of the mind. The effect of the medium is made strong and intense just because it is given another medium as "content." The content of a movie is a novel or a play or an opera. The effect of the movie form is not related to its program content. The "content" of writing or print is speech, but the reader is almost entirely unaware either of print or of speech.

Arnold Toynbee is innocent of any understanding of media as they have shaped history, but he is full of examples that the student of media can use. At one moment he can seri-ously suggest that adult education, such as the Workers Educational Association in Britain, is a useful counterforce to the popular press. Toynbee considers that although all of the oriental societies have in our time accepted the industrial technology and its political consequences: "On the cultural plane, however, there is no uniform corresponding tendency." (Somervell, I.267) This is like the voice of the literate man, floundering in a milieu of ads, who boasts, "Personally, I pay no attention to ads." The spiritual and cultural reservations that the oriental peoples may have toward our technology will avail them not at all. The effects of technology do not occur at the level of opinions or concepts, but alter sense ratios or patterns of perception steadily and without any resistance. The serious artist is the only person able to encounter tech-nology with impunity, just because he is an expert aware of the changes in sense perception.

The operation of the money medium in seventeenth cen-tury Japan had effects not unlike the operation of typography in the West. The penetration of the money economy, wrote G. B. Sansom (in *Japan*, Cresset Press, London, 1931) "caused a

slow but irresistible revolution, culminating in the breakdown
of feudal government and the resumption of intercourse
with foreign countries after more than two hundred years of
seclusion." Money has reorganized the sense life of peoples
just because it is an *extension* of our sense lives. This change
does not depend upon approval or disapproval of those living
in the society.

Arnold Toynbee made one approach to the transforming
power of media in his concept of "etherialization," which
he holds to be the principle of progressive simplification
and efficiency in any organization or technology. Typically,
he is ignoring the *effect* of the challenge of these forms upon
the response of our senses. He imagines that it is the response
of our opinions that is relevant to the effect of media and
technology in society, a "point of view" that is plainly the
result of the typographic spell. For the man in a literate
and homogenized society ceases to be sensitive to the
diverse and discontinuous life of forms. He acquires the
illusion of the third dimension and the "private point of
view" as part of his Narcissus fixation, and is quite shut
off from Blake's awareness or that of the Psalmist, that we
become what we behold.

Today when we want to get our bearings in our own culture, and have need to stand aside from the bias and pressure
exerted by any technical form of human expression, we have
only to visit a society where that particular form has not been
felt, or a historical period in which it was unknown. Professor Wilbur Schramm made such a tactical move in studying
Television in the Lives of Our Children. He found areas where
TV had not penetrated at all and ran some tests. Since he
had made no study of the peculiar nature of the TV image,
his tests were of "content" preferences, viewing time, and
vocabulary counts. In a word, his approach to the problem
was a literary one, albeit unconsciously so. Consequently, he
had nothing to report. Had his methods been employed in

1500 A.D. to discover the effects of the printed book in the lives of children or adults, he could have found out nothing of the changes in human and social psychology resulting from typography. Print created individualism and nationalism in the sixteenth century. Program and "content" analysis offer no clues to the magic of these media or to their subliminal charge.

Leonard Doob, in his report *Communication in Africa*, tells of one African who took great pains to listen each evening to the BBC news, even though he could understand nothing of it. Just to be in the presence of those sounds at 7 P.M. each day was important for him. His attitude to speech was like ours to melody — the resonant intonation was meaning enough. In the seventeenth century our ancestors still shared this native's attitude to the forms of the media, as is plain in the following sentiment of the Frenchman Bernard Lam expressed in *The Art of Speaking* (London, 1696):

> 'Tis an effect of the wisdom of God, who created Man to be happy, that whatever is useful to his conversation (way of life) is agreeable to him … because all victual that conduces to nourishment is relishable, whereas other things that cannot be assimilated and be turned into our substance are insipid. A Discourse cannot be pleasant to the Hearer that is not easie to the Speaker; nor can it be easily pronounced unless it be heard with delight.

Here is an equilibrium theory of human diet and expression such as even now we are only striving to work out again for media after centuries of fragmentation and specialism.

Pope Pius XII was deeply concerned that there be serious study of the media today. On February 17, 1950, he said:

> It is not an exaggeration to say that the future of modern society and the stability of its inner life depend in large part on the maintenance of an

33

equilibrium between the strength of the techniques of communication and the capacity of the individual's own reaction.

Failure in this respect has for centuries been typical and total for mankind. Subliminal and docile acceptance of media impact has made them prisons without walls for their human users. As A. J. Liebling remarked in his book *The Press*, a man is not free if he cannot see where he is going, even if he has a gun to help him get there. For each of the media is also a powerful weapon with which to clobber other media and other groups. The result is that the present age has been one of multiple civil wars that are not limited to the world of art and entertainment. In *War and Human Progress,* Professor J. U. Nef declared: "The total wars of our time have been the result of a series of intellectual mistakes ..."

If the formative power in the media are the media themselves, that raises a host of large matters that can only be mentioned here, although they deserve volumes. Namely, that technological media are staples or natural resources, exactly as are coal and cotton and oil. Anybody will concede that a society whose economy is dependent upon one or two major staples like cotton, or grain, or lumber, or fish, or cattle is going to have some obvious social patterns of organization as a result. Stress on a few major staples creates extreme instability in the economy but great endurance in the population. The pathos and humor of the American South are embedded in such an economy of limited staples. For a society configured by reliance on a few commodities accepts them as a social bond quite as much as the metropolis does the press. Cotton and oil, like radio and TV, become "fixed charges" on the entire psychic life of the community. And this pervasive fact creates the unique cultural flavor of any society. It pays through the nose and all its other senses for each staple that shapes its life.

That our human senses, of which all media are extensions,

are also fixed charges on our personal energies, and that they also configure the awareness and experience of each one of us, may be perceived in another connection mentioned by the psychologist C. G. Jung:

> Every Roman was surrounded by slaves. The slave and his psychology flooded ancient Italy, and every Roman became inwardly, and of course unwittingly, a slave. Because living constantly in the atmosphere of slaves, he became infected through the unconscious with their psychology. No one can shield himself from such an influence (*Contributions to Analytical Psychology*, London, 1928).

CAPACITY OF INDIVIDUALS REACTION

STRENGTH OF TECHNIQUE

2

Media Hot and Cold

This chapter contains the first of many references to the writings of Lewis Mumford, but as often as not, Mumford serves McLuhan as a negative model for media analysis. Jack Paar, given pride of place in the introduction to the second edition of the book, ranks a prominent mention here as a bellwether of the clash between hot and cool media. He will vanish for over two hundred pages and then reappear, both anecdotally and to illuminate a fundamental point about the television medium and its personalities. A rich, embedded probe occurs where McLuhan observes that "It is play that cools off the hot situations of actual life by miming them." Play is collective and archetypal, while the situations of actual life are plural and cliché. Together they constitute the principle that McLuhan would later refer to as "reamalgamergence." The probe in question illustrates why Marcel Marceau can sit comfortably side by side with mass media in McLuhan's analysis. It remains a probe, because the reader is challenged to continue applying and reapplying it, in order to achieve a full understanding of how media operate. Probes, or aphorisms, indispensable tools for McLuhan, are explicitly mentioned here in connection with the writing style of Francis Bacon as an example of a cool medium compelling participation in depth.

— (Editor)

HIGH DEFINITION   RELIGION /POLITICS

HOT - RADIO   MOVIES   LOW IN PARTICIPATION

COOL - TELEPHONE   T.V.   HIGH IN PARTICIPATION

LOW DEFINITION

"The rise of the waltz," explained Curt Sachs in the *World History of Dance*, "was the result of that longing for truth, simplicity, closeness to nature, and primitivism, which the last two-thirds of the eighteenth century fulfilled." In the century of jazz we are likely to overlook the emergence of the waltz as a hot and explosive human expression that broke through the formal feudal barriers of courtly and choral dance styles.

There is a basic principle that distinguishes a hot medium like radio from a cool one like the telephone, or a hot medium like the movie from a cool one like TV. A hot medium is one that extends one single sense in "high definition." High definition is the state of being well filled with data. A photograph is, visually, "high definition." A cartoon is "low definition," simply because very little visual information is provided. Telephone is a cool medium, or one of low definition, because the ear is given a meager amount of information. And speech is a cool medium of low definition, because so little is given and so much has to be filled in by the listener. On the other hand, hot media do not leave so much to be filled in or completed by the audience. Hot media are, therefore, low in participation, and cool media are high in participation or completion by the audience. Naturally, therefore, a hot medium like radio has very different effects on the user from a cool medium like the telephone.

A cool medium like hieroglyphic or ideogrammic written characters has very different effects from the hot and explosive medium of the phonetic alphabet. The alphabet, when pushed to a high degree of abstract visual intensity, became typography. The printed word with its specialist intensity burst the bonds of medieval corporate guilds and monasteries, creating extreme individualist patterns of enter prise and monopoly. But the typical reversal occurred when extremes of monopoly brought back the corporation, with its impersonal empire over many lives. The hotting-up of the medium of writing to repeatable print intensity led to nationalism and the religious wars of the sixteenth century. The heavy and unwieldy media, such as stone, are time binders. Used for writing, they are very cool indeed, and serve to unify the ages; whereas paper is a hot medium that serves to unify spaces horizontally, both in political and entertainment empires.

Any hot medium allows of less participation than a cool one, as a lecture makes for less participation than a seminar, and a book for less than dialogue. With print many earlier forms were excluded from life and art, and many were given strange new intensity. But our own time is crowded with examples of the principle that the hot form excludes, and the cool one includes. When ballerinas began to dance on their toes a century ago, it was felt that the art of the ballet had acquired a new "spirituality." With this new intensity, male figures were excluded from ballet. The role of women had also become fragmented with the advent of industrial specialism and the explosion of home functions into laundries, bakeries and hospitals on the periphery of the community. Intensity or high definition engenders specialism and fragmentation in living as in entertainment, which explains why any intense experience must be "forgotten," "censored," and reduced to a very cool state before it can be "learned" or assimilated. The Freudian "censor" is less of a moral function than an indispensable condition of learning. Were we to accept fully and directly every shock to our various structures of awareness, we would soon be nervous wrecks, doing double-takes and pressing panic buttons every minute. The "censor" protects our central system of values, as it does our physical nervous system by simply cooling off the onset of experience a great deal. For many people, this cooling system brings on a life-long state of psychic *rigor mortis,* or of somnambulism, particularly observable in periods of new technology.

An example of the disruptive impact of a hot technology succeeding a cool one is given by Robert Theobald in *The Rich and the Poor.* When Australian natives were given steel axes by the missionaries, their culture, based on the stone axe, collapsed. The stone axe had not only been scarce but had always been a basic status symbol of male importance. The missionaries provided quantities of sharp steel axes and gave them to women and children. The men had even to borrow these from the women, causing a collapse of male dignity. A tribal and feudal hierarchy of traditional kind collapses quickly when it meets any hot medium of the mechanical, uniform, and repetitive kind. The medium of money or wheel or writing, or any other form of specialist speedup of exchange and information, will serve to fragment a tribal structure. Similarly, a very much greater speedup, such as occurs with electricity, may serve to restore a tribal pattern of intense involvement such as took place with the introduction of radio in Europe, and is now tending to happen as a result of TV in America. Specialist technologies detribalize. The nonspecialist electric technology retribalizes. The process of upset resulting from a new distribution of skills is accompanied by much culture lag in which people feel compelled to look at new situations as if they were old ones, and come up with ideas of "population explosion" in an age of implosion. Newton, in an age of clocks, managed to present the physical universe in the image of a clock. But poets like Blake were far ahead of Newton in their response to the challenge of the clock. Blake spoke of the need to be delivered "from single vision and Newton's sleep," knowing very well that Newton's response to the challenge of the new mechanism was itself merely a mechanical repetition of the challenge. Blake saw Newton and Locke and others as hypnotized Narcissus types quite unable to meet the challenge of mechanism. W. B. Yeats gave the full Blakean version of Newton and Locke in a famous epigram:

41

> Locke sank into a swoon;
> The garden died;
> God took the spinning jenny
> Out of his side.

Yeats presents Locke, the philosopher of mechanical and lineal associationism, as hypnotized by his own image. The "garden," or unified consciousness, ended. Eighteenth century man got an extension of himself in the form of the spinning machine that Yeats endows with its full sexual significance. Woman, herself, is thus seen as a technological extension of man's being.

Blake's counterstrategy for his age was to meet mechanism with organic myth. Today, deep in the electric age, organic myth is itself a simple and automatic response capable of mathematical formulation and expression, without any of the imaginative perception of Blake about it. Had he encountered the electric age, Blake would not have met its challenge with a mere repetition of electric form. For myth *is* the instant vision of a complex process that ordinarily extends over a long period. Myth is contraction or implosion of any process, and the instant speed of electricity confers the mythic dimension on ordinary industrial and social action today. We *live* mythically but continue to think fragmentarily and on single planes.

Scholars today are acutely aware of a discrepancy between their ways of treating subjects and the subject itself. Scriptural scholars of both the Old and New Testaments frequently say that while their treatment must be linear, the subject is not. The subject treats of the relations between God and man, and between God and the world, and of the relations between man and his neighbor — all these subsist together, and act and react upon one another at the same time. The Hebrew and Eastern mode of thought tackles problem and resolution, at the outset of a discussion, in a way typical of oral societies in general. The entire message is then traced and retraced, again and again, on the rounds of a concentric spiral with seeming redundancy. One can stop anywhere after the first few sentences and have the full message, if one is prepared to "dig" it. This kind of plan seems to have inspired Frank Lloyd Wright in designing the Guggenheim Art Gallery on a spiral, concentric basis. It is a redundant form inevitable to the electric age, in which the concentric pattern is imposed by the instant quality, and overlay in depth, of electric speed. But the concentric with its endless intersection of planes is necessary for insight. In fact, it is the technique of insight, and as such is necessary for media study, since no medium has its meaning or existence alone, but only in constant interplay with other media.

The new electric structuring and configuring of life more and more encounters the old lineal and fragmentary procedures and tools of analysis from the mechanical age. More and more we turn from the content of messages to study total effect. Kenneth Boulding put this matter in *The Image* by saying, "The meaning of a message is the change which it produces in the image." Concern with *effect* rather than *meaning* is a basic change of our electric time, for effect involves the total situation, and not a single level of information movement. Strangely, there is recognition of this matter of effect rather than information in the British idea of libel: "The greater the truth, the greater the libel."

The effect of electric technology had at first been anxiety. Now it appears to create boredom. We have been through the three stages of alarm, resistance, and exhaustion that occur in every disease or stress of life, whether individual or collective. At least our exhausted slump after the first encounter with the electric has inclined us to expect new problems. However, backward countries that have experienced little permeation with our own mechanical and specialist culture are much better able to confront and to understand electric technology. Not only have backward and nonindustrial

43

cultures no specialist habits to overcome in their encounter with electromagnetism, but they have still much of their traditional oral culture that has the total, unified "field" character of our new electromagnetism. Our old industrialized areas, having eroded their oral traditions automatically, are in the position of having to rediscover them in order to cope with the electric age.

In terms of the theme of media hot and cold, backward countries are cool, and we are hot. The "city slicker" is hot, and the rustic is cool. But in terms of the reversal of procedures and values in the electric age, the past mechanical time was hot, and we of the TV age are cool. The waltz was a hot, fast mechanical dance suited to the industrial time in its moods of pomp and circumstance. In contrast, the Twist is a cool, involved and chatty form of improvised gesture. The jazz of the period of the hot new media of movie and radio was hot jazz. Yet jazz of itself tends to be a casual dialogue form of dance quite lacking in the repetitive and mechanical forms of the waltz. Cool jazz came in quite naturally after the first impact of radio and movie had been absorbed.

In the special Russian issue of *Life* magazine for September 13, 1963, it is mentioned that in Russian restaurants and night clubs, "though the Charleston is tolerated, the Twist is taboo." All this is to say that a country in the process of industrialization is inclined to regard hot jazz as consistent with its developing programs. The cool and involved form of the Twist, on the other hand, would strike such a culture at once as retrograde and incompatible with its new mechanical stress. The Charleston, with its aspect of a mechanical doll agitated by strings, appears in Russia as an *avant-garde* form. We, on the other hand, find the *avant-garde* in the cool and the primitive, with its promise of depth involvement and integral expression.

The "hard" sell and the "hot" line become mere comedy in the TV age, and the death of all the salesmen at one stroke of the TV axe has turned the hot American culture into a cool one that is quite unacquainted with itself. America, in fact, would seem to be living through the reverse process that Margaret Mead described in *Time* magazine (September 4, 1954): "There are too many complaints about society having to move too fast to keep up with the machine. There is great advantage in moving fast if you move completely, if social, educational, and recreational changes keep pace. You must change the whole pattern at once and the whole group together — and the people themselves must decide to move."

Margaret Mead is thinking here of change as uniform speedup of motion or a uniform hotting-up of temperatures in backward societies. We are certainly coming within conceivable range of a world automatically controlled to the point where we could say, "Six hours less radio in Indonesia next week or there will be a great falling off in literary attention." Or, "We can program twenty more hours of TV in South Africa next week to cool down the tribal temperature raised by radio last week." Whole cultures could now be programmed to keep their emotional climate stable in the same way that we have begun to know something about maintaining equilibrium in the commercial economies of the world.

In the merely personal and private sphere we are often reminded of how changes of tone and attitude are demanded of different times and seasons in order to keep situations in hand. British clubmen, for the sake of companionship and amiability have long excluded the hot topics of religion and politics from mention inside the highly participational club. In the same vein, W. H. Auden wrote, "… this season the man of goodwill will wear his heart up his sleeve, not on it … the honest manly style is today suited only to Iago" (Introduction to John Betjeman's *Slick But Not Streamlined*). In the Renaissance, as print technology hotted up the social *milieu* to a very high point, the gentleman and the courtier (Hamlet-Mercutio style) adopted, in contrast, the casual and cool nonchalance of the playful and superior being. The Iago allusion of Auden reminds us that Iago was the *alter ego* and assistant of the intensely earnest and very non-nonchalant General Othello. In imitation of the earnest and forthright general, Iago hotted up his own image and wore his heart on his sleeve, until General Othello read him loud and clear as "honest Iago," a man after his own grimly earnest heart.

Throughout *The City in History*, Lewis Mumford favors the cool or casually structured towns over the hot and intensely filled-in cities. The great period of Athens, he feels, was one during which most of the democratic habits of village life and participation still obtained. Then burst forth the full variety of human expression and exploration such as was later impossible in highly developed urban centers. For the highly developed situation is, by definition, low in opportunities of participation, and rigorous in its demands of specialist fragmentation from those who would control it. For example, what is known as "job enlargement" today in business and in management consists in allowing the employee more freedom to discover and define his function. Likewise, in reading a detective story the reader participates as co-author simply because so much has been left out of the narrative. The open-mesh silk stocking is far more sensuous than the smooth nylon, just because the eye must act as hand in filling in and completing the image, exactly as in the mosaic of the TV image.

Douglas Cater in *The Fourth Branch of Government* tells how the men of the Washington press bureaus delighted to complete or fill in the blank of Calvin Coolidge's personality. Because he was so like a mere cartoon, they felt the urge to complete his image for him and his public. It is instructive that the press applied the word "cool" to Cal. In the very sense of a cool medium, Calvin Coolidge was so lacking in any articulation of data in his public image that there was only one word for him. He was real cool. In the hot 1920s, the hot press medium found Cal very cool and rejoiced in his lack of image, since it compelled the participation of the press in filling in an image of him for the public. By contrast, F.D.R. was a hot press agent, himself a rival of the newspaper medium and one who delighted in scoring off the press on the rival hot medium of radio. Quite in contrast, Jack Paar ran a cool show for the cool TV medium, and became a rival for the patrons of the night spots and their allies in the gossip columns. Jack Paar's war with the gossip columnists was a weird example of clash between a hot and cold medium such as had occurred with the "scandal of the rigged TV quiz shows." The rivalry between the hot press and radio media, on one hand, and TV on the other, for the hot ad buck, served to confuse and to overheat the issues in the affair that pointlessly involved Charles Van Doren.

An Associated Press story from Santa Monica, California, August 9, 1962, reported how

> Nearly 100 traffic violators watched a police traffic accident film today to atone for their violations. Two had to be treated for nausea and shock …
>
> Viewers were offered a $5.00 reduction in fines if they agreed to see the movie, *Signal 30*, made by Ohio State police.
>
> It showed twisted wreckage and mangled bodies and recorded the screams of accident victims.

Whether the hot film medium using hot content would cool off the hot drivers is a moot point. But it does concern any understanding of media. The effect of hot media treatment cannot include much empathy or participation at any time. In this connection an insurance ad that featured Dad in an iron lung surrounded by a joyful family group did more to strike terror into the reader than all the warning wisdom in the world. It is a question that arises in connection with capital punishment. Is a severe penalty the best deterrent to serious crime? With regard to the bomb and the cold war, is the threat of massive retaliation the most effective means to peace? Is it not evident in every human situation that is pushed to a point of saturation that some precipitation occurs? When all the available resources and energies have been played up in an organism or in any structure there is some kind of reversal of pattern. The spectacle of brutality used as deterrent can brutalize. Brutality used in sports may humanize under some conditions, at least. But with regard to the bomb and retaliation as deterrent, it is obvious that numbness is the result of any prolonged terror, a fact that was discovered when the fallout shelter program was broached. The price of eternal vigilance is indifference.

Nevertheless, it makes all the difference whether a hot medium is used in a hot or a cool culture. The hot radio medium used in cool or nonliterate cultures has a violent effect, quite unlike its effect, say in England or America, where radio is felt as entertainment. A cool or low literacy culture cannot accept hot media like movies or radio as entertainment. They are, at least, as radically upsetting for them as the cool TV medium has proved to be for our high literacy world.

And as for the cool war and the hot bomb scare, the cultural strategy that is desperately needed is humor and play. It is play that cools off the hot situations of actual life by miming them. Competitive sports between Russia and the West will hardly serve that purpose of relaxation. Such sports are inflammatory, it is plain. And what we consider entertainment or fun in our media inevitably appears as violent political agitation to a cool culture.

One way to spot the basic difference between hot and cold media uses is to compare and contrast a broadcast of a symphony performance with a broadcast of a symphony rehearsal. Two of the finest shows ever released by the CBC were of Glenn Gould's procedure in recording piano recitals, and Igor Stravinsky's rehearsing the Toronto symphony in some of his new work. A cool medium like TV, when really used, demands this involvement in process. The neat tight package is suited to hot media, like radio and gramophone. Francis Bacon never tired of contrasting hot and cool prose. Writing in "methods" or complete packages, he contrasted with writing in aphorisms, or single observations such as "Revenge is a kind of wild justice." The passive consumer wants packages, but those, he suggested, who are concerned in pursuing knowledge and in seeking causes will resort to aphorisms, just because they are incomplete and require participation in depth.

The principle that distinguishes hot and cold media is perfectly embodied in the folk wisdom: "Men seldom make passes at girls who wear glasses." Glasses intensify the outward-going vision, and fill in the feminine image exceedingly, Marion the Librarian notwithstanding. Dark glasses, on the other hand, create the inscrutable and inaccessible image that invites a great deal of participation and completion.

Again, in a visual and highly literate culture, when we meet a person for the first time his visual appearance dims out the sound of the name, so that in self-defense we add: "How do you spell your name?" Whereas, in an ear culture, the *sound* of a man's name is the overwhelming fact, as Joyce knew when he said in *Finnegans Wake*, "Who gave you that numb?" For the name of a man is a numbing blow from which he never recovers.

Another vantage point from which to test the difference between hot and cold media is the practical joke. The hot literary medium excludes the practical and participant aspect of the joke so completely that Constance Rourke, in her *American Humor*, considers it as no joke at all. To literary people, the practical joke with its total physical involvement is as distasteful as the pun that derails us from the smooth and uniform progress that is typographic order. Indeed, to the literary person who is quite unaware of the intensely abstract nature of the typographic medium, it is the grosser and participant forms of art that seem "hot," and the abstract and intensely literary form that seems "cool." "You may perceive, Madam," said Dr. Johnson, with a pugilistic smile, "that I am well-bred to a degree of needless scrupulosity." And Dr. Johnson was right in supposing that "well-bred" had come to mean a white-shirted stress on attire that rivaled the rigor of the printed page. "Comfort" consists in abandoning a visual arrangement in favor of one that permits casual participation of the senses, a state that is excluded when any one sense, but especially the visual sense, is hotted up to the point of dominant command of a situation.

On the other hand, in experiments in which all outer sensation is withdrawn, the subject begins a furious fill-in or completion of senses that is sheer hallucination. So the hotting-up of one sense tends to effect hypnosis, and the cooling of all sense tends to result in hallucination.

3

Reversal of the Overheated Medium

*McLuhan announces one of his most important themes: elec-
tricity decentralizes. This pervasive and inescapable effect of
electric technology reverses the dominant trend to expansion
that marked thousands of years of social organization under
mechanical technology. Here, in the first of many such passages,
the scope of McLuhan's terms, and an important clue to full
understanding of his ideas, emerges from the word or, in the
phrase "medium or structure." (This was already implicit in
Chapter 1: "The medium is socially the message.") He continues to cite the writings of those who showed him the way by
missing it themselves: "Had Benda known his history ..." and
ridicules "the Angries," (Britain's Angry Young Men) for
their loss of nerve. But Kenneth Boulding, cited already in the
previous chapter, supplies McLuhan with the important notion
of* break boundaries, *the essence of the type of reversal cited
in the chapter title. In exploring the endless and endlessly rich
facets of media reversal, McLuhan links the notion of retribalization (see previous chapter and Glossary) to what he views
as the inevitable transformation of the* global *village into the*
global *city (the latter term only evoked here but occurring in
his private writings): "The new magnetic or world city will be
static and iconic or inclusive." It is useful, at one and the same
time, to consolidate the lessons of this chapter and anticipate
those of later chapters on the printed word and television, by
reflecting that a talking head on television is an overheating
of a cool medium, but a one-word sentence on a page is not an
overcooling of a hot medium.*

— *(Editor)*

A headline for June 21, 1963, read:

WASHINGTON-MOSCOW HOT LINE TO OPEN IN 60 DAYS

The *Times* of London Service, Geneva:
The agreement to establish a direct communication link between Washington and Moscow for emergencies was signed here yesterday by Charles Stelle of the United States and Semyon Tsarapkin of the Soviet Union....

The link, known as the hot line, will be opened within sixty days, according to U.S. officials. It will make use of leased commercial circuits, one cable and the other wireless, using teleprinter equipment.

The decision to use the hot printed medium in place of the cool, participational, telephone medium is unfortunate in the extreme. No doubt the decision was prompted by the literary bias of the West for the printed form, on the ground that it is more impersonal than the telephone. The printed form has quite different implications in Moscow from what it has in Washington. So with the telephone. The Russians' love of this instrument, so congenial to their oral traditions, is owing to the rich nonvisual involvement it affords. The Russian uses the telephone for the sort of effects we associate with the eager conversation of the lapel-gripper whose face is twelve inches away.

Both telephone and teleprinter as amplifications of the unconscious cultural bias of Moscow, on one hand, and of Washington, on the other, are invitations to monstrous misunderstandings. The Russian bugs rooms and spies by ear, finding this quite natural. He is outraged by our visual spying, however, finding this quite unnatural.

The principle that during the stages of their development

53

all things appear under forms opposite to those that they finally present is an ancient doctrine. Interest in the power of things to reverse themselves by evolution is evident in a great diversity of observations, sage and jocular. Alexander Pope wrote:

> Vice is a monster of such frightful mien
> As to be hated needs but to be seen;
> But seen too oft, familiar with its face,
> We first endure, then pity, then embrace.

A caterpillar gazing at the butterfly is supposed to have remarked, "Waal, you'll never catch me in one of those durn things."

At another level we have seen in this century the change-over from the debunking of traditional myths and legends to their reverent study. As we begin to react in depth to the social life and problems of our global village, we become reactionaries. Involvement that goes with our instant technologies transforms the most "socially conscious" people into conservatives. When Sputnik had first gone into orbit a schoolteacher asked her second-graders to write some verse on the subject. One child wrote:

> The stars are so big,
> The earth is so small,
> Stay as you are.

With man his knowledge and the process of obtaining knowledge are of equal magnitude. Our ability to apprehend galaxies and subatomic structures, as well, is a movement of faculties that include and transcend them. The second-grader who wrote the words above *lives* in a world much vaster than any which a scientist today has instruments to measure, or concepts to describe. As W. B. Yeats wrote of this reversal, "The visible world is no longer a reality and the unseen world is no longer a dream."

Associated with this transformation of the real world

into science fiction is the reversal now proceeding apace, by which the Western world is going Eastern, even as the East goes Western. Joyce encoded this reciprocal reverse in his cryptic phrase:

> The West shall shake the East awake
> While ye have the night for morn.

The title of his *Finnegans Wake* is a set of multi-leveled puns on the reversal by which Western man enters his tribal, or Finn, cycle once more, following the track of the old Finn, but wide awake this time as we re-enter the tribal night. It is like our contemporary consciousness of the Unconscious.

The stepping-up of speed from the mechanical to the instant electric form reverses explosion into implosion. In our present electric age the imploding or contracting energies of our world now clash with the old expansionist and traditional patterns of organization. Until recently our institutions and arrangements, social, political, and economic, had shared a one-way pattern. We still think of it as "explosive," or expansive; and though it no longer obtains, we still talk about the population explosion and the explosion in learning. In fact, it is not the increase of numbers in the world that creates our concern with population. Rather, it is the fact that everybody in the world has to live in the utmost proximity created by our electric involvement in one another's lives. In education, likewise, it is not the increase in numbers of those seeking to learn that creates the crisis. Our new concern with education follows upon the changeover to an interrelation in knowledge, where before the separate subjects of the curriculum had stood apart from each other. Departmental sovereignties have melted away as rapidly as national sovereignties under conditions of electric speed. Obsession with the older patterns of mechanical, one-way expansion from centers to margins is no longer relevant to our electric world. Electricity does not centralize, but decentralizes. It is like the difference between a railway system and an electric grid system: the one requires railheads and big urban centers. Electric power, equally available in the farmhouse and the Executive Suite, permits any place to be a center, and does not require large aggregations. This reverse pattern appeared quite early in electrical "labor-saving" devices, whether a toaster or washing machine or vacuum cleaner. Instead of saving work, these devices permit everybody to do his own work. What the nineteenth century had delegated to servants and housemaids we now do for ourselves. This principle applies *in toto* in the electric age. In politics, it permits Castro to exist as independent nucleus or center. It would permit Quebec to leave the Canadian union in a way quite inconceivable under the regime of the railways. The railways require a uniform political and economic space. On the other hand, airplane and radio permit the utmost discontinuity and diversity in spatial organization.

Today the great principle of classical physics and economics and political science, namely that of the divisibility of each process, has reversed itself by sheer extension into the unified field theory; and automation in industry replaces the divisibility of process with the organic interlacing of all functions in the complex. The electric tape succeeds the assembly line.

In the new electric Age of Information and programmed production, commodities themselves assume more and more the character of information, although this trend appears mainly in the increasing advertising budget. Significantly, it is those commodities that are most used in social communication, cigarettes, cosmetics, and soap (cosmetic removers), that bear much of the burden of the upkeep of the media in general. As electric information levels rise, almost any kind of material will serve any kind of need or function, forcing the intellectual more and more into the role of social command and into the service of production.

It was Julien Benda's *Great Betrayal* that helped to clarify the new situation in which the intellectual suddenly holds the whip hand in society. Benda saw that the artists and intellectuals who had long been alienated from power, and who since Voltaire had been in opposition, had now been drafted for service in the highest echelons of decision-making. Their great betrayal was that they had surrendered their autonomy and had become the flunkies of power, as the atomic physicist at the present moment is the flunky of the war lords.

Had Benda known his history, he would have been less angry and less surprised. For it has always been the role of intelligentsia to act as liaison and as mediators between old and new power groups. Most familiar of such groups is the case of the Greek slaves, who were for long the educators and confidential clerks of the Roman power. And it is precisely this servile role of the confidential clerk to the tycoon — commercial, military, or political — that the educator has continued to play in the Western world until the present moment. In England "the Angries" were a group of such clerks who had suddenly emerged from the lower echelons by the educational escape hatch. As they emerged into the upper world of power they found that the air was not at all fresh or bracing. But they lost their nerve even quicker than Bernard Shaw lost his. Like Shaw, they quickly settled down to whimsy and to the cultivation of entertainment values.

In his *Study of History*, Toynbee notes a great many reversals of form and dynamic, as when, in the middle of the fourth century A.D., the Germans in the Roman service began abruptly to be proud of their tribal names and to retain them. Such a moment marked new confidence born of saturation with Roman values, and it was a moment marked by the complementary Roman swing toward primitive values. (As Americans saturate with European values, especially since TV, they begin to insist upon American coach lamps, hitching posts, and colonial kitchenware as cultural objects.) Just as the barbarians got to the top of the Roman social ladder, the Romans themselves were disposed to assume the dress and manners of tribesmen out of the same frivolous and snobbish spirit that attached the French court of Louis XVI to the world of shepherds and shepherdesses. It would have seemed a natural moment for the intellectuals to have taken over while the governing class was touring Disneyland, as it were. So it must have appeared to Marx and his followers. But they reckoned without understanding the dynamics of the new media of communication. Marx based his analysis most untimely on the machine, just as the telegraph and other implosive forms began to reverse the mechanical dynamic.

The present chapter is concerned with showing that in any medium or structure there is what Kenneth Boulding calls a "break boundary at which the system suddenly changes into another or passes some point of no return in its dynamic processes." Several such "break boundaries" will be discussed later, including the one from stasis to motion, and from the mechanical to the organic in the pictorial world. One effect of the static photo had been to suppress the conspicuous consumption of the rich, but the effect of the speedup of the photo had been to provide fantasy riches for the poor of the entire globe.

Today the road beyond its break boundary turns cities into highways, and the highway proper takes on a continuous urban character. Another characteristic reversal after passing a road break boundary is that the country ceases to be the center of all work, and the city ceases to be the center of leisure. In fact, improved roads and transport have reversed the ancient pattern and made cities the centers of work and the country the place of leisure and recreation.

Earlier, the increase of traffic that came with money and roads had ended the static tribal state (as Toynbee calls the nomadic food-gathering culture). Typical of the reversing that occurs at break boundaries is the paradox that nomadic mobile man, the hunter and food-gatherer, is socially static. On the other hand, sedentary, specialist man is dynamic, explosive, progressive. The new magnetic or world city will be static and iconic or inclusive.

In the ancient world the intuitive awareness of break boundaries as points of reversal and of no return was embodied in the Greek idea of *hubris*, which Toynbee presents in his *Study of History*, under the head of "The Nemesis of Creativity" and "The Reversal of Roles." The Greek dramatists presented the idea of creativity as creating, also, its own kind of blindness, as in the case of Oedipus Rex, who solved the riddle of the Sphinx. It was as if the Greeks felt that the penalty for one break-through was a general sealing-off of awareness to the total field. In a Chinese work—*The Way and Its Power* (A. Waley translation)—there is a series of instances of the overheated medium, the overextended man or culture, and the peripety or reversal that inevitably follows:

> He who stands on tiptoe does not stand firm;
> He who takes the longest strides does not walk the
>   fastest ...
> He who boasts of what he will do succeeds in nothing;
> He who is proud of his work achieves nothing that
>   endures.

One of the most common causes of breaks in any system is the cross-fertilization with another system, such as happened to print with the steam press, or with radio and movies (that yielded the talkies). Today with microfilm and micro-cards, not to mention electric memories, the printed word assumes again much of the handicraft character of a manuscript. But printing from movable type was, itself, the major break boundary in the history of phonetic literacy, just as the phonetic alphabet had been the break boundary between tribal and individualist man.

The endless reversals or break boundaries passed in the

interplay of the structures of bureaucracy and enterprise include the point at which individuals began to be held responsible and accountable for their "private actions." That was the moment of the collapse of tribal collective authority. Centuries later, when further explosion and expansion had exhausted the powers of private action, corporate enterprise invented the idea of Public Debt, making the individual privately accountable for group action.

As the nineteenth century heated up the mechanical and dissociative procedures of technical fragmentation, the entire attention of men turned to the associative and the corporate. In the first great age of the substitution of machine for human toil Carlyle and the Pre-Raphaelites promulgated the doctrine of Work as a mystical social communion, and millionaires like Ruskin and Morris toiled like navvies for esthetic reasons. Marx was an impressionable recipient of these doctrines. Most bizarre of all the reversals in the great Victorian age of mechanization and high moral tone is the counter-strategy of Lewis Carroll and Edward Lear, whose nonsense has proved exceedingly durable. While the Lord Cardigans were taking their blood baths in the Valley of Death, Gilbert and Sullivan were announcing that the boundary break had been passed.

4

The Gadget Lover:
Narcissus as Narcosis

*McLuhan continues to encapsulate his themes in probes or
aphorisms, handed to the reader as tools to continue exploring
those themes: "Self-amputation forbids self-recognition." Later,
the typical amputating effect of a medium will be re-expressed
in terms of* obsolescence *(see Subject Index and Glossary).
And* or *appears again as a guide to the meaning of a key term:
"closure or equilibrium-seeking," "closure or displacement of
perception." In Chapter 7, Challenge and Collapse, the variant
will become " 'closure' or psychic consequence," in an echo of
Echo's unheeded words to Narcissus, no less urgent for warning of dangers more sublime and subliminal than those to life
and limb.*

— *(Editor)*

The Greek myth of Narcissus is directly concerned with a fact of human experience, as the word *Narcissus* indicates. It is from the Greek word *narcosis*, or numbness. The youth Narcissus mistook his own reflection in the water for another person. This extension of himself by mirror numbed his perceptions until he became the servomechanism of his own extended or repeated image. The nymph Echo tried to win his love with fragments of his own speech, but in vain. He was numb. He had adapted to his extension of himself and had become a closed system.

Now the point of this myth is the fact that men at once become fascinated by any extension of themselves in any material other than themselves. There have been cynics who insisted that men fall deepest in love with women who give them back their own image. Be that as it may, the wisdom of the Narcissus myth does not convey any idea that Narcissus fell in love with anything he regarded as himself. Obviously he would have had very different feelings about the image had he known it was an extension or repetition of himself. It is, perhaps, indicative of the bias of our intensely technological and, therefore, narcotic culture that we have long interpreted the Narcissus story to mean that he fell in love with himself, that he imagined the reflection to be Narcissus!

Physiologically there are abundant reasons for an extension of ourselves involving us in a state of numbness. Medical researchers like Hans Selye and Adolphe Jonas hold that all extensions of ourselves, in sickness or in health, are attempts to maintain equilibrium. Any extension of ourselves they regard as "autoamputation," and they find that the autoamputative power or strategy is resorted to by the body when the perceptual power cannot locate or avoid the cause of irritation. Our language has many expressions that indicate this self-amputation that is imposed by various pressures. We speak of "wa=nting to jump out of my skin" or of "going out of my mind," being "driven batty" or "flipping my lid." And we often create artificial situations that rival the irritations and stresses of real life under controlled conditions of sport and play.

While it was no part of the intention of Jonas and Selye to provide an explanation of human invention and technology, they have given us a theory of disease (discomfort) that goes far to explain why man is impelled to extend various parts of his body by a kind of autoamputation. In the physical stress of superstimulation of various kinds, the central nervous system acts to protect itself by a strategy of amputation or isolation of the offending organ, sense, or function. Thus, the stimulus to new invention is the stress of acceleration of pace and increase of load. For example, in the case of the wheel as an extension of the foot, the pressure of new burdens resulting from the acceleration of exchange by written and monetary media was the immediate occasion of the extension or "amputation" of this function from our bodies. The wheel as a counter-irritant to increased burdens, in turn, brings about a new intensity of action by its amplification of a separate or isolated function (the feet in rotation). Such amplification is bearable by the nervous system only through numbness or blocking of perception. This is the sense of the Narcissus myth. The young man's image is a self-amputation or extension induced by irritating pressures. As counter-irritant, the image produces a generalized numbness or shock that declines recognition. Self-amputation forbids self-recognition.

The principle of self-amputation as an immediate relief of strain on the central nervous system applies very readily to the origin of the media of communication from speech to computer.

Physiologically, the central nervous system, that electric network that coordinates the various media of our senses, plays the chief role. Whatever threatens its function must be contained, localized, or cut off, even to the total removal of the offending organ. The function of the body, as a group of sustaining and protective organs for the central nervous system, is to act as buffers against sudden variations of stimulus in the physical and social environment. Sudden social failure or shame is a shock that some may "take to heart" or that may cause muscular disturbance in general, signaling for the person to withdraw from the threatening situation.

Therapy, whether physical or social, is a counter-irritant that aids in that equilibrium of the physical organs which protect the central nervous system. Whereas pleasure is a counter-irritant (e.g., sports, entertainment, and alcohol), comfort is the removal of irritants. Both pleasure and comfort are strategies of equilibrium for the central nervous system.

With the arrival of electric technology, man extended, or set outside himself, a live model of the central nervous system itself. To the degree that this is so, it is a development that suggests a desperate and suicidal autoamputation, as if the central nervous system could no longer depend on the physical organs to be protective buffers against the slings and arrows of outrageous mechanism. It could well be that the successive mechanizations of the various physical organs since the invention of printing have made too violent and superstimulated a social experience for the central nervous system to endure.

In relation to that only too plausible cause of such development, we can return to the Narcissus theme. For if Narcissus is numbed by his self-amputated image, there is a very good reason for the numbness. There is a close parallel of response between the patterns of physical and psychic trauma or shock. A person suddenly deprived of loved ones and a person who drops a few feet unexpectedly will both register shock. Both the loss of family and a physical fall are extreme instances of amputations of the self. Shock induces a generalized numbness or an increased threshold to all types of perception. The victim seems immune to pain or sense.

Battle shock created by violent noise has been adapted for dental use in the device known as *audiac*. The patient puts on headphones and turns a dial raising the noise level to the point that he feels no pain from the drill. The selection of a *single* sense for intense stimulus, or of a single extended, isolated, or "amputated" sense in technology, is in part the reason for the numbing effect that technology as such has on its makers and users. For the central nervous system rallies a response of general numbness to the challenge of specialized irritation.

The person who falls suddenly experiences immunity to all pain or sensory stimuli because the central nervous system has to be protected from any intense thrust of sensation. Only gradually does he regain normal sensitivity to sights and sounds, at which time he may begin to tremble and perspire and to react as he would have done if the central nervous system had been prepared in advance for the fall that occurred unexpectedly.

Depending on which sense or faculty is extended technologically, or "autoamputated," the "closure" or equilibrium-seeking among the other senses is fairly predictable. It is with the senses as it is with color. Sensation is always 100 per cent, and a color is always 100 per cent color. But the ratio among the components in the sensation or the color can differ infinitely. Yet if sound, for example, is intensified, touch and taste and sight are affected at once. The effect of radio on literate or visual man was to reawaken his tribal memories, and the effect of sound added to motion pictures was to diminish the role of mime, tactility, and kinesthesis. Similarly, when nomadic man turned to sedentary and specialist ways, the senses specialized too. The development of writing and the visual organization of life made possible the discovery of individualism, introspection and so on.

Any invention or technology is an extension or self-amputation of our physical bodies, and such extension also demands new ratios or new equilibriums among the other organs and extensions of the body. There is, for example, no way of refusing to comply with the new sense ratios or sense "closure" evoked by the TV image. But the effect of the entry of the TV image will vary from culture to culture in accordance with the existing sense ratios in each culture. In audile-tactile Europe TV has intensified the visual sense, spurring them toward American styles of packaging and dressing. In America, the intensely visual culture, TV has opened the doors of audile-tactile perception to the non-visual world of spoken languages and food and the plastic arts. As an extension and expediter of the sense life, any medium at once affects the entire field of the senses, as the Psalmist explained long ago in the 113th Psalm:

> Their idols are silver and gold,
> The work of men's hands.
> They have mouths, but they speak not;
> Eyes they have, but they see not;
> They have ears, but they hear not;
> Noses have they, but they smell not;
> They have hands, but they handle not;
> Feet have they, but they walk not;
> Neither speak they through their throat.
> They that make them shall be like unto them;
> Yea, every one that trusteth in them.

The concept of "idol" for the Hebrew Psalmist is much like that of Narcissus for the Greek mythmaker. And the Psalmist insists that the *beholding* of idols, or the use of technology, conforms men to them. "They that make them shall be like unto them." This is a simple fact of sense "closure." The poet Blake developed the Psalmist's ideas into an entire theory of communication and social change. It is in his long poem of *Jerusalem* that he explains why men have become what they have beheld. What they have, says Blake, is "the spectre of the Reasoning Power in Man" that has become fragmented and "separated from Imagination and enclosing itself as in steel." Blake, in a word, sees man as fragmented by his technologies. But he insists that these technologies are self-amputations of our own organs. When so amputated, each organ becomes a closed system of great new intensity that hurls man into "martyrdoms and wars." Moreover, Blake announces as his theme in *Jerusalem* the organs of perception:

> If Perceptive Organs vary, Objects of Perception
> seem to vary:
> If Perceptive Organs close, their Objects seem to
> close also.

To behold, use or perceive any extension of ourselves in technological form is necessarily to embrace it. To listen to radio or to read the printed page is to accept these extensions of ourselves into our personal system and to undergo the "closure" or displacement of perception that follows automatically. It is this continuous embrace of our own technology in daily use that puts us in the Narcissus role of subliminal awareness and numbness in relation to these images of ourselves. By continuously embracing technologies, we relate ourselves to them as servomechanisms. That is why we must, to use them at all, serve these objects, these extensions of ourselves, as gods or minor religions. An Indian is the servomechanism of his canoe, as the cowboy of his horse or the executive of his clock.

Physiologically, man in the normal use of technology (or his variously extended body) is perpetually modified by it and in turn finds ever new ways of modifying his technology. Man becomes, as it were, the sex organs of the machine world, as the bee of the plant world, enabling it to fecundate and to evolve ever new forms. The machine world reciprocates man's love by expediting his wishes and desires, namely, in providing him with wealth. One of the merits of motivation research has been the revelation of man's sex relation to the motor car.

Socially, it is the accumulation of group pressures and irritations that prompt invention and innovation as counter-irritants. War and the fear of war have always been considered the main incentives to technological extension of our bodies. Indeed, Lewis Mumford, in his *The City in History*, considers the walled city itself an extension of our skins, as much as housing and clothing. More even than the preparation for war, the aftermath of invasion is a rich technological period; because the subject culture has to adjust all its sense ratios to accommodate the impact of the invading culture. It is from such intensive hybrid exchange and strife of ideas and forms that the greatest social energies are released, and from which arise the greatest technologies. Buckminster Fuller estimates that since 1910 the governments of the world have spent 3.5 trillion dollars on airplanes. That is 62 times the existing gold supply of the world.

The principle of numbness comes into play with electric technology, as with any other. We have to numb our central nervous system when it is extended and exposed, or we will die. Thus the age of anxiety and of electric media is also the age of the unconscious and of apathy. But it is strikingly the age of consciousness of the unconscious, in addition. With our central nervous system strategically numbed, the tasks of conscious awareness and order are transferred to the physical life of man, so that for the first time he has become aware of technology as an extension of his physical body. Apparently this could not have happened before the electric age gave us the means of instant, total field-awareness. With such awareness, the subliminal life, private and social, has been hoicked up into full view, with the result that we have "social consciousness" presented to us as a cause of guilt-feelings.

69

Existentialism offers a philosophy of structures, rather than categories, and of total social involvement instead of the bourgeois spirit of individual separateness or points of view. In the electric age we wear all mankind as our skin.

5

Hybrid Energy
les liaisons dangereuses

In the closing pages of the preceding chapter, and even more so here, McLuhan begins to focus his attention on the media of communication (references to McWhinnie, Eisenstein). Inevitably, he relates these to "the medium of language itself as shaping the arrangements of daily existence." This is an oblique reference to the Sapir-Whorf hypothesis about language as a sort of grid or filter for reality. And it is the starting point for McLuhan's unique development, beginning in the following chapter (the title Media as Translators is itself a hint), of the entire and vast subject of language (too seldom recognized for its central place in his work). It is integrated with the very foundation of his media analysis when he observes: "The spoken word was the first technology by which man was able to let go of his environment in order to grasp it in a new way." The production of probes continues in the present chapter, also with an emphasis on media as creators of new environments and the necessity of understanding and controlling media: "We can, if we choose, think things out before we put them out." An important lesson on the eclipse of narrative/descriptive technique produces a pantheon of Klee, Picasso, Braque, Eisenstein, the Marx brothers, and James Joyce. Never averse to letting the sublime and the absurd sit cheek by jowl, McLuhan offers a rare historical footnote, intended to tweak the noses of the city fathers of Toronto the Good: "In print-oriented Toronto, poetry-reading in the public parks is a public offense. Religion and politics are permitted, but not poetry, as many young poets recently discovered."

— (Editor)

"For most of our lifetime civil war has been raging in the world of art and entertainment.... Moving pictures, gramophone records, radio, talking pictures...." This is the view of Donald McWhinnie, analyst of the radio medium. Most of this civil war affects us in the depths of our psychic lives, as well, since the war is conducted by forces that are extensions and amplifications of our own beings. Indeed, the interplay among media is only another name for this "civil war" that rages in our society and our psyches alike. "To the blind all things are sudden," it has been said. The crossings or hybridizations of the media release great new force and energy as by fission or fusion. There need be no blindness in these matters once we have been notified that there is anything to observe.

It has now been explained that media, or the extensions of man, are "make happen" agents, but not "make aware" agents. The hybridizing or compounding of these agents offers an especially favorable opportunity to notice their structural components and properties. "As the silent film cried out for sound, so does the sound film cry out for color," wrote Sergei Eisenstein in his *Notes of a Film Director*. This type of observation can be extended systematically to all media: "As the printing press cried out for nationalism, so did the radio cry out for tribalism." These media, being extensions of ourselves, also depend upon us for their interplay and their evolution. The fact that they do interact and spawn new progeny has been a source of wonder over the ages. It need baffle us no longer if we trouble to scrutinize their action. We can, if we choose, think things out before we put them out.

Plato, in all his striving to imagine an ideal training school, failed to notice that Athens was a greater school than any university even he could dream up. In other words, the greatest school had been put out for human use before it had been thought out. Now, this is especially true of our media. They are put out long before they are thought out. In fact, their being put outside us tends to cancel the possibility of their being thought of at all.

Everybody notices how coal and steel and cars affect the arrangements of daily existence. In our time, study has finally turned to the medium of language itself as shaping the arrangements of daily life, so that society begins to look like a linguistic echo or repeat of language norms, a fact that has disturbed the Russian Communist party very deeply. Wedded as they are to nineteenth-century industrial technology as the basis of class liberation, nothing could be more subversive of the Marxian dialectic than the idea that linguistic media shape social development, as much as do the means of production.

In fact, of all the great hybrid unions that breed furious release of energy and change, there is none to surpass the meeting of literate and oral cultures. The giving to man of an eye for an ear by phonetic literacy is, socially and politically, probably the most radical explosion that can occur in any social structure. This explosion of the eye, frequently repeated in "backward areas," we call Westernization. With literacy now about to hybridize the cultures of the Chinese, the Indians, and the Africans, we are about to experience such a release of human power and aggressive violence as makes the previous history of phonetic alphabet technology seem quite tame.

That is only the East-side story, for the electric implosion now brings oral and tribal ear-culture to the literate West. Not only does the visual, specialist, and fragmented Westerner have now to live in closest daily association with all the ancient oral cultures of the earth, but his own electric technology now begins to translate the visual or eye man back into the tribal and oral pattern with its seamless web of kinship and interdependence.

We know from our own past the kind of energy that is released, as by fission, when literacy explodes the tribal or family unit. What do we know about the social and psychic energies that develop by electric fusion or implosion when literate individuals are suddenly gripped by an electromagnetic field, such as occurs in the new Common Market pressure in Europe? Make no mistake, the fusion of people who have known individualism and nationalism is not the same process as the fission of "backward" and oral cultures that are just coming to individualism and nationalism. It is the difference between the A-bomb and the H-bomb. The latter is more violent, by far. Moreover, the products of electric fusion are immensely complex, while the products of fission are simple. Literacy creates very much simpler kinds of people than those that develop in the complex web of ordinary tribal and oral societies. For the fragmented man creates the homogenized Western world, while oral societies are made up of people differentiated, not by their specialist skills or visible marks, but by their unique emotional mixes. The oral man's inner world is a tangle of complex emotions and feelings that the Western practical man has long ago eroded or suppressed within himself in the interest of efficiency and practicality.

The immediate prospect for literate, fragmented Western man encountering the electric implosion within his own culture is his steady and rapid transformation into a complex and depth-structured person emotionally aware of his total interdependence with the rest of human society. Representatives of the older Western individualism are even now assuming the appearance, for good or ill, of Al Capp's General Bull Moose or of the John Birchers, tribally dedicated to opposing the tribal. Fragmented, literate, and visual individualism is not possible in an electrically patterned and imploded society. So what is to be done? Do we dare to confront such facts at the conscious level, or is it best to becloud and repress such matters until some violence releases us from the entire burden? For the fate of implosion and interdependence is more terrible for Western man than the fate of explosion and independence for tribal man. It may be merely temperament in my own case, but I find some easing of the burden in just understanding and clarifying the issues. On the other hand, since consciousness and awareness seem to be a human privilege, may it not be desirable to extend this condition to our hidden conflicts, both private and social?

The present book, in seeking to understand many media, the conflicts from which they spring, and the even greater conflicts to which they give rise, holds out the promise of reducing these conflicts by an increase of human autonomy. Let us now note a few of the effects of media hybrids, or of the interpenetration of one medium by another.

Life at the Pentagon has been greatly complicated by jet travel, for example. Every few minutes an assembly gong rings to summon many specialists from their desks to hear a personal report from an expert from some remote part of the world. Meantime, the undone paper work mounts on each desk. And each department daily dispatches personnel by jet to remote areas for more data and reports. Such is the speed of this process of the meeting of the jet plane, the oral report, and the typewriter that those going forth to the ends of the earth often arrive unable to spell the name of the spot to which they have been sent as experts. Lewis Carroll pointed out that as large-scale maps got more and more detailed and extensive, they would tend to blanket agriculture and rouse the protest of farmers. So why not use the actual earth as a map of itself? We have reached a similar point of data gathering when each stick of chewing gum we reach for is acutely noted by some computer that translates our least gesture into a new probability curve or some parameter of social science. Our private and corporate lives have become information processes just because we have put our central nervous systems outside us in electric technology. That is the key to Professor Boorstin's bewilderment in *The Image, or What Happened to the American Dream.*

The electric light ended the regime of night and day, of indoors and out-of-doors. But it is when the light encounters already existing patterns of human organization that the hybrid energy is released. Cars can travel all night, ball players can play all night, and windows can be left out of buildings. In a word, the message of the electric light is total change. It is pure information without any content to restrict its transforming and informing power.

If the student of media will but meditate on the power of this medium of electric light to transform every structure of time and space and work and society that it penetrates or contacts, he will have the key to the form of the power that is in all media to reshape any lives that they touch. Except for light, all other media come in pairs, with one acting as the "content" of the other, obscuring the operation of both.

It is a peculiar bias of those who operate media for the owners that they be concerned about the program content of radio, or press, or film. The owners themselves are concerned more about the media as such, and are not inclined to go beyond "what the public wants" or some vague formula. Owners are aware of the media as power, and they know that this power has little to do with "content" or the media within the media.

When the press opened up the "human interest" keyboard after the telegraph had restructured the press medium, the newspaper killed the theater, just as TV hit the movies and the night clubs very hard. George Bernard Shaw had the wit and imagination to fight back. He put the press into the theater, taking over the controversies and the human interest world of the press for the stage, as Dickens had done for the novel. The movie took over the novel and the newspaper and the stage, all at once. Then TV pervaded the movie and gave the theater-in-the-round back to the public.

What I am saying is that media as extensions of our senses institute new ratios, not only among our private senses, but among themselves, when they interact among themselves. Radio changed the form of the news story as much as it altered the film image in the talkies. TV caused drastic changes in radio programming, and in the form of the *thing* or documentary novel.

It is the poets and painters who react instantly to a new medium like radio or TV. Radio and gramophone and tape recorder gave us back the poet's voice as an important dimension of the poetic experience. Words became a kind of painting with light, again. But TV, with its deep-participation mode, caused young poets suddenly to present their poems in cafés, in public parks, anywhere. After TV, they suddenly felt the need for personal contact with their public. (In print-oriented Toronto, poetry-reading in the public parks is a public offense. Religion and politics are permitted, but not poetry, as many young poets recently discovered.)

John O'Hara, the novelist, wrote in *The New York Times Book Review* of November 27, 1955:

> You get a great satisfaction from a book. You know your reader is captive inside those covers, but as novelist you have to imagine the satisfaction he's getting. Now, in the theater — well, I used to drop in during both productions of Pal Joey and watch, not imagine, the people enjoy it. I'd willingly start my next novel — about a small town — right now, but I need the diversion of a play.

In our age artists are able to mix their media diet as easily as their book diet. A poet like Yeats made the fullest use of oral peasant culture in creating his literary effects. Quite early, Eliot made a great impact by the careful use of jazz and film form. *The Love Song of J. Alfred Prufrock* gets much of its power from an interpenetration of film form and jazz idiom. But this mix reached its greatest power in *The Waste Land* and

Sweeney Agonistes. *Prufrock* uses not only film form but the film theme of Charlie Chaplin, as did James Joyce in *Ulysses*. Joyce's Bloom is a deliberate takeover from Chaplin ("Chorney Choplain," as he called him in *Finnegans Wake*). And Chaplin, just as Chopin had adapted the pianoforte to the style of the ballet, hit upon the wondrous media mix of ballet and film in developing his Pavlova-like alternation of ecstasy and waddle. He adapted the classical steps of ballet to a movie mime that converged exactly the right blend of the lyric and the ironic that is found also in *Prufrock* and *Ulysses*. Artists in various fields are always the first to discover how to enable one medium to use or to release the power of another. In a simpler form, it is the technique employed by Charles Boyer in his kind of French-English blend of urbane, throaty delirium.

The printed book had encouraged artists to reduce all forms of expression as much as possible to the single descriptive and narrative plane of the printed word. The advent of electric media released art from this straitjacket at once, creating the world of Paul Klee, Picasso, Braque, Eisenstein, the Marx Brothers, and James Joyce.

A headline in *The New York Times Book Review* (September 16, 1962) trills: THERE'S NOTHING LIKE A BEST SELLER TO SET HOLLYWOOD A-TINGLE.

Of course, nowadays, movie stars can only be lured from the beaches or science-fiction or some self-improvement course by the cultural lure of a role in a famous book. That is the way that the interplay of media now affects many in the movie colony. They have no more understanding of their media problems than does Madison Avenue. But from the point of view of the owners of the film and related media, the best seller is a form of insurance that some massive new *gestalt* or pattern has been isolated in the public psyche. It is an oil strike or a gold mine that can be depended on to yield a fair amount of boodle to the careful and canny processor.

Hollywood bankers, that is, are smarter than literary historians, for the latter despise popular taste except when it has been filtered down from lecture course to literary handbook.

Lillian Ross in *Picture* wrote a snide account of the filming of *The Red Badge of Courage*. She got a good deal of easy kudos for a foolish book about a great film by simply *assuming* the superiority of the literary medium to the film medium. Her book got much attention as a hybrid.

Agatha Christie wrote far above her usual good level in a group of twelve short stories about Hercule Poirot, called *The Labours of Hercules*. By adjusting the classical themes to make reasonable modern parallels, she was able to lift the detective form to extraordinary intensity.

Such was, also, the method of James Joyce in *Dubliners* and *Ulysses*, when the precise classical parallels created the true hybrid energy. Baudelaire, said Mr. Eliot, "taught us how to raise the imagery of common life to first intensity." It is done, not by any direct heave-ho of poetic strength, but by a simple adjustment of situations from one culture in hybrid form with those of another. It is precisely in this way that during wars and migrations new cultural mix is the norm of ordinary daily life. Operations Research programs the hybrid principle as a technique of creative discovery.

When the movie scenario or picture story was applied to the *idea* article, the magazine world had discovered a hybrid that ended the supremacy of the short story. When wheels were put in tandem form, the wheel principle combined with the lineal typographic principle to create aerodynamic balance. The wheel crossed with industrial, lineal form released the new form of the airplane.

The hybrid or the meeting of two media is a moment of truth and revelation from which new form is born. For the parallel between two media holds us on the frontiers between forms that snap us out of the Narcissus-narcosis.

The moment of the meeting of media is a moment of free-dom and release from the ordinary trance and numbness imposed by them on our senses.

6

Media as Translators

Important working definitions to be developed in later chapters are established: all technologies as techniques for transfer of knowledge from one mode to another (an expansion of the title phrase), automation as organic patterns, words as information retrieval. In the definition of prayer, cited from George Herbert, language is the thread linking it to the theme of the chapter and the first inkling that McLuhan's media analysis can accommodate a spiritual dimension. Even phrased as a question, the closing paragraph of this chapter was enough for commentators to brand McLuhan from the outset as a technological utopian. Rephrased as a statement, McLuhan's question compels no optimism but issues an implicit warning in anticipation of the following chapter: meet the challenge or face collapse.

— (Editor)

The tendency of neurotic children to lose neurotic traits when telephoning has been a puzzle to psychiatrists. Some stutterers lose their stutter when they switch to a foreign language. That technologies are ways of translating one kind of knowledge into another mode has been expressed by Lyman Bryson in the phrase "technology is explicitness." Translation is thus a "spelling-out" of forms of knowledge. What we call "mechanization" is a translation of nature, and of our own natures, into amplified and specialized forms. Thus the quip in *Finnegans Wake*, "What bird has done yesterday man may do next year," is a strictly literal observation of the courses of technology. The power of technology as dependent on alternately grasping and letting go in order to enlarge the scope of action has been observed as the power of the higher arboreal apes as compared with those that are on the ground. Elias Canetti made the proper association of this power of the higher apes to grasp and let go, with the strategy of the stock market speculators. It is all capsulated in the popular variant on Robert Browning: "A man's reach must exceed his grasp or what's a metaphor." All media are active metaphors in their power to translate experience into new forms. The spoken word was the first technology by which man was able to let go of his environment in order to grasp it in a new way. Words are a kind of information retrieval that can range over the total environment and experience at high speed. Words are complex systems of metaphors and symbols that translate experience into our uttered or outered senses. They are a technology of explicitness. By means of translation of immediate sense experience into vocal symbols the entire world can be evoked and retrieved at any instant.

In this electric age we see ourselves being translated more and more into the form of information, moving toward the technological extension of consciousness. That is what is meant when we say that we daily know more and more about man. We mean that we can translate more and more of ourselves into other forms of expression that exceed ourselves. Man is a form of expression who is traditionally expected to repeat himself and to echo the praise of his Creator. "Prayer," said George Herbert, "is reversed thunder." Man has the power to reverberate the Divine thunder, by verbal translation.

By putting our physical bodies inside our extended nervous systems, by means of electric media, we set up a dynamic by which all previous technologies that are mere extensions of hands and feet and teeth and bodily heat-controls — all such extensions of our bodies, including cities — will be translated into information systems. Electromagnetic technology requires utter human docility and quiescence of meditation such as befits an organism that now wears its brain outside its skull and its nerves outside its hide. Man must serve his electric technology with the same servo-mechanistic fidelity with which he served his coracle, his canoe, his typography, and all other extensions of his physical organs. But there is this difference, that previous technologies were partial and fragmentary, and the electric is total and inclusive. An external consensus or conscience is now as necessary as private consciousness. With the new media, however, it is also possible to store and to translate everything; and, as for speed, that is no problem. No further acceleration is possible this side of the light barrier.

Just as when information levels rise in physics and chemistry, it is possible to use anything for fuel or fabric or building material, so with electric technology all solid goods can be summoned to appear as solid commodities by means of information circuits set up in the organic patterns that we call "automation" and information retrieval. Under electric technology the entire business of man becomes learning and knowing. In terms of what we still consider an "economy" (the Greek word for a household), this means that all forms of employment become "paid learning," and all forms of wealth result from the movement of information. The problem of discovering occupations or employment may prove as difficult as wealth is easy.

The long revolution by which men have sought to translate nature into art we have long referred to as "applied knowledge." "Applied" means translated or carried across from one kind of material form into another. For those who care to consider this amazing process of applied knowledge in Western civilization, Shakespeare's *As You Like It* provides a good deal to think about. His Forest of Arden is just such a golden world of translated benefits and joblessness as we are now entering via the gate of electric automation.

It is no more than one would expect that Shakespeare should have understood the Forest of Arden as an advance model of the age of automation when all things are translatable into anything else that is desired:

> And this our life, exempt from public haunt,
> Finds tongues in trees, books in running brooks,
> Sermons in stones, and good in every thing.
> I would not change it.
>
> AMIENS: Happy is your Grace,
> That can translate the stubbornness of fortune
> Into so quiet and so sweet a style.
>
> (*As You Like It*, II, i. 15–21)

Shakespeare speaks of a world into which, by programming, as it were, one can play back the materials of the natural world in a variety of levels and intensities of style. We are close to doing just this on a massive scale at the present time electronically. Here is the image of the golden age as one of complete metamorphoses or translations of nature into human art, that stands ready of access to our electric age. The poet Stéphane Mallarmé thought "the world exists to end in a

87

book." We are now in a position to go beyond that and to transfer the entire show to the memory of a computer. For man, as Julian Huxley observes, unlike merely biological creatures, possesses an apparatus of transmission and transformation based on his power to store experience. And his power to store, as in a language itself, is also a means of transformation of experience:

Those pearls that were his eyes.

Our dilemma may become like that of the listener who phoned the radio station: "Are you the station that gives twice as much weather? Well, turn it off. I'm drowning."

Or we might return to the state of tribal man, for whom magic rituals are his means of "applied knowledge." Instead of translating nature into art, the native nonliterate attempts to invest nature with spiritual energy.

Perhaps there is a key to some of these problems in the Freudian idea that when we fail to translate some natural event or experience into conscious art we "repress" it. It is this mechanism that also serves to numb us in the presence of those extensions of ourselves that are the media studied in this book. For just as a metaphor transforms and transmits experience, so do the media. When we say, "I'll take a rain check on that," we translate a social invitation into a sporting event, stepping up the conventional regret to an image of spontaneous disappointment: "Your invitation is not just one of those casual gestures that I must brush off. It makes me feel all the frustration of an interrupted ball game that I can't get with it." As in all metaphors, there are complex ratios among four parts: "Your invitation is to ordinary invitations as ball games are to conventional social life." It is in this way that by seeing one set of relations through another set that we store and amplify experience in such forms as money. For money is also a metaphor. And all media as extensions of ourselves serve to provide new transforming vision and awareness. "It is an excellent invention," Bacon says, "that Pan or the world is said to make choice of Echo only (above all other speeches or voices) for his wife, for that alone is true philosophy which doth faithfully render the very words of the world ..."

Today Mark II stands by to render the masterpieces of literature from any language into any other language, giving as follows, the words of a Russian critic of Tolstoy about "War and World (peace ... But nonetheless culture not stands) costs on place. Something translate. Something print." (Boorstin, 141)

Our very word "grasp" or "apprehension" points to the process of getting at one thing through another, of handling and sensing many facets at a time through more than one sense at a time. It begins to be evident that "touch" is not skin but the interplay of the senses, and "keeping in touch" or "getting in touch" is a matter of a fruitful meeting of the senses, of sight translated into sound and sound into movement, and taste and smell. The "common sense" was for many centuries held to be the peculiar human power of translating one kind of experience of one sense into all the senses, and presenting the result continuously as a unified image to the mind. In fact, this image of a unified ratio among the senses was long held to be the mark of our *ratio*-nality, and may in the computer age easily become so again. For it is now possible to program ratios among the senses that approach the condition of consciousness. Yet such a condition would necessarily be an extension of our own consciousness as much as wheel is an extension of feet in rotation. Having extended or translated our central nervous system into the electromagnetic technology, it is but a further stage to transfer our consciousness to the computer world as well. Then, at least, we shall be able to program consciousness in such wise that it cannot be numbed nor distracted by the Narcissus illusions of the entertainment world that beset mankind when he encounters himself extended in his own gimmickry.

If the work of the city is the remaking or translating of man into a more suitable form than his nomadic ancestors achieved, then might not our current translation of our entire lives into the spiritual form of information seem to make of the entire globe, and of the human family, a single consciousness?

7

Challenge and Collapse:
The Nemesis of Creativity

*The title is rich in echoes harking back to the first page of the
book ("the creative process of knowing") and very explicitly
anticipating a much later passage: "Humpty-Dumpty met the
challenge of the wall with a spectacular collapse." (p. [249-50])
Even without that fatal wall, Humpty-Dumpty may have met
his fate performing a banana-skin pirouette of the kind evoked
here. Less whimsical is McLuhan's warning and his own challenge inviting collapse with the mention of another wall: "Every
American home has its Berlin Wall." Important new dimensions
of media effects emerge: "The effect of radio is visual; the effect
of the photo is auditory." As do definitions ("specialism as a
counter-irritant") and metaphors ("the alphabet was the greatest
processor of men for homogenized military life that was known
to antiquity"). One passage affords a relatively rare glimpse of
McLuhan's attitude to big business ("We have leased these 'places
to stand' [our physical senses] to private corporations.") and
another, rarely discussed by commentators, deals with the power
of will vs. perception of situations (p.[102]). But it is a passage
on the medieval Schoolmen that delivers a message of one of the
greatest failures of all time, a collapse in the face of an educational challenge, the very issue that first impelled McLuhan to
undertake media analysis: "Had the Schoolmen with their complex oral culture understood the Gutenberg technology, they
could have created a new synthesis of written and oral education,
instead of bowing out of the picture and allowing the merely
visual page to take over the educational enterprise." Chastized
again in Chapter 20, The Photograph, the Schoolmen find
themselves there in the company of the modern educator who
"intensifies his ineptness by a defensive arrogance and condescension to 'pop kulch.' "*

— *(Editor)*

It was Bertrand Russell who declared that the great discovery of the twentieth century was the technique of the suspended judgment. A. N. Whitehead, on the other hand, explained how the great discovery of the nineteenth century was the discovery of the technique of discovery. Namely, the technique of starting with the thing to be discovered and working back, step by step, as on an assembly line, to the point at which it is necessary to start in order to reach the desired object. In the arts this meant starting with the *effect* and then inventing a poem, painting, or building that would have just that effect and no other.

But the "technique of the suspended judgment" goes further. It anticipates the effect of, say, an unhappy childhood on an adult, and offsets the effect before it happens. In psychiatry, it is the technique of total permissiveness extended as an anesthetic for the mind, while various adhesions and moral effects of false judgments are systematically eliminated.

This is a very different thing from the numbing or narcotic effect of new technology that lulls attention while the new form slams the gates of judgment and perception. For massive social surgery is needed to insert new technology into the group mind, and this is achieved by the built-in numbing apparatus discussed earlier. Now the "technique of the suspended judgment" presents the possibility of rejecting the narcotic and of postponing indefinitely the operation of inserting the new technology in the social psyche. A new stasis is in prospect.

Werner Heisenberg, in *The Physicist's Conception of Nature*, is an example of the new quantum physicist whose overall awareness of forms suggests to him that we would do well to stand aside from most of them. He points out that technical change alters not only habits of life, but patterns of thought and valuation, citing with approval the outlook of the Chinese sage:

> As Tzu-Gung was travelling through the regions north of the river Han, he saw an old man working in his vegetable garden. He had dug an irrigation ditch. The man would descend into a well, fetch up a vessel of water in his arms and pour it out into the ditch. While his efforts were tremendous the results appeared to be very meager.
>
> Tzu-Gung said, "There is a way whereby you can irrigate a hundred ditches in one day, and whereby you can do much with little effort. Would you not like to hear of it?"
>
> Then the gardener stood up, looked at him and said, "And what would that be?"
>
> Tzu-Gung replied, "You take a wooden lever, weighted at the back and light in front. In this way you can bring up water so quickly that it just gushes out. This is called a draw-well."
>
> Then anger rose up in the old man's face, and he said, "I have heard my teacher say that whoever uses machines does all his work like a machine. He who does his work like a machine grows a heart like a machine, and he who carries the heart of a machine in his breast loses his simplicity. He who has lost his simplicity becomes unsure in the strivings of his soul. Uncertainty in the strivings of the soul is something which does not agree with honest sense. It is not that I do not know of such things; I am ashamed to use them."

Perhaps the most interesting point about this anecdote is that it appeals to a modern physicist. It would not have appealed to Newton or to Adam Smith, for they were great experts and advocates of the fragmentary and the specialist approaches. It is by means quite in accord with the outlook of the Chinese sage that Hans Selye works at his "stress" idea of illness. In the 1920s he had been baffled at why physicians always seemed to concentrate on the recognition of individual diseases and specific remedies for such isolated causes, while never paying any attention to the "syndrome of just being sick." Those who are concerned with the program "content" of media and not with the medium proper, appear to be in the position of physicians who ignore the "syndrome of just being sick." Hans Selye, in tackling a total, inclusive approach to the field of sickness, began what Adolphe Jonas has continued in *Irritation and Counter-Irritation*; namely, a quest for the response to injury as such, or to novel impact of any kind. Today we have anesthetics that enable us to perform the most frightful physical operations on one another.

The new media and technologies by which we amplify and extend ourselves constitute huge collective surgery carried out on the social body with complete disregard for antiseptics. If the operations are needed, the inevitability of infecting the whole system during the operation has to be considered. For in operating on society with a new technology, it is not the incised area that is most affected. The area of impact and incision is numb. It is the entire system that is changed. The effect of radio is visual, the effect of the photo is auditory. Each new impact shifts the ratios among all the senses. What we seek today is either a means of controlling these shifts in the sense-ratios of the psychic and social outlook, or a means of avoiding them altogether. To have a disease without its symptoms is to be immune. No society has ever known enough about its actions to have developed immunity to its new extensions or technologies. Today we have begun to sense that art may be able to provide such immunity.

In the history of human culture there is no example of a conscious adjustment of the various factors of personal and social life to new extensions except in the puny and peripheral efforts of artists. The artist picks up the message of cultural and technological challenge decades before its transforming impact occurs. He, then, builds models or Noah's arks for facing the change that is at hand. "The war of 1870 need never have been fought had people read my *Sentimental Education*," said Gustave Flaubert.

It is this aspect of *new* art that Kenneth Galbraith recommends to the careful study of businessmen who want to stay in business. For in the electric age there is no longer any sense in talking about the artist's being ahead of his time. Our technology is, also, ahead of its time, if we reckon by the ability to recognize it for what it is. To prevent undue wreckage in society, the artist tends now to move from the ivory tower to the control tower of society. Just as higher education is no longer a frill or luxury but a stark need of production and operational design in the electric age, so the artist is indispensable in the shaping and analysis and understanding of the life of forms, and structures created by electric technology.

The percussed victims of the new technology have invariably muttered clichés about the impracticality of artists and their fanciful preferences. But in the past century it has come to be generally acknowledged that, in the words of Wyndham Lewis, "The artist is always engaged in writing a detailed history of the future because he is the only person aware of the nature of the present." Knowledge of this simple fact is now needed for human survival. The ability of the artist to sidestep the bully blow of new technology of any age, and to parry such violence with full awareness, is age-old. Equally age-old is the inability of the percussed victims, who cannot sidestep the new violence, to recognize their need of the artist. To reward and to make celebrities of artists can, also, be a way of ignoring their prophetic work, and preventing its timely use for survival. The artist is the man in any field, scientific or humanistic, who grasps the implications of his actions and of new knowledge in his own time. He is the man of integral awareness.

The artist can correct the sense ratios before the blow of

new technology has numbed conscious procedures. He can correct them before numbness and subliminal groping and reaction begin. If this is true, how is it possible to present the matter to those who are in a position to do something about it? If there were even a remote likelihood of this analysis being true, it would warrant a global armistice and period of stock-taking. If it is true that the artist possesses the means of anticipating and avoiding the consequences of technological trauma, then what are we to think of the world and bureau-cracy of "art appreciation?" Would it not seem suddenly to be a conspiracy to make the artist a frill, a fribble, or a Milltown?[1] If men were able to be convinced that art is precise advance knowledge of how to cope with the psychic and social consequences of the next technology, would they all become artists? Or would they begin a careful translation of new art forms into social navigation charts? I am curious to know what would happen if art were suddenly seen for what it is, namely, exact information of how to rearrange one's psyche in order to anticipate the next blow from our own extended faculties. Would we, then, cease to look at works of art as an explorer might regard the gold and gems used as the ornaments of simple nonliterates?

At any rate, in experimental art, men are given the exact specifications of coming violence to their own psyches from their own counter-irritants or technology. For those parts of ourselves that we thrust out in the form of new invention are attempts to counter or neutralize collective pressures and irritations. But the counter-irritant usually proves a greater plague than the initial irritant, like a drug habit. And it is here that the artist can show us how to "ride with the punch," instead of "taking it on the chin." It can only be repeated that

---

1. This is a reference to the anti-anxiety agent meprobamate, marketed under various names, including Miltown (*sic*).

human history is a record of "taking it on the chin."

Emile Durkheim long ago expressed the idea that the specialized task always escaped the action of the social conscience. In this regard, it would appear that the artist is the social conscience and is treated accordingly! "We have no art," say the Balinese, "we do everything as well as possible."

The modern metropolis is now sprawling helplessly after the impact of the motorcar. As a response to the challenge of railway speeds the suburb and the garden city arrived too late, or just in time to become a motorcar disaster. For an arrangement of functions adjusted to one set of intensities becomes unbearable at another intensity. And a technological extension of our bodies designed to alleviate physical stress can bring on psychic stress that may be much worse. Western specialist technology transferred to the Arab world in late Roman times released a furious discharge of tribal energy.

The somewhat devious means of diagnosis that have to be used to pin down the actual form and impact of a new medium are not unlike those indicated in detective fiction by Peter Cheyney. In *You Can't Keep the Change* (Collins, London, 1956) he wrote:

> A case to Callaghan was merely a collection of people, some of whom — all of whom — were giving incorrect information, or telling lies, because circumstances either forced them or led them into the process.
>
> But the fact that they *had* to tell lies, *had* to give false impressions, necessitated a reorientation of their own viewpoints and their own lives. Sooner or later they became exhausted or careless. Then, and not until then, was an investigator able to put his finger on the one fact that would lead him to a possible logical solution.

It is interesting to note that success in keeping up a respectable front of the customary kind can only be done by a frantic scramble back of the façade. After the crime, after the blow has fallen, the façade of custom can only be held up by swift rearrangement of the props. So it is in our social lives when a new technology strikes, or in our private life when some intense and, therefore, indigestible experience occurs, and the censor acts at once to numb us from the blow and to ready the faculties to assimilate the intruder. Peter Cheyney's observations of a mode of detective fiction is another instance of a popular form of entertainment functioning as mimic model of the real thing.

Perhaps the most obvious "closure" or psychic consequence of any new technology is just the demand for it. Nobody wants a motorcar till there are motorcars, and nobody is interested in TV until there are TV programs. This power of technology to create its own world of demand is not independent of technology being first an extension of our own bodies and senses. When we are deprived of our sense of sight, the other senses take up the role of sight in some degree. But the need to use the senses that are available is as insistent as breathing—a fact that makes sense of the urge to keep radio and TV going more or less continuously. The urge to continuous use is quite independent of the "content" of public programs or of the private sense life, being testimony to the fact that technology is part of our bodies. Electric technology is directly related to our central nervous systems, so it is ridiculous to talk of "what the public wants" played over its own nerves. This question would be like asking people what sort of sights and sounds they would prefer around them in an urban metropolis! Once we have surrendered our senses and nervous systems to the private manipulation of those who would try to benefit from taking a lease on our eyes and ears and nerves, we don't really have any rights left. Leasing our eyes and ears and nerves to commercial interests is like handing over the common speech to a private corporation, or like giving the earth's atmosphere to a company as a monopoly. Something like this has already happened with outer space, for the same reasons that we have leased our central nervous systems to various corporations. As long as we adopt the Narcissus attitude of regarding the extensions of our own bodies as really *out there* and really independent of us, we will meet all technological challenges with the same sort of banana-skin pirouette and collapse.

Archimedes once said, "Give me a place to stand and I will move the world." Today he would have pointed to our electric media and said, "I will stand on your eyes, your ears, your nerves, and your brain, and the world will move in any tempo or pattern I choose." We have leased these "places to stand" to private corporations.

Arnold Toynbee has devoted much of his *A Study of History* to analyzing the kinds of challenge faced by a variety of cultures during many centuries. Highly relevant to Western man is Toynbee's explanation of how the lame and the crippled respond to their handicaps in a society of active warriors. They become specialists like Vulcan, the smith and armorer. And how do whole communities act when conquered and enslaved? The same strategy serves them as it does the lame individual in a society of warriors. They specialize and become indispensable to their masters. It is probably the long human history of enslavement, and the collapse into specialism as a counter-irritant, that have put the stigma of servitude and pusillanimity on the figure of the specialist, even in modern times. The capitulation of Western man to his technology, with its crescendo of specialized demands, has always appeared to many observers of our world as a kind of enslavement. But the resulting fragmentation has been voluntary and enthusiastic, unlike the conscious strategy of specialism on the part of the captives of military conquest.

It is plain that fragmentation or specialism as a technique of achieving security under tyranny and oppression of any kind has an attendant danger. Perfect adaptation to any environment is achieved by a total channeling of energies and vital force that amounts to a kind of static terminus for a creature. Even slight changes in the environment of the very well adjusted find them without any resource to meet new challenge. Such is the plight of the representatives of "conventional wisdom" in any society. Their entire stake of security and status is in a single form of acquired knowledge, so that innovation is for them not novelty but annihilation.

A related form of challenge that has always faced cultures is the simple fact of a frontier or a wall, on the other side of which exists another kind of society. Mere existence side by side of any two forms of organization generates a great deal of tension. Such, indeed, has been the principle of symbolist artistic structures in the past century. Toynbee observes that the challenge of a civilization set side by side with a tribal society has over and over demonstrated that the simpler society finds its integral economy and institutions "disintegrated by a rain of psychic energy generated by the civilization" of the more complex culture. When two societies exist side by side, the psychic challenge of the more complex one acts as an explosive release of energy in the simpler one. For prolific evidence of this kind of problem it is not necessary to look beyond the life of the teenager lived daily in the midst of a complex urban center. As the barbarian was driven to furious restlessness by the civilized contact, collapsing into mass migration, so the teenager, compelled to share the life of a city that cannot accept him as an adult, collapses into "rebellion without a cause." Earlier the adolescent had been provided with a rain check. He was prepared to wait it out. But since TV, the drive to participation has ended adolescence, and every American home has its Berlin wall.

Toynbee is very generous in providing examples of widely varied challenge and collapse, and is especially apt in pointing to the frequent and futile resort to futurism and archaism as strategies of encountering radical change. But to point back to the day of the horse or to look forward to the coming of anti-gravitational vehicles is not an adequate response to the challenge of the motorcar. Yet these two uniform ways of backward and forward looking are habitual ways of avoiding the discontinuities of present experience with their demand for sensitive inspection and appraisal. Only the dedicated artist seems to have the power for encountering the present actuality.

Toynbee urges again and again the cultural strategy of the imitation of the example of great men. This, of course, is to locate cultural safety in the power of the *will*, rather than in the power of adequate *perception* of situations. Anybody could quip that this is the British trust in character as opposed to intellect. In view of the endless power of men to hypnotize themselves into unawareness in the presence of challenge, it may be argued that willpower is as useful as intelligence for survival. Today we need also the will to be exceedingly informed and aware.

Arnold Toynbee gives an example of Renaissance technology being effectively encountered and creatively controlled when he shows how the revival of the decentralized medieval parliament saved English society from the monopoly of centralism that seized the continent. Lewis Mumford in *The City in History* tells the strange tale of how the New England town was able to carry out the pattern of the medieval ideal city because it was able to dispense with walls and to mix town and country. When the technology of a time is powerfully thrusting in one direction, wisdom may well call for a countervailing thrust. The implosion of electric energy in our century cannot be met by explosion or expansion, but it can be met by decentralism and the flexibility of multiple small centers. For example, the rush of students into our universities is not explosion but implosion. And the needful strategy to encounter this force is not to enlarge the university, but to create numerous groups of autonomous colleges in place of our centralized university plant that grew up on the lines of European government and nineteenth-century industry.

In the same way the excessive tactile effects of the TV image cannot be met by mere program changes. Imaginative strategy based on adequate diagnosis would prescribe a corresponding depth or structural approach to the existing literary and visual world. If we persist in a conventional approach to these developments our traditional culture will be swept aside as scholasticism was in the sixteenth century. Had the Schoolmen with their complex oral culture understood the Gutenberg technology, they could have created a new synthesis of written and oral education, instead of bowing out of the picture and allowing the merely visual page to take over the educational enterprise. The oral Schoolmen did not meet the new visual challenge of print, and the resulting expansion or explosion of Gutenberg technology was in many respects an impoverishment of the culture, as historians like Mumford are now beginning to explain. Arnold Toynbee, in *A Study of History*, in considering "the nature of growths of civilizations," not only abandons the concept of enlargement as a criterion of real growth of society, but states: "More often geographical expansion is a concomitant of real decline and coincides with a 'time of troubles' or a universal state—both of them stages of decline and disintegration."

Toynbee expounds the principle that times of trouble or rapid change produce militarism, and it is militarism that produces empire and expansion. The old Greek myth which taught that the alphabet produced militarism ("King Cadmus sowed the dragon's teeth, and they sprang up armed men") really goes much deeper than Toynbee's story. In fact, "militarism" is just vague description, not analysis of causality at all. Militarism is a kind of visual organization of social energies that is both specialist and explosive, so that it is merely repetitive to say, as Toynbee does, that it both creates large empires and causes social breakdown. But militarism is a form of industrialism or the concentration of large amounts of homogenized energies into a few kinds of production. The Roman soldier was a man with a spade. He was an expert workman and builder who processed and packaged the resources of many societies and sent them home. Before machinery, the only massive work forces available for processing material were soldiers or slaves. As the Greek myth of Cadmus points out, the phonetic alphabet was the greatest processor of men for homogenized military life that was known to antiquity. The age of Greek society that Herodotus acknowledges to have been "overwhelmed by more troubles than in the twenty preceding generations" was the time that to our literary retrospect appears as one of the greatest of human centuries. It was Macaulay who remarked that it was not pleasant to live in times about which it was exciting to read. The succeeding age of Alexander saw Hellenism expand into Asia and prepare the course of the later Roman expansion. These, however, were the very centuries in which Greek civilization obviously fell apart.

Toynbee points to the strange falsification of history by archeology, insofar as the survival of many material objects of the past does not indicate the quality of ordinary life and experience at any particular time. Continuous technical improvement in the means of warfare occurs over the entire period of Hellenic and Roman decline. Toynbee checks out his hypothesis by testing it with the developments in Greek agriculture. When the enterprise of Solon weaned the Greeks from mixed farming to a program of specialized products for export, there were happy consequences and a glorious manifestation of energy in Greek life. When the next phase of the same specialist stress involved much reliance on slave labor, there was spectacular increase of production. But the armies of technologically specialized slaves working the land blighted the social existence of the independent yeomen and small farmers, and led to the strange world of the Roman towns and cities crowded with rootless parasites.

To a much greater degree than Roman slavery, the specialism of mechanized industry and market organization has faced Western man with the challenge of manufacture by mono-fracture, or the tackling of all things and operations one-bit-at-a-time. This is the challenge that has permeated all aspects of our lives and enabled us to expand so triumphantly in all directions and in all spheres.

Part II

Part II

8

The Spoken Word:
Flower of Evil?

The subtitle is inspired by Les fleurs du mal, *the work of nineteenth-century poet Charles Baudelaire, father of French symbolism, mentioned earlier in the company of Joyce and Eliot. Baudelaire must wait until Chapter 11, Number, for a discussion of his intuitive grasp of media and their effects. Here it is his countryman, philosopher Henri Bergson, whose views on language are discussed. And it is not till Chapter 13, Housing, that McLuhan explains Baudelaire's phrase: "Our letting-go of ourselves, self-alienations, as it were, in order to amplify or increase the power of various functions, Baudelaire considered to be flowers of growths of evil." (p. [169]) Perhaps as a concession to readers disconcerted by seven chapters in which McLuhan relentlessly rubs ideas together in unexpected combinations, he states plainly the rationale for considering both language and the wheel under the heading of media: "Language does for intelligence what the wheel does for the feet and the body. It enables them to move from thing to thing with greater ease and speed and ever less involvement."*

— *(Editor)*

A few seconds from a popular disk-jockey show were typed out as follows:

> That's Patty Baby and that's the girl with the dancing feet and that's Freddy Cannon there on the David Mickie Show in the night time ooohbah scuba-doo how are you booboo. Next we'll be Swinging on a Star and ssshhhwwoooo and sliding on a moonbeam.
>
> Waaaaaaa how about that … one of the goodest guys with you … this is lovable kissable D.M. in the P.M. at 22 minutes past nine o'clock there, aahrightie, we're gonna have a Hitline, all you have to do is call WAlnut 5-1151, WAlnut 5-1151, tell them what number it is on the Hitline.

Dave Mickie alternately soars, groans, swings, sings, solos, intones, and scampers, always reacting to his own actions. He moves entirely in the spoken rather than the written area of experience. It is in this way that audience participation is created. The spoken word involves all of the senses dramatically, though highly literate people tend to speak as connectedly and casually as possible. The sensuous involvement natural to cultures in which literacy is not the ruling form of experience is sometimes indicated in travel guides, as in this item from a guide to Greece:

> You will notice that many Greek men seem to spend a lot of time counting the beads of what appear to be amber rosaries. But these have no religious significance. They are *komboloia* or "worry beads," a legacy from the Turks, and Greeks click them on land, on the sea, in the air to ward off that insupportable silence which threatens to reign whenever conversation lags. Shepherds do it, cops do it, stevedores and merchants in their shops do it. And if you wonder why so few Greek women wear

beads, you'll know it's because their husbands have pre-empted them for the simple pleasure of clicking. More aesthetic than thumb-twiddling, less expensive than smoking, this Queeg-like obsession indicates a tactile sensuousness characteristic of a race which has produced the western world's greatest sculpture ...

Where the heavy visual stress of literacy is lacking in a culture, there occurs another form of sensuous involvement and cultural appreciation that our Greek guide explains whimsically:

... do not be surprised at the frequency with which you are patted, petted and prodded in Greece. You may end up feeling like the family dog ... in an affectionate family. This propensity to pat seems to us a tactile extension of the avid Greek curiosity noted before. It's as though your hosts are trying to find out what you are made of.

The widely separate characters of the spoken and written words are easy to study today when there is ever closer touch with nonliterate societies. One native, the only literate member of his group, told of acting as reader for the others when they received letters. He said he felt impelled to put his fingers to his ears while reading aloud, so as not to violate the privacy of their letters. This is interesting testimony to the values of privacy fostered by the visual stress of phonetic writing. Such separation of the senses, and of the individual from the group, can scarcely occur without the influence of phonetic writing. The spoken word does not afford the extension and amplification of the visual power needed for habits of individualism and privacy.

It helps to appreciate the nature of the spoken word to contrast it with the written form. Although phonetic writing separates and extends the visual power of words, it is comparatively crude and slow. There are not many ways of writing "tonight," but Stanislavsky used to ask his young actors to pronounce and stress it fifty different ways while the audience wrote down the different shades of feeling and meaning expressed. Many a page of prose and many a narrative has been devoted to expressing what was, in effect, a sob, a moan, a laugh, or a piercing scream. The written word spells out in sequence what is quick and implicit in the spoken word.

Again, in speech we tend to react to each situation that occurs, reacting in tone and gesture even to our own act of speaking. But writing tends to be a kind of separate or specialist action in which there is little opportunity or call for reaction. The literate man or society develops the tremendous power of acting in any matter with considerable detachment from the feelings or emotional involvement that a nonliterate man or society would experience.

Henri Bergson, the French philosopher, lived and wrote in a tradition of thought in which it was and is considered that language is a human technology that has impaired and diminished the values of the collective unconscious. It is the extension of man in speech that enables the intellect to detach itself from the vastly wider reality. Without language, Bergson suggests, human intelligence would have remained totally involved in the objects of its attention. Language does for intelligence what the wheel does for the feet and the body. It enables them to move from thing to thing with greater ease and speed and ever less involvement. Language extends and amplifies man but it also divides his faculties. His collective consciousness or intuitive awareness is diminished by this technical extension of consciousness that is speech.

Bergson argues in *Creative Evolution* that even consciousness is an extension of man that dims the bliss of union in the collective unconscious. Speech acts to separate man from man, and mankind from the cosmic unconscious. As an extension or uttering (outering) of all our senses at once, language has always been held to be man's richest art form, that which distinguishes him from the animal creation.

If the human ear can be compared to a radio receiver that is able to decode electromagnetic waves and recode them as sound, the human voice may be compared to the radio transmitter in being able to translate sound into electromagnetic waves. The power of the voice to shape air and space into verbal patterns may well have been preceded by a less specialized expression of cries, grunts, gestures, and commands, of song and dance. The patterns of the senses that are extended in the various languages of men are as varied as styles of dress and art. Each mother tongue teaches its users a way of seeing and feeling the world, and of acting in the world, that is quite unique.

Our new electric technology that extends our senses and nerves in a global embrace has large implications for the future of language. Electric technology does not need words any more than the digital computer needs numbers. Electricity points the way to an extension of the process of consciousness itself, on a world scale, and without any verbalization whatever. Such a state of collective awareness may have been the preverbal condition of men. Language as the technology of human extension, whose powers of division and separation we know so well, may have been the "Tower of Babel" by which men sought to scale the highest heavens. Today computers hold out the promise of a means of instant translation of any code or language into any other code or language. The computer, in short, promises by technology a Pentecostal condition of universal understanding and unity. The next logical step would seem to be, not to translate, but to by-pass languages in favor of a general cosmic consciousness which might be very like the collective unconscious dreamt of by Bergson. The condition of "weightlessness" that biologists say promises a physical immortality, may be paralleled by the condition of speechlessness that could confer a perpetuity of collective harmony and peace.

9

The Written Word:
An Eye for an Ear

The development of the phonetic alphabet opens the floodgates for the effects of this powerful medium to inundate Western civilization for centuries. McLuhan's account of the phonetic alphabet as a unique technology emphasizes that it was achieved at the expense of "worlds of meaning and perception" which inhere in non-alphabetic writing systems such as hieroglyphs and ideograms. He pinpoints a paradox in that the latter are discontinuous (not systematically related to each other) but integrated (the shape of an ideogram carries meaning and so does the way it is said aloud), whereas the alphabet is a uniform system without being integrated, because neither its letters nor the way they are pronounced has any meaning in itself. This is what McLuhan means by the alphabet's "stark division and parallelism between a visual and an auditory world." He is very explicit here about the sense in which the medium is the message: "It is in its power to extend patterns of visual uniformity that the 'message' of the alphabet is felt by cultures." This observation leads him to follow up the present chapter not with an account of the next major step in the evolution of the alphabet—this will come in Chapter 18, The Printed Word—but by showing the effects and consequences of the alphabet technology on the social, cultural, political, economic, and military organization of the ancient world.

— (Editor)

Prince Modupe[1] wrote of his encounter with the written word in his West African days:

> The one crowded space in Father Perry's house was his bookshelves. I gradually came to understand that the marks on the pages were *trapped words*. Anyone could learn to decipher the symbols and turn the trapped words loose again into speech. The ink of the print trapped the thoughts; they could no more get away than a *doomboo* could get out of a pit. When the full realization of what this meant flooded over me, I experienced the same thrill and amazement as when I had my first glimpse of the bright lights of Konakry. I shivered with the intensity of my desire to learn to do this wondrous thing myself.

In striking contrast to the native's eagerness, there are the current anxieties of civilized man concerning the written word. To some Westerners the written or printed word has become a very touchy subject. It is true that there is more material written and printed and read today than ever before, but there is also a new electric technology that threatens this ancient technology of literacy built on the phonetic alphabet. Because of its action in extending our central nervous system, electric technology seems to favor the inclusive and participational spoken word over the specialist written word. Our Western values, built on the written word, have already been considerably affected by the electric media of telephone, radio, and TV. Perhaps that is the reason why many highly literate people in our time find it difficult to examine this question without getting into a moral panic. There is the

1. The title of the cited work is given in Chapter 16, p. 215.

further circumstance that, during his more than two thousand years of literacy, Western man has done little to study or to understand the effects of the phonetic alphabet in creating many of his basic patterns of culture. To begin now to examine the question may, therefore, seem too late.

Suppose that, instead of displaying the Stars and Stripes, we were to write the words "American flag" across a piece of cloth and to display that. While the symbols would convey the same meaning, the effect would be quite different. To translate the rich visual mosaic of the Stars and Stripes into written form would be to deprive it of most of its qualities of corporate image and of experience, yet the abstract literal bond would remain much the same. Perhaps this illustration will serve to suggest the change the tribal man experiences when he becomes literate. Nearly all the emotional and corporate family feeling is eliminated from his relationship with his social group. He is emotionally free to separate from the tribe and to become a civilized individual, a man of visual organization who has uniform attitudes, habits, and rights with all other civilized individuals.

The Greek myth about the alphabet was that Cadmus, reputedly the king who introduced the phonetic letters into Greece, sowed the dragon's teeth, and they sprang up armed men. Like any other myth, this one capsulates a prolonged process into a flashing insight. The alphabet meant power and authority and control of military structures at a distance. When combined with papyrus, the alphabet spelled the end of the stationary temple bureaucracies and the priestly monopolies of knowledge and power. Unlike pre-alphabetic writing, which with its innumerable signs was difficult to master, the alphabet could be learned in a few hours. The acquisition of so extensive a knowledge and so complex a skill as pre-alphabetic writing represented, when applied to such unwieldy materials as brick and stone, insured for the scribal caste a monopoly of priestly power. The easier alphabet and the light, cheap, transportable papyrus together effected the transfer of power from the priestly to the military class. All this is implied in the myth about Cadmus and the dragon's teeth, including the fall of the city states, the rise of empires and military bureaucracies.

In terms of the extensions of man, the theme of the dragon's teeth in the Cadmus myth is of the utmost importance. Elias Canetti in *Crowds and Power* reminds us that the teeth are an obvious agent of power in man, and especially in many animals. Languages are filled with testimony to the grasping, devouring power and precision of teeth. That the power of letters as agents of aggressive order and precision should be expressed as extensions of the dragon's teeth is natural and fitting. Teeth are emphatically visual in their lineal order. Letters are not only like teeth visually, but their power to put teeth into the business of empire-building is manifest in our Western history.

The phonetic alphabet is a unique technology. There have been many kinds of writing, pictographic and syllabic, but there is only one phonetic alphabet in which semantically meaningless letters are used to correspond to semantically meaningless sounds. This stark division and parallelism between a visual and an auditory world was both crude and ruthless, culturally speaking. The phonetically written word sacrifices worlds of meaning and perception that were secured by forms like the hieroglyph and the Chinese ideogram. These culturally richer forms of writing, however, offered men no means of sudden transfer from the magically discontinuous and traditional world of the tribal word into the hot and uniform visual medium. Many centuries of ideogrammic use have not threatened the seamless web of family and tribal subtleties of Chinese society. On the other hand, a single generation of alphabetic literacy suffices in Africa today, as in Gaul two thousand years ago, to release the individual initially, at least, from the tribal web. This fact has nothing to do with the *content* of the alphabetized words; it is the result of the sudden breach between the auditory and the visual experience of man. Only the phonetic alphabet makes such a sharp division in experience, giving to its user an eye for an ear, and freeing him from the tribal trance of resonating word magic and the web of kinship.

It can be argued, then, that the phonetic alphabet, alone, is the technology that has been the means of creating "civilized man" — the separate individuals equal before a written code of law. Separateness of the individual, continuity of space and of time, and uniformity of codes are the prime marks of literate and civilized societies. Tribal cultures like those of the Indian and the Chinese may be greatly superior to the Western cultures, in the range and delicacy of their perceptions and expression. However, we are not here concerned with the question of values, but with the configurations of societies. Tribal cultures cannot entertain the possibility of the individual or of the separate citizen. Their ideas of spaces and times are neither continuous nor uniform, but compassional and compressional in their intensity. It is in its power to extend patterns of visual uniformity and continuity that the "message" of the alphabet is felt by cultures.

As an intensification and extension of the visual function, the phonetic alphabet diminishes the role of the other senses of sound and touch and taste in any literate culture. The fact that this does not happen in cultures such as the Chinese, which use nonphonetic scripts, enables them to retain a rich store of inclusive perception in depth of experience that tends to become eroded in civilized cultures of the phonetic alphabet. For the ideogram is an inclusive *gestalt*, not an analytic dissociation of senses and functions like phonetic writing.

The achievements of the Western world, it is obvious, are testimony to the tremendous values of literacy. But many people are also disposed to object that we have purchased our structure of specialist technology and values at too high a price. Certainly the lineal structuring of rational life by phonetic literacy has involved us in an interlocking set of consistencies that are striking enough to justify a much more extensive inquiry than that of the present chapter. Perhaps there are better approaches along quite different lines; for example, consciousness is regarded as the mark of a rational being, yet there is nothing lineal or sequential about the total field of awareness that exists in any moment of consciousness. Consciousness is not a verbal process. Yet during all our centuries of phonetic literacy we have favored the chain of inference as the mark of logic and reason. Chinese writing, in contrast, invests each ideogram with a total intuition of being and reason that allows only a small role to visual sequence as a mark of mental effort and organization. In Western literate society it is still plausible and acceptable to say that something "follows" from something, as if there were some cause at work that makes such a sequence. It was David Hume who, in the eighteenth century, demonstrated that there is no causality indicated in any sequence, natural or logical. The sequential is merely additive, not causative. Hume's argument, said Immanuel Kant, "awoke me from my dogmatic slumber." Neither Hume nor Kant, however, detected the hidden cause of our Western bias toward sequence as "logic" in the all-pervasive technology of the alphabet. Today in the electric age we feel as free to invent nonlineal logics as we do to make non-Euclidean geometries. Even the assembly line, as the method of analytic sequence for mechanizing every kind of making and production, is nowadays yielding to new forms.

Only alphabetic cultures have ever mastered connected lineal sequences as pervasive forms of psychic and social organization. The breaking up of every kind of experience into uniform units in order to produce faster action and change of form (applied knowledge) has been the secret of Western power over man and nature alike. That is the reason why our Western industrial programs have quite involuntarily been so militant, and our military programs have been so industrial. Both are shaped by the alphabet in their technique of trans-formation and control by making all situations uniform and continuous. This procedure, manifest even in the Graeco-Roman phase, became more intense with the uniformity and repeata-bility of the Gutenberg development.

Civilization is built on literacy because literacy is a uni-form processing of a culture by a visual sense extended in space and time by the alphabet. In tribal cultures, experience is arranged by a dominant auditory sense-life that represses visual values. The auditory sense, unlike the cool and neu-tral eye, is hyper-esthetic and delicate and all-inclusive. Oral cultures act and react at the same time. Phonetic culture endows men with the means of repressing their feelings and emotions when engaged in action. To act without reacting, without involvement, is the peculiar advantage of Western literate man.

The story of *The Ugly American* describes the endless succession of blunders achieved by visual and civilized Americans when confronted with the tribal and auditory cultures of the East. As a civilized UNESCO experiment, running water — with its lineal organization of pipes — was installed recently in some Indian villages. Soon the villagers requested that the pipes be removed, for it seemed to them that the whole social life of the village had been impover-ished when it was no longer necessary for all to visit the communal well. To us the pipe is a convenience. We do not think of it as culture or as a product of literacy, any more than we think of literacy as changing our habits, our emo-tions, or our perceptions. To nonliterate people, it is perfectly obvious that the most commonplace conveniences represent total changes in culture.

The Russians, less permeated with the patterns of literate

culture than Americans, have much less difficulty in perceiving and accommodating the Asiatic attitudes. For the West, literacy has long been pipes and taps and streets and assembly lines and inventories. Perhaps most potent of all as an expression of literacy is our system of uniform pricing that penetrates distant markets and speeds the turnover of commodities. Even our ideas of cause and effect in the literate West have long been in the form of things in sequence and succession, an idea that strikes any tribal or auditory culture as quite ridiculous, and one that has lost its prime place in our own new physics and biology.

All the alphabets is use in the Western world, from that of Russia to that of the Basques, from that of Portugal to that of Peru, are derivatives of the Graeco-Roman letters. Their unique separation of sight and sound from semantic and verbal content made them a most radical technology for the translation and homogenization of cultures. All other forms of writing had served merely one culture, and had served to separate that culture from others. The phonetic letters alone could be used to translate, albeit crudely, the sounds of any language into one-and-the-same visual code. Today, the effort of the Chinese to use our phonetic letters to translate their language has run into special problems in the wide tonal variations and meanings of similar sounds. This has led to the practice of fragmenting Chinese monosyllables into polysyllables in order to eliminate tonal ambiguity. The Western phonetic alphabet is now at work transforming the central auditory features of the Chinese language and culture in order that China can also develop the lineal and visual patterns that give central unity and aggregate uniform power to Western work and organization. As we move out of the Gutenberg era of our own culture, we can more readily discern its primary features of homogeneity, uniformity, and continuity. These were the characteristics that gave the Greeks and Romans their easy ascendancy over the nonliterate barbarians. The barbarian or tribal man, then as now, was hampered by cultural pluralism, uniqueness, and discontinuity.

To sum up, pictographic and hieroglyphic writing as used in Babylonian, Mayan, and Chinese cultures represents an extension of the visual sense for storing and expediting access to human experience. All of these forms give pictorial expression to oral meanings. As such, they approximate the animated cartoon and are extremely unwieldy, requiring many signs for the infinity of data and operations of social action. In contrast, the phonetic alphabet, by a few letters only, was able to encompass all languages. Such an achievement, however, involved the separation of both signs and sounds from their semantic and dramatic meanings. No other system of writing had accomplished this feat.

The same separation of sight and sound and meaning that is peculiar to the phonetic alphabet also extends to its social and psychological effects. Literate man undergoes much separation of his imaginative, emotional, and sense life, as Rousseau (and later the Romantic poets and philosophers) proclaimed long ago. Today the mere mention of D. H. Lawrence will serve to recall the twentieth-century efforts made to by-pass literate man in order to recover human "wholeness." If Western literate man undergoes much dissociation of inner sensibility from his use of the alphabet, he also wins his personal freedom to dissociate himself from clan and family. This freedom to shape an individual career manifested itself in the ancient world in military life. Careers were open to talents in Republican Rome, as much as in Napoleonic France, and for the same reasons. The new literacy had created an homogeneous and malleable milieu in which the mobility of armed groups and of ambitious individuals, equally, was as novel as it was practical.

10

Roads and Paper Routes

The shadow of Canadian political economist and pioneer of com-munication studies, Harold Innis, looms large on the opening page of this chapter. At the same time, there is a crucial link to both Innis's ideas and those developed in the earlier chapters: "Each form of transport not only carries, but translates and transforms the sender, the receiver, and the message." The global village and the tribal village are set in a four-term comparison: "The speedup of the electronic age is as disruptive for literate, lineal and Western man as the Roman paper routes were for tribal villagers." The paradox of media effects is linked to the topic of center-margin dynamics (itself extensively developed in the following pages): "Paradoxically, the effect of the wheel and of paper in organizing new power structures was not to decentralize but to centralize." And the profound coherence of McLuhan's percepts emerges clearly, as he presents the favorable conditions for the development of center-margin dynamics in terms of village life as an intermediary stage in social organiza-tion between that of food-gathering hunters and the formation of cities and city-states, for this is yet another instance of the medium being the message.

— (Editor)

It was not until the advent of the telegraph that messages could travel faster than a messenger. Before this, roads and the written word were closely interrelated. It is only since the telegraph that information has detached itself from such solid commodities as stone and papyrus, much as money had earlier detached itself from hides, bullion, and metals, and has ended as paper. The term "communication" has had an extensive use in connection with roads and bridges, sea routes, rivers, and canals, even before it became transformed into "information movement" in the electric age. Perhaps there is no more suitable way of defining the character of the electric age than by first studying the rise of the idea of transportation as communication, and then the transition of the idea from transport to information by means of electricity. The word "metaphor" is from the Greek *meta* plus *pherein*, to carry across or transport. In this book, we are concerned with all forms of transport of goods and information, both as metaphor and exchange. Each form of transport not only carries, but translates and transforms the sender, the receiver, and the message. The use of any kind of medium or extension of man alters the patterns of interdependence among people, as it alters the ratios among our senses.

It is a persistent theme of this book that all technologies are extensions of our physical and nervous systems to increase power and speed. Again, unless there were such increases of power and speed, new extensions of ourselves would not occur or would be discarded. For an increase of power or speed in any kind of grouping of any components whatever is itself a disruption that causes a change of organization. The alteration of social groupings, and the formation of new communities, occur with the increased speed of information movement by means of paper messages and road transport. Such speedup means much more control at much greater distances. Historically, it meant the formation of the Roman Empire and the disruption of the previous city-states of the Greek world. Before the use of papyrus and alphabet created the incentives for building fast, hard-surface roads, the walled town and the city-state were natural forms that could endure.

Village and city-state essentially are forms that include all human needs and functions. With greater speed and, therefore, greater military control at a distance, the city-state collapsed. Once inclusive and self-contained, its needs and functions were extended in the specialist activities of an empire. Speedup tends to separate functions, both commercial and political, and acceleration beyond a point in any system becomes disruption and breakdown. So when Arnold Toynbee turns, in *A Study of History*, to a massive documentation of "the breakdowns of civilizations," he begins by saying: "One of the most conspicuous marks of disintegration, as we have already noticed, is … when a disintegrating civilisation purchases a reprieve by submitting to forcible political unification in a universal state." Disintegration and reprieve, alike, are the consequence of ever faster movement of information by couriers on excellent roads.

Speedup creates what some economists refer to as a *center-margin* structure. When this becomes too extensive for the generating and control center, pieces begin to detach themselves and to set up new center-margin systems of their own. The most familiar example is the story of the American colonies of Great Britain. When the thirteen colonies began to develop a considerable social and economic life of their own, they felt the need to become centers themselves, with their own margins. This is the time when the original center may make a more rigorous effort of centralized control of the margins, as, indeed, Great Britain did. The slowness of sea travel proved altogether inadequate to the maintenance of so extensive an empire on a mere center-margin basis. Land powers can more easily attain a unified center-margin pattern than sea powers. It is the relative slowness of sea travel that inspires sea powers to foster multiple centers by a kind of seeding process. Sea powers thus tend to create centers without margins, and land empires favor the center-margin structure. Electric speeds create centers everywhere. Margins cease to exist on this planet.

Lack of homogeneity in speed of information movement creates diversity of patterns in organization. It is quite predictable, then, that any new means of moving information will alter any power structure whatever. So long as the new means is everywhere available at the same time, there is a possibility that the structure may be changed without breakdown. Where there are great discrepancies in speeds of movement, as between air and road travel or between telephone and typewriter, serious conflicts occur within organizations. The metropolis of our time has become a test case for such discrepancies. If homogeneity of speeds were total, there would be no rebellion and no breakdown. With print, political unity via homogeneity became feasible for the first time. In ancient Rome, however, there was only the light paper manuscript to pierce the opacity, or to reduce the discontinuity, of the tribal villages; and when the paper supplies failed, the roads were vacated, as they were in our own age during gas-rationing. Thus the old city-state returned, and feudalism replaced republicanism.

It seems obvious enough that technical means of speedup should wipe out the independence of villages and city-states. Whenever speedup has occurred, the new centralist power always takes action to homogenize as many marginal areas as possible. The process that Rome effected by the phonetic alphabet geared to its paper routes has been occurring in Russia for the last century. Again, from the current example of Africa we can observe how very much visual processing of the human psyche by alphabetic means will be needed before any appreciable degree of homogenized social organization is possible. Much of this visual processing was done in the ancient world by nonliterate technologies, as in Assyria. The phonetic alphabet has no rival, however, as a translator of man out of the closed tribal echo-chamber into the neutral visual world of lineal organization.

The situation of Africa today is complicated by the new electronic technology. Western man is himself being de-Westernized by his own new speedup, as much as the Africans are being detribalized by our old print and industrial technology. If we understood our own media old and new, these confusions and disruptions could be programmed and synchronized. The very success we enjoy in specializing and separating functions in order to have speedup, however, is at the same time the cause of inattention and unawareness of the situation. It has ever been thus in the Western world at least. Self-consciousness of the causes and limits of one's own culture seems to threaten the ego structure and is, therefore, avoided. Nietzsche said understanding stops action, and men of action seem to have an intuition of the fact in their shunning the dangers of comprehension.

The point of the matter of speedup by wheel, road and paper is the extension of power in an ever more homogenous and uniform space. This the real potential of the Roman technology was not realized until printing had given road and wheel a much greater speed than that of the Roman vortex. Yet the speedup of the electronic age is as disrupting for literate, lineal, and Western man as the Roman paper routes were for tribal villagers. Our speedup today is not a slow explosion outward from center to margins but an instant implosion and an interfusion of space and functions. Our specialist and fragmented civilization of center-margin structure is suddenly experiencing an instantaneous reassembling of all its mechanized bits into an organic whole. This is the new world of the global village. The village, as Mumford explains in *The City in History*, had achieved a social and institutional extension of all human faculties. Speedup and city aggregates only served to separate these from one another in more specialist forms. The electronic age cannot sustain the very low gear of a center-margin structure such as we associate with the past two thousand years of the Western world. Nor is this a question of values. If we understood our older media, such as roads and the written word, and if we valued their human effects sufficiently, we could reduce or even eliminate the electronic factor from our lives. Is there an instance of any culture that understood the technology that sustained its structure and was prepared to keep it that way? If so, that would be an instance of values or reasoned preference. The values or preferences that arise from the mere automatic operation of this or that technology in our social lives are not capable of being perpetuated.

In the chapter on the wheel it will be shown that transport without wheels had played a big role before the wheel, some of which was by sledge, over both snow and bogs. Much of it was by pack animal — woman being the first pack animal. Most wheelless transport in the past, however, was by river and by sea, a fact that is today as richly expressed as ever in the location and form of the great cities of the world. Some writers have observed that man's oldest beast of burden was woman, because the male had to be free to run interference for the woman, as ball-carrier, as it were. But that phase belonged to the pre-wheel stage of transport, when there was only the tractless waste of man the hunter and food-gatherer. Today, when the greatest volume of transport consists in the moving of information, the wheel and the road are undergoing recession and obsolescence; but in the first instance, given the pressure for, and from, wheels, there had to be roads to accommodate them. Settlements had created the impulse for exchange and for the increasing movement of raw material and produce from countryside to processing centers, where there was division of labor and specialist craft skills. Improvement of wheel and road more and more brought the town to the country in a reciprocal spongelike action of give-and-take. It is a process we have seen in this century with the motorcar. Great improvements in roads brought the city more and more to the country. The road became a substitute for the country by the time people began to talk about "taking a spin in the country." With superhighways the road became a wall between man and the country. Then came the stage of the highway as city, a city stretching continuously across the continent, dissolving all earlier cities into the sprawling aggregates that desolate their populations today.

With air transport comes a further disruption of the old town-country complex that had occurred with wheel and road. With the plane the cities began to have the same slender relation to human needs that museums do. They became corridors of showcases echoing the departing forms of industrial assembly lines. The road is, then, used less and less for travel, and more and more for recreation. The traveler now turns to the airways, and thereby ceases to experience the act of traveling. As people used to say that an ocean liner might as well be a hotel in a big city, the jet traveler, whether he is over Tokyo or New York, might just as well be in a cocktail lounge so far as travel experience is concerned. He will begin to travel only after he lands.

Meantime, the countryside, as oriented and fashioned by plane, by highway, and by electric information-gathering, tends to become once more the nomadic trackless area that preceded the wheel. The beatniks gather on the sands to meditate *haiku*.

The principal factors in media impact on existing social forms are acceleration and disruption. Today the acceleration tends to be total, and thus ends space as the main factor in social arrangements. Toynbee sees the acceleration factor as translating the physical into moral problems, pointing to the antique road crowded with dog carts, wagons, and rickshaws as full of minor nuisance, but also minor dangers. Further, as the forces impelling traffic mount in power, there is no more problem of hauling and carrying, but the physical problem is translated into a psychological one as the annihilation of space permits easy annihilation of travelers as well. This principle applies to all media study. All means of interchange and of human interassociation tend to improve by acceleration. Speed, in turn, accentuates problems of form and structure. The older arrangements had not been made with a view to such speeds, and people begin to sense a draining-away of life values as they try to make the old physical forms adjust to the new and speedier movement. These problems, however, are not new. Julius Caesar's first act upon assuming power was to restrict the night movement of wheeled vehicles in the city of Rome in order to permit sleep. Improved transport in the Renaissance turned the medieval walled towns into slums.

Prior to the considerable diffusion of power through alphabet and papyrus, even the attempts of kings to extend their rule in spatial terms were opposed at home by the priestly bureaucracies. Their complex and unwieldy media of stone inscription made wide-ranging empires appear very dangerous to such static monopolies. The struggles between those who exercised power over the hearts of men and those who sought to control the physical resources of nations were not of one time and place. In the Old Testament, just this kind of struggle is reported in the Book of Samuel (I, viii) when the children of Israel besought Samuel to give them a king. Samuel explained to them the nature of kingly, as opposed to priestly, rule:

> This will be the manner of the King that shall reign over you: he will take your sons, and appoint them unto him for his chariots; and they shall run before

133

his chariots: and he will appoint them unto him for captains of thousands, and captains of fifties; and he will set some to plough his ground, and to reap his harvest, and to make his instruments of war, and the instruments of his chariots. And he will take your daughters to be confectionaries, and to be cooks and to be bakers. And he will take your fields, and your vineyards, and your oliveyards, even the best of them, and give them to his servants.

Paradoxically, the effect of the wheel and of paper in organizing new power structures was not to decentralize but to centralize. A speedup in communications always enables a central authority to extend its operations to more distant margins. The introduction of alphabet and papyrus meant that many more people had to be trained as scribes and administrators. However, the resulting extension of homogenization and of uniform training did not come into play in the ancient or medieval world to any great degree. It was not really until the mechanization of writing in the Renaissance that intensely unified and centralized power was possible. Since this process is still occurring, it should be easy for us to see that it was in the armies of Egypt and Rome that a kind of democratization by uniform technological education occurred. Careers were then open to talents for those with literate training. In the chapter on the written word we saw how phonetic writing translated tribal man into a visual world and invited him to undertake the visual organization of space. The priestly groups in the temples had been more concerned with the records of the past and with the control of the inner space of the unseen than with outward military conquest. Hence, there was a clash between the priestly monopolizers of knowledge and those who wished to apply it abroad as new conquest and power. (This same clash now recurs between the university and the business world.) It was this kind of rivalry that inspired Ptolemy II to establish the great library at Alexandria as a center of imperial power. The huge staff of civil servants and scribes assigned to many specialist tasks was an antithetic and countervailing force to the Egyptian priesthood. The library could serve the political organization of empire in a way that did not interest the priesthood at all. A not-dissimilar rivalry is developing today between the atomic scientists and those who are mainly concerned with power.

If we realize that the city as center was in the first instance an aggregate of threatened villagers, it is then easier for us to grasp how such harassed companies of refugees might fan out into an empire. The city-state as a form was not a response to peaceful commercial development, but a huddling for security amidst anarchy and dissolution. Thus the Greek city-state was a tribal form of inclusive and integral community, quite unlike the specialist cities that grew up as extensions of Roman military expansion. The Greek city-states eventually disintegrated by the usual action of specialist trading and the separation of functions that Mumford portrays in *The City in History*. The Roman cities began that way—as specialist operations of the central power. The Greek cities ended that way.

If a city undertakes rural trade, it sets up at once a center-margin relation with the rural area in question. That relation involves taking staples and raw produce from the country in exchange for specialist products of the craftsman. If, on the other hand, the same city attempts to engage in overseas trade, it is more natural to "seed" another city center, as the Greeks did, rather than to deal with the overseas area as a specialized margin or raw material supply.

A brief review of the structural changes in the organization of space as they resulted from wheel, road, and papyrus could go as follows: There was first the village, which lacked all of these group extensions of the private physical body.

The village, however, was already a form or community different from that of food-gathering hunters and fishers, for villagers may be sedentary and may begin a division of labor and functions. Their being congregated is, itself, a form of acceleration of human activities which provides momentum for further separation and specialization of action. Such are the conditions for the extension of feet-as-wheel to speed production and exchange. These are, also, the conditions that intensify communal conflicts and ruptures that send men huddling into ever larger aggregates, in order to resist the accelerated activities of other communities. The villages are swept up into the city-state by way of resistance and for the purpose of security and protection.

The village had institutionalized all human functions in forms of low intensity. In this mild form everyone could play many roles. Participation was high, and organization was low. This is the formula for stability in any type of organization. Nevertheless, the enlargement of village forms in the city-state called for greater intensity and the inevitable separation of functions to cope with this intensity and competition. The villagers had all participated in the seasonal rituals that in the city became the specialized Greek drama. Mumford feels that "The village measure prevailed in the development of the Greek cities, down to the fourth century ..." (*The City in History*). It is this extension and translation of the human organs into the village model without loss of corporal unity that Mumford uses as a criterion of excellence for city forms in any time or locale. This biological approach to the man-made environment is sought today once more in the electric age. How strange that the idea of the "human scale" should have seemed quite without appeal during the mechanical centuries.

The natural tendency of the enlarged community of the city is to increase the intensity and accelerate functions of every sort, whether of speech, or crafts, or currency and exchange. This, in turn, implies an inevitable extension of these actions by subdivision or, what is the same thing, new invention. So that even though the city was formed as a kind of protective hide or shield for man, this protective layer was purchased at the cost of maximized struggle within the walls. War games such as those described by Herodotus began as ritual blood baths between the citizenry. Rostrum, law courts, and marketplace all acquired the intense image of divisive competition that is nowadays called "the rat race." Nevertheless, it was amidst such irritations that man produced his greatest inventions as counter-irritants. These inventions were extensions of himself by means of concentrated toil, by which he hoped to neutralize distress. The Greek word *ponos*, or "toil," was a term used by Hippocrates, the father of medicine, to describe the fight of the body in disease. Today this idea is called *homeostasis*, or equilibrium as a strategy of the staying power of any body. All organizations, but especially biological ones, struggle to remain constant in their inner condition amidst the variations of outer shock and change. The man-made social environment as an extension of man's physical body is no exception. The city, as a form of the body politic, responds to new pressures and irritations by resourceful new extensions—always in the effort to exert staying power, constancy, equilibrium, and *homeostasis*.

The city, having been formed for protection, unexpectedly generated fierce intensities and new hybrid energies from accelerated interplay of functions and knowledge. It burst forth into aggression. The alarm of the village, followed by the resistance of the city, expanded into the exhaustion and inertia of empire. These three stages of the disease and irritation syndrome were felt, by those living through them, as normal physical expressions of counter-irritant recovery from disease.

The third stage of struggle for equilibrium among the forces within the city took the form of empire, or a universal state, that generated the extension of human senses in wheel, road, and alphabet. We can sympathize with those who first saw in these tools a providential means of bringing order to distant areas of turbulence and anarchy. These tools would have seemed a glorious form of "foreign aid," extending the blessings of the center to the barbarian margins. At this moment, for example, we are quite in the dark about the political implications of Telstar. By outering these satellites as extensions of our nervous system, there is an automatic response in all the organs of the body politic of mankind. Such new intensity of proximity imposed by Telstar calls for radical rearrangement of all organs in order to maintain staying power and equilibrium. The teaching and learning process for every child will be affected sooner rather than later. The time factor in every decision of business and finance will acquire new patterns. Among the peoples of the world strange new vortices of power will appear unexpectedly.

The full-blown city coincides with the development of writing — especially of phonetic writing, the specialist form of writing that makes a division between sight and sound. It was with this instrument that Rome was able to reduce the tribal areas to some visual order. The effects of phonetic literacy do not depend upon persuasion or cajolery for their acceptance. This technology for translating the resonating tribal world into Euclidean lineality and visuality is automatic. Roman roads and Roman streets were uniform and repeatable wherever they occurred. There was no adaptation to the contours of local hill or custom. With the decline of papyrus supplies, the wheeled traffic stopped on these roads, too. Deprivation of papyrus, resulting from the Roman loss of Egypt, meant the decline of bureaucracy, and of army organization as well. Thus the medieval world grew up without uniform roads or cities or bureaucracies, and it fought the wheel, as later city forms fought the railways; and as we, today, fight the automobile. For new speed and power are never compatible with existing spatial and social arrangements.

Writing about the new straight avenues of the seventeenth-century cities, Mumford points to a factor that was also present in the Roman city with its wheeled traffic; namely, the need for broad straight avenues to speed military movements, and to express the pomp and circumstance of power. In the Roman world the army was the work force of a mechanized wealth-creating process. By means of soldiers as uniform and replaceable parts, the Roman military machine made and delivered the goods, very much in the manner of industry during the early phases of the industrial revolution. Trade followed the legions. More than that, the legions were the industrial machine, itself; and numerous new cities were like new factories manned by uniformly trained army personnel. With the spread of literacy after printing, the bond between the uniformed soldier and the wealth-making factory hand became less visible. It was obvious enough in Napoleon's armies. Napoleon, with his citizen-armies, was the industrial revolution itself, as it reached areas long protected from it.

The Roman army as a mobile, industrial wealth-making force created in addition a vast consumer public in the Roman towns. Division of labor always creates a separation between producer and consumer, even as it tends to separate the place of work and the living space. Before Roman literate bureaucracy, nothing comparable to the Roman consumer specialists had been seen in the world. This fact was institutionalized in the individual known as "parasite," and in the social institution of the gladiatorial games. (*Panem et circenses*.) The private sponge and the collective sponge, both reaching out for their rations of sensation, achieved a horrible distinctness and clarity that matched the raw power of the predatory army machine.

With the cutting-off of the supplies of papyrus by the

Mohammedans, the Mediterranean, long a Roman lake, became a Muslim lake, and the Roman center collapsed. What had been the margins of this center-margin structure became independent centers on a new feudal, structural base. The Roman center collapsed by the fifth century A.D. as wheel, road, and paper dwindled into a ghostly paradigm of former power.

Papyrus never returned. Byzantium, like the medieval centers, relied heavily on parchment, but this was too expensive and scarce a material to speed commerce or even education. It was paper from China, gradually making its way through the Near East to Europe, that accelerated education and commerce steadily from the eleventh century, and provided the basis for "the Renaissance of the twelfth century," popularizing prints and, finally, making printing possible by the fifteenth century.

With the moving of information in printed form, the wheel and the road came into play again after having been in abeyance for a thousand years. In England, pressure from the press brought about hard-surface roads in the eighteenth century, with all the population and industrial rearrangement that entailed. Print, or mechanized writing, introduced a separation and extension of human functions unimaginable even in Roman times. It was only natural, therefore, that greatly increased wheel speeds, both on road and in factory, should be related to the alphabet that had once done a similar job of speedup and specialization in the ancient world. Speed, at least in its lower reaches of the mechanical order, always operates to separate, to extend, and to amplify functions of the body. Even specialist learning in higher education proceeds by ignoring interrelationships; for such complex awareness slows down the achieving of expertness.

The post roads of England were, for the most part, paid for by the newspapers. The rapid increase of traffic brought in the railway, that accommodated a more specialized form of wheel than the road. The story of modern America that began with the discovery of the white man by the Indians, as a wag has truly said, quickly passed from exploration by canoe to development by railway. For three centuries Europe invested in America for its fish and its furs. The fishing schooner and the canoe preceded the road and the postal route as marks of our North American spatial organization. The European investors in the fur trade naturally did not want the trapping lines overrun by Tom Sawyers and Huck Finns. They fought land surveyors and settlers, like Washington and Jefferson, who simply would not think in terms of mink. Thus the War of Independence was deeply involved in media and staple rivalries. Any new medium, by its acceleration, disrupts the lives and investments of whole communities. It was the railway that raised the art of war to unheard-of intensity, making the American Civil War the first major conflict fought by rail, and causing it to be studied and admired by all European general staffs, who had not yet had an opportunity to use railways for a general blood-letting.

War is never anything less than accelerated technological change. It begins when some notable disequilibrium among existing structures has been brought about by inequality of rates of growth. The very late industrialization and unification of Germany had left her out of the race for staples and colonies for many years. As the Napoleonic wars were technologically a sort of catching-up of France with England, the First World War was itself a major phase of the final industrialization of Germany and America. As Rome had not shown before, and Russia has shown today, militarism is itself the main route of technological education and acceleration for lagging areas.

Almost unanimous enthusiasm for improved routes of land transportation followed the War of 1812. Furthermore, the British blockade of the Atlantic coast had compelled an unprecedented amount of land carriage, thus emphasizing the unsatisfactory character of the highways. War is certainly a form of emphasis that delivers many a telling touch to lagging social attention. However, in the very Hot Peace since the Second War, it is the highways of the mind that have been found inadequate. Many have felt dissatisfaction with our educational methods since Sputnik, in exactly the same spirit that many complained about the highways during the War of 1812.

Now that man has extended his central nervous system by electric technology, the field of battle has shifted to mental image-making-and-breaking, both in war and in business. Until the electric age, higher education had been a privilege and a luxury for the leisured classes; today it has become a necessity for production and survival. Now, when information itself is the main traffic, the need for advanced knowledge presses on the spirits of the most routine-ridden minds. So sudden an upsurge of academic training into the marketplace has in it the quality of classical peripety or reversal, and the result has been a wild guffaw from the gallery and the campus. The hilarity, however, will die down as the Executive Suites are taken over by the Ph.D.s.

For an insight into the ways in which the acceleration of wheel and road and paper rescramble population and settlement patterns, let us glance at some instances provided by Oscar Handlin in his study *Boston's Immigrants*. In 1790, he tells us, Boston was a compact unit with all workers and traders living in sight of each other, so that there was no tendency to section residential areas on a class basis: "But as the town grew, as the outlying districts became more accessible, the people spread out and at the same time were localized in distinctive areas." That one sentence capsulates the theme of this chapter. The sentence can be generalized to include the art of writing: "As knowledge was spread out visually and as it became more accessible in alphabetic form, it was localized and divided into specialties." Up to the point just short of electrification, increase of speed produces division of function, and of social classes, and of knowledge.

At electric speed, however, all that is reversed. Implosion and contraction then replace mechanical explosion and expansion. If the Handlin formula is extended to power, it becomes: "As power grew, and as outlying areas became accessible to power, it was localized in distinctive delegated jobs and functions." This formula is a principle of acceleration at all levels of human organization. It concerns especially those extensions of our physical bodies that appear in wheel and road and paper messages. Now that we have extended not just our physical organs but the nervous system, itself, in electric technology, the principle of specialism and division as a factor of speed no longer applies. When information moves at the speed of signals in the central nervous system, man is confronted with the obsolescence of all earlier forms of acceleration, such as road and rail. What emerges is a total field of inclusive awareness. The old patterns of psychic and social adjustment become irrelevant.

Until the 1820s, Handlin tells us, Bostonians walked to and fro, or used private conveyances. Horse-drawn buses were introduced in 1826, and these speeded up and extended business a great deal. Meantime the speedup of industry in England had extended business into the rural areas, dislodging many from the land and increasing the rate of immigration. Sea transport of immigrants became lucrative and encouraged a great speedup of ocean transport. Then the Cunard Line was subsidized by the British government in order to ensure swift contact with the colonies. The railways soon linked into this Cunard service, to convey mail and immigrants inland.

Although America developed a massive service of inland canals and river steamboats, they were not geared to the speeding wheels of the new industrial production. The railroad was needed to cope with mechanized production, as much as to span the great distances of the continent. The steam railroad as an accelerator proved to be one of the most revolutionary of all extensions of our physical bodies, creating a new political centralism and a new kind of urban shape and size. It is to the railroad that the American city owes its abstract grid layout, and the nonorganic separation of production, consumption, and residence. It is the motorcar that scrambled the abstract shape of the industrial town, mixing up its separated functions to a degree that has frustrated and baffled both planner and citizen. It remained for the airplane to complete the confusion by amplifying the mobility of the citizen to the point where urban space as such was irrelevant. Metropolitan space is equally irrelevant for the telephone, the telegraph, the radio, and television. What the town planners call "the human scale" in discussing ideal urban spaces is equally unrelated to these electric forms. Our electric extensions of ourselves simply by-pass space and time, and create problems of human involvement and organization for which there is no precedent. We may yet yearn for the simple days of the automobile and the superhighway.

11

Number:
Profile of the Crowd

It is a measure of McLuhan's originality that he does not confine himself to inevitable sources such as Tobias Dantzig's Number: The Language of Science, *but puts the writings of sociologist Elias Canetti, the work of Bauhaus artists, the symbolist poet Baudelaire, and the observations of Oswald Spengler to the exploration of a topic readers are unlikely to have anticipated finding. But McLuhan is very explicit about why number must occupy a central role in a full understanding of media as technological extensions: it extends the most intimate and interrelating of the physical senses—touch. Asides on the paradox of abstract art, Manhattan as tactile jazz, and gangster slang all add rich insights to the central topic. This chapter could well be titled "Number Magic;" explicitly contrasting writing and number, McLuhan develops a subtle counterpoint to the lessons on Word Magic (see pp. 120, 183, 189) that he absorbed from his Cambridge mentor I. A. Richards, inviting readers at the same time to reflect on the thread of semiotic principles that runs through the study of media viewed as any and all technologies.*

— *(Editor)*

Hitler made a special horror of the Versailles Treaty because it had deflated the German army. After 1870 the heel-clicking members of the German army had become the new symbol of tribal unity and power. In England and America the same sense of numerical grandeur from sheer numbers was associated with the mounting output of industry, and the statistics of wealth and production: "tanks a million." The power of sheer numbers, in wealth or in crowds, to set up a dynamic drive toward growth and aggrandizement is mysterious. Elias Canetti in *Crowds and Power* illustrates the profound tie between monetary inflation and crowd behavior. He is baffled by our failure to study inflation as a crowd phenomenon, since its effects on our modern world are pervasive. The drive toward unlimited growth inherent in any kind of crowd, heap, or horde would seem to link economic and population inflation.

In the theater, at a ball, at a ball game, in church, every individual enjoys all those others present. The pleasure of being among the masses is the sense of the joy in the multiplication of numbers, which has long been suspect among the literate members of Western society.

In such society, the separation of the individual from the group in space (privacy), and in thought ("point of view"), and in work (specialism), has had the cultural and technological support of literacy, and its attendant galaxy of fragmented industrial and political institutions. But the power of the printed word to create the homogenized social man grew steadily until our time, creating the paradox of the "mass mind" and the mass militarism of citizen armies. Pushed to the mechanized extreme, letters have often seemed to produce effects opposite to civilization, just as numbering in earlier times seemed to break tribal unity, as the Old Testament declares ("And Satan rose up against Israel, and moved

David to number Israel"). Phonetic letters and numbers were the first means of fragmenting and detribalizing man.

Throughout Western history we have traditionally and rightly regarded letters as the source of civilization, and looked to our literatures as the hallmark of civilized attainment. Yet all along, there has been with us a shadow of number, the language of science. In isolation, number is as mysterious as writing. Seen as an extension of our physical bodies, it becomes quite intelligible. Just as writing is an extension and separation of our most neutral and objective sense, the sense of sight, number is an extension and separation of our most intimate and interrelating activity, our sense of touch.

This faculty of touch, called the "haptic" sense by the Greeks, was popularized as such by the Bauhaus program of sensuous education, through the work of Paul Klee, Walter Gropius, and many others in the Germany of the 1920s. The sense of touch, as offering a kind of nervous system or organic unity in the work of art, has obsessed the minds of the artists since the time of Cézanne. For more than a century now artists have tried to meet the challenge of the electric age by investing the tactile sense with the role of a nervous system for unifying all the others. Paradoxically, this has been achieved by "abstract art," which offers a central nervous system for a work of art, rather than the conventional husk of the old pictorial image. More and more it has occurred to people that the sense of touch is necessary to integral existence. The weightless occupant of the space capsule has to fight to retain the integrating sense of touch. Our mechanical technologies for extending and separating the functions of our physical beings have brought us near to a state of disintegration by putting us out of touch with ourselves. It may very well be that in our conscious inner lives the interplay among our senses is what constitutes the sense of touch. Perhaps *touch* is not just skin contact with *things*, but the very life of

things in the *mind*? The Greeks had the notion of a consensus
or a faculty of "common sense" that translated each sense into
each other sense, and conferred consciousness on man. Today,
when we have extended all parts of our bodies and senses by
technology, we are haunted by the need for an outer consensus
of technology and experience that would raise our communal
lives to the level of a world-wide consensus. When we have
achieved a world-wide fragmentation, it is not unnatural to
think about a world-wide integration. Such a universality of
conscious being for mankind was dreamt of by Dante, who
believed that men would remain mere broken fragments
until they should be united in an inclusive consciousness.
What we have today, instead of a social consciousness electrically ordered, however, is a private subconsciousness or
individual "point of view" rigorously imposed by older
mechanical technology. This is a perfectly natural result of
"culture lag" or conflict, in a world suspended between two
technologies.

The ancient world associated number magically with the
properties of physical things, and with the necessary causes of
things, much as science has tended until recent times to reduce
all objects to numerical quantities. In any and all of its manifestations, however, number seems to have both auditory and
repetitive resonance, and a tactile dimension as well.

It is the quality of number that explains its power to
create the effect of an icon or an inclusive compressed image.
Such is its use in newspaper and magazine reporting, as:
"Cyclist John Jameson, 12, Collides with Bus," or "William
Samson, 51, New Vice-President in Charge of Brooms." By
rule of thumb the journalists have discovered the iconic
power of number.

Since Henri Bergson and the Bauhaus group of artists, to
say nothing of Jung and Freud, the nonliterate and even anti-
literate values of tribal man have in general received enthusiastic study and promotion. For many European artists and

149

intellectuals, jazz became one of the rallying points in their quest for the integral Romantic Image. The uncritical enthusiasm of the European intellectual for tribal culture appears in the exclamation of the architect Le Corbusier on first seeing Manhattan: "It is hot-jazz in stone." It appears again in the artist Moholy-Nagy's account of his visit to a San Francisco night club in 1940. A Negro band was playing with zest and laughter. Suddenly a player intoned, "*One million and three*," and was answered: "*One million and seven and a half*." Then another sang, "*Eleven*," and another, "*Twenty-one*." Then amidst "happy laughter and shrill singing the numbers took over the place."

Moholy-Nagy notes how, to Europeans, America seems to be the land of abstractions, where numbers have taken on an existence of their own in phrases like "57 Varieties," "the 5 and 10," or "7 Up" and "behind the 8-ball." It figures. Perhaps this is a kind of echo of an industrial culture that depends heavily on prices, charts, and figures. Take 36-24-36. Numbers cannot become more sensuously tactile than when mumbled as the magic formula for the female figure while the haptic hand sweeps the air.

Baudelaire had the true intuition of number as a tactile hand or nervous system for interrelating separate units, when he said that "number is within the individual. Intoxication is a number." That explains why "the pleasure of being in a crowd is a mysterious expression of delight in the multiplication of number." Number, that is to say, is not only auditory and resonant, like the spoken word, but originates in the sense of touch, of which it is an extension. The statistical aggregation or crowding of numbers yields the current cave-drawings or finger-paintings of the statisticians' charts. In every sense, the amassing of numbers statistically gives man a new influx of primitive intuition and magically subconscious awareness, whether of public taste or feeling: "You feel better satisfied when you use well-known brands."

Like money and clocks and all other forms of measurement, numbers acquired a separate life and intensity with the growth of literacy. Nonliterate societies had small use for numbers, and today the nonliterate digital computer substitutes "yes" and "no" for numbers. The computer is strong on contours, weak on digits. In effect, then, the electric age brings number back into unity with visual and auditory experience, for good or ill.

Oswald Spengler's *The Decline of the West* originated in large part from his concern with the new mathematics. Non-Euclidean geometries, on one hand, and the rise of Functions in number theory, on the other, seemed to Spengler to spell the end of Western man. He had not grasped the fact that the invention of Euclidean space is, itself, a direct result of the action of the phonetic alphabet on the human senses. Nor had he realized that number is an extension of the physical body of man, an extension of our sense of touch. The "infinity of functional processes," into which Spengler gloomily saw traditional number and geometry dissolving, is, also, the extension of our central nervous system in electrical technologies. We need not feel grateful to apocalyptic writers like Spengler, who see our technologies as cosmic visitors from outer space. The Spenglers are tribally entranced men who crave the swoon back into collective unconsciousness and all the intoxication of number. In India the idea of *darshan* — of the mystical experience of being in very large gatherings — stands at the opposite end of the spectrum from the Western idea of conscious values.

The most primitive tribes of Australia and Africa, like the Eskimos of today, have not yet reached finger-counting, nor do they have numbers in series. Instead they have a binary system of independent numbers for *one* and *two*, with composite numbers up to *six*. After *six*, they perceive only "heap." Lacking the sense of series, they will scarcely notice when two pins have been removed from a row of seven. They become aware at once, however, if *one* pin is missing. Tobias Dantzig, who investigated these matters, points out (in *Number: The Language of Science*) that the parity or kinesthetic sense of these people is stronger than their number sense. It is certainly an indication of a developing visual stress in a culture when number appears. A closely integrated tribal culture will not easily yield to the separatist visual and individualistic pressures that lead to the division of labor, and then to such accelerated forms as writing and money. On the other hand, Western man, were he determined to cling to the fragmented and individualist ways that he has derived from the printed word in particular, would be well advised to scrap all his electric technology since the telegraph. The implosive (compressional) character of the electric technology plays the disk or film of Western man backward, into the heart of tribal darkness, or into what Joseph Conrad called "the Africa within." The instant character of electric information movement does not enlarge, but involves, the family of man in the cohesive state of village living.

It seems contradictory that the fragmenting and divisive power of our analytic Western world should derive from an accentuation of the visual faculty. This same visual sense is, also, responsible for the habit of seeing all things as continuous and connected. Fragmentation by means of visual stress occurs in that isolation of moment in time, or of aspect in space, that is beyond the power of touch, or hearing, or smell, or movement. By imposing unvisualizable relationships that are the result of instant speed, electric technology dethrones the visual sense and restores us to the dominion of synesthesia, and the close interinvolvement of the other senses.

Spengler was plunged into a Slough of Despond by what he saw as the Western retreat from Numerical Magnitude into a Faery Land of Functions and abstract relations. "The most valuable thing in classical mathematic," he wrote, "is its proposition that number is the essence of all things *perceptible to the senses.* Defining number as a measure, it contains the whole world-feeling of a soul passionately devoted to the 'here' and 'now.' Measurement in this sense means the measurement of something near and corporeal."

The ecstatic tribal man emanates from every page of Spengler. It never occurred to him that the *ratio* among corporeal things could never be less than rational. That is to say, rationality or consciousness is itself a ratio or proportion among the sensuous components of experience, and is not something *added* to such sense experience. Subrational beings have no means of achieving such a ratio or proportion in their sense lives but are wired for fixed wave lengths, as it were, having infallibility in their own area of experience. Consciousness, complex and subtle, can be impaired or ended by a mere stepping-up or dimming-down of any one sense intensity, which is the procedure in hypnosis. And the intensification of one sense by a new medium can hypnotize an entire community. Thus, when he thought he saw modern mathematics and science abandoning visual relations and constructions for a nonvisual theory of relations and functions, Spengler pronounced the demise of the West.

Had Spengler taken the time to discover the origins of both number and Euclidean space in the psychological effects of the phonetic alphabet, *The Decline of the West* might never have been written. That work is based on the assumption that classical man, Apollonian man, was not the product of a technological bias in Greek culture (namely, the early impact of literacy on a tribal society), but rather the result of a special tremor in the soul stuff that embosomed the Greek world. This is a striking instance of how easily men of any one particular culture will panic when some familiar pattern or landmark gets smudged or shifted because of the indirect pressure of new media. Spengler, as much as Hitler, had derived from radio a subconscious mandate to announce the end of all "rational" or visual values. He was acting like Pip

in Dickens' *Great Expectations*. Pip was a poor boy who had a
hidden benefactor who wanted to raise Pip to the status of a
gentleman. Pip was ready and willing until he found that his
benefactor was an escaped convict. Spengler and Hitler and
many more of the would-be "irrationalists" of our century
are like singing-telegram delivery boys, who are quite innocent of any understanding of the medium that prompts the
song they sing.

So far as Tobias Dantzig is concerned in his *Number: The
Language of Science*, the progress from the tactile fingering of
toes and fingers to "the homogeneous number concept, which
made mathematics possible" is the result of visual abstraction
from the operation of tactile manipulation. We have both
extremes of this process in our daily speech. The gangster
term "to put the finger on" says that somebody's "number"
has come up. At the extreme of the graph profiles of the statisticians there is the frankly expressed object of manipulation
of population for varieties of power purposes. For example,
in any large stockbroker's office there is a modern medicine
man known as "Mr. Odd Lots." His magical function is to
study the daily purchases and sales of the small buyers on
the big exchanges. Long experience has revealed that these
small buyers are wrong 80 per cent of the time. A statistical
profile of the failure of the little man to be in touch enables
the big operators to be about 80 per cent right. Thus from error
comes truth; and from poverty, riches, thanks to numbers.
This is the modern magic of numbers. The more primitive
attitude toward the magical power of numbers appeared
in the dread of the English when William the Conqueror
numbered them and their chattels in what the folk called the
Doomsday Book.

To turn again briefly to the question of number in its
more limited manifestation, Dantzig, having made clear that
the idea of homogeneity had to come before primitive numbers could be advanced to the level of mathematics, points to

another literate and visual factor in the older mathematics. "Correspondence and succession, the two principles which permeate all mathematics—nay, all realms of exact thought —are woven into the very fabric of our number system," he observes. So, indeed, are they woven into the very fabric of Western logic and philosophy. We have already seen how the phonetic technology fostered visual continuity and individual point of view, and how these contributed to the rise of uniform Euclidean space. Dantzig says that it is the idea of correspondence which gives us cardinal numbers. Both of these spatial ideas—lineality and point of view—come with writing, especially with phonetic writing; but neither is necessary in our new mathematics and physics. Nor is writing necessary to an electric technology. Of course, writing and conventional arithmetic may long continue to be of the utmost use to man, for all that. Even Einstein could not face the new quantum physics with comfort. Too visual a Newtonian for the new task, he said that *quanta* could not be handled mathematically. That is as much as to say that poetry cannot be properly translated into merely visual form on the printed page.

Dantzig develops his points about number by saying that a literate population soon departs from the abacus and from finger enumeration, though arithmetic manuals in the Renaissance continued to give elaborate rules for calculating on the hands. It could be true that numbers preceded literacy in some cultures, but so did visual stress precede writing. For writing is only the principal manifestation of the extension of our visual sense, as the photograph and the movie today may well remind us. And long before literate technology, the binary factors of hands and feet sufficed to launch man on the path of counting. Indeed, the mathematical Leibniz saw in the mystic elegance of the binary system of *zero* and *1* the image of Creation. The unity of the Supreme Being operating in the void by binary function would, he felt, suffice to make all beings from the void.

Dantzig reminds us also that in the age of manuscript there was a chaotic variety of signs for numerals, and that they did not assume a stable form until printing. Although this was one of the least of the cultural effects of printing, it should serve to recall that one of the big factors in the Greek adoption of the letters of the phonetic alphabet was the prestige and currency of the number system of the Phoenician traders. The Romans got the Phoenician letters from the Greeks but retained a number system that was much more ancient. Wayne and Shuster, the comedian team, never fail to get a good laugh when they line up a group of ancient Roman cops in togas and have them number themselves from left to right, uttering Roman numerals. This joke demonstrates how the pressure of numbers caused men to seek ever more streamlined methods of numeration. Before the advent of ordinal, successive, or positional numbers, rulers had to count large bodies of soldiery by displacement methods. Sometimes they were herded by groups into spaces of approximately known area. The method of having them march in file and of dropping pebbles into containers was another method not unrelated to the abacus and the counting board. Eventually the method of the counting board gave rise to the great discovery of the principle of position in the early centuries of our era. By simply putting 3 and 4 and 2 in position on the board, one after another, it was possible to step up the speed and potential of calculation fantastically. The discovery of calculation by positional numbers rather than by merely additive numbers led, also, to the discovery of zero. Mere positions for 3 and 2 on the board created ambiguities about whether the number was 32 or 302. The need was to have a sign for the gaps between numbers. It was not till the thirteenth century that *sifr*, the Arab word for "gap" or "empty," was Latinized and added to our culture as "cipher" (*ziphrium*) and finally became the Italian *zero*. Zero really meant a positional gap. It did not acquire the indispensable quality of "infinity" until the rise of perspective and "vanishing point" in Renaissance painting. The new visual space of Renaissance painting affected number as much as lineal waiting had done centuries earlier.

A main fact about numbers has now been reached, with the link between the medieval positional zero and the Renaissance vanishing point. That both vanishing point and infinity were unknown in the Greek and Roman cultures can be explained as by-products of literacy. It was not until printing extended the visual faculty into very high precision, uniformity, and intensity of special order that the other senses could be restrained or depressed sufficiently to create the new awareness of infinity. As one aspect of perspective and printing, mathematical or numerical infinity serves as an instance of how our various physical extensions or media act upon one another through the agency of our own senses. It is in this mode that man appears as the reproductive organ of the technological world, a fact that Samuel Butler bizarrely announced in *Erewhon*.

The effect of any kind of technology engenders a new equilibrium in us that brings quite new technologies to birth, as we have just seen in the interplay of number (the tactile and quantitative form), and the more abstract forms of written or visual culture. Print technology transformed the medieval zero into the Renaissance infinity, not only by convergence — perspective and vanishing point — but by bringing into play for the first time in human history the factor of exact repeatability. Print gave to men the concept of indefinite repetition so necessary to the mathematical concept of infinity.

The same Gutenberg fact of uniform, continuous, and indefinitely repeatable bits inspired also the related concept of the infinitesimal calculus, by which it became possible to translate any kind of tricky space into the straight, the flat, the uniform, and the "rational." This concept of infinity was not imposed upon us by logic. It was the gift of Gutenberg. So, also, later on, was the industrial assembly line. The power to translate knowledge into mechanical production by the breaking up of any process into fragmented aspects to be placed in a lineal sequence of movable, yet uniform, parts was the formal essence of the printing press. This amazing technique of spatial analysis duplicating itself at once, by a kind of echo, invaded the world of number and touch.

Here, then, is merely one familiar, if unrecognized, instance of the power of one medium to translate itself into another medium. Since all media are extensions of our own bodies and senses, and since we habitually translate one sense into another in our own experience, it need not surprise us that our extended senses or technologies should repeat the process of translation and assimilation of one form into another. This process may well be inseparable from the character of touch, and from the abrasively interfaced action of surfaces, whether in chemistry or crowds or technologies. The mysterious need of crowds to grow and to reach out, equally characteristic of large accumulations of wealth, can be understood if money and numbers are, indeed, technologies that extend the power of touch and the grasp of the hand. For numbers, whether of people or of digits, and units of money would seem to possess the same factual magic for seizing and incorporating.

The Greeks ran head-on into the problem of translating their own new media when they tried to apply rational arithmetic to a problem in geometry. Up arose the specter of Achilles and the tortoise. These attempts resulted in the first crisis in the history of our Western mathematics. Such a crisis concerned the problems of determining the diagonal of a square and the circumference of a circle: a clear case of number, the tactile sense, trying to cope with visual and pictorial space by reduction of the visual space to itself.

For the Renaissance, it was the infinitesimal calculus that enabled arithmetic to take over mechanics, physics, and geometry. The idea of an infinite but continuous and uniform process, so basic to the Gutenberg technology of movable types, gave rise to the calculus. Banish the infinite process and mathematics, pure and applied, is reduced to the state known to the pre-Pythagoreans. This is to say, banish the new medium of print with its fragmented technology of uniform, lineal repeatability, and modern mathematics disappears. Apply, however, this infinite uniform process to finding the length of an arc, and all that need be done is to inscribe in the arc a sequence of rectilinear contours of an increasing number of sides. When these contours approach a limit, the length of the arc becomes the limit of this sequence. The older method of determining volumes by liquid displacement is thus translated into abstract visual terms by calculus. The principles regarding the concept of length apply also to notions of areas, volumes, masses, moments, pressures, forces, stresses and strains, velocities and accelerations.

The miracle-maker, the sheer function of the infinitely fragmented and repeatable, became the means of making visually flat, straight, and uniform all that was nonvisual: the skew, the curved, and the bumpy. In the same way, the phonetic alphabet had, centuries before, invaded the discontinuous cultures of the barbarians, and translated their sinuosities and obtusities into the uniformities of the visual culture of the Western world. It is this uniform, connected, and visual order that we still use as the norm of "rational" living. In our electric age of instant and non-visual forms of interrelation, therefore, we find ourselves at a loss to define the "rational," if only because we never noticed whence it came in the first place.

12

Clothing:
Our Extended Skin

McLuhan opens this chapter by explicitly anticipating the following one ("clothing and housing are near twins") and immediately moves on to expand his integrated explanations of media effects ("the sewing machine, for example, created the long straight line in clothes, as much as the linotype flattened the human vocal style"). This integration extends to commenting on the potent transformation of non-verbal into verbal communication (Mrs. Khrushchev's plain cotton dress, simple clothing in revolutionary times, etc). And the McLuhan technique of anchoring media studies in the economy of the sensorium allows him to range here over the tactility of four-letter words, the obsolescence of strip-tease, and Alfred Kinsey's blind spot.

— (Editor)

Economists have estimated that an unclad society eats 40 per cent more than one in Western attire. Clothing as an extension of our skin helps to store and to channel energy, so that if the Westerner needs less food, he may also demand more sex. Yet neither clothing nor sex can be understood as separate isolated factors, and many sociologists have noted that sex can become a compensation for crowded living. Privacy, like individualism, is unknown in tribal societies, a fact that Westerners need to keep in mind when estimating the attractions of our way of life to nonliterate peoples.

Clothing, as an extension of the skin, can be seen both as a heat-control mechanism and as a means of defining the self socially. In these respects, clothing and housing are near twins, though clothing is both nearer and elder; for housing extends the inner heat-control mechanisms of our organism, while clothing is a more direct extension of the outer surface of the body. Today Europeans have begun to dress for the eye, American-style, just at the moment when Americans have begun to abandon their traditional visual style. The media analyst knows why these opposite styles suddenly transfer their locations. The European, since the Second War, has begun to stress visual values; his economy, not coincidentally, now supports a large amount of uniform consumer goods. Americans, on the other hand, have begun to rebel against uniform consumer values for the first time. In cars, in clothes, in paperback books; in beards, babies, and beehive hairdos, the American has declared for stress on touch, on participation, involvement, and sculptural values. America, once the land of an abstractly visual order, is profoundly "in touch" again with European traditions of food and life and art. What was an *avant-garde* program for the 1920 expatriates is now the teenagers' norm.

The Europeans, however, underwent a sort of consumer

revolution at the end of the eighteenth century. When industrialism was a novelty, it became fashionable among the upper classes to abandon rich, courtly attire in favor of simpler materials. That was the time when men first donned the trousers of the common foot soldier (or *pioneer*, the original French usage), but it was done at that time as a kind of brash gesture of social "integration." Up until then, the feudal system had inclined the upper classes to dress as they spoke, in a courtly style quite removed from that of ordinary people. Dress and speech were accorded a degree of splendor and richness of texture that universal literacy and mass production were eventually to eliminate completely. The sewing machine, for example, created the long straight line in clothes, as much as the linotype flattened the human vocal style.

A recent ad for C-E-I-R Computer Services pictured a plain cotton dress and the headline: "Why does Mrs. 'K' dress that way?"—referring to the wife of Nikita Khrushchev. Some of the copy of this very ingenious ad continued: "It is an icon. To its own underprivileged population and to the uncommitted of the East and South, it says: 'We are thrif-ty, simple, hon-est; peaceful, home-y, go-od.' To the free nations of the West it says: 'We will bury you.'"

This is precisely the message that the new simple clothing of our forefathers had for the feudal classes at the time of the French Revolution. Clothing was then a nonverbal manifesto of political upset.

Today in America there is a revolutionary attitude expressed as much in our attire as in our patios and small cars. For a decade and more, women's dress and hair styles have abandoned visual for iconic—or sculptural and tactual—stress. Like toreador pants and gaiter stockings, the beehive hairdo is also iconic and sensuously inclusive, rather than abstractly visual. In a word, the American woman for the first time presents herself as a person to be touched and handled, not just to be looked at. While the Russians are groping vaguely

toward visual consumer values, North Americans are frolick-
ing amidst newly discovered tactile, sculptural spaces in cars,
clothes, and housing. For this reason, it is relatively easy for us
now to recognize clothing as an extension of the skin. In the
age of the bikini and of skin-diving, we begin to understand
"the castle of our skin" as a space and world of its own. Gone
are the thrills of strip-tease. Nudity could be naughty excitement only for a visual culture that had divorced itself from the
audile-tactile values of less abstract societies. As late as 1930,
four-letter words made visual on the printed page seemed
portentous. Words that most people used every hour of the
day became as frantic as nudity, when printed. Most "four-
letter words" are heavy with tactile-involving stress. For this
reason they seem earthy and vigorous to visual man. So it is
with nudity. To backward cultures still embedded in the full
gamut of sense-life, not yet abstracted by literacy and industrial visual order, nudity is merely pathetic. The Kinsey
Report on the sex life of the male expressed bafflement that
peasants and backward peoples did not relish marital or
boudoir nudity. Khrushchev did not enjoy the can-can dance
provided for his entertainment in Hollywood. Naturally
not. That sort of mime of sense involvement is meaningful
only to long-literate societies. Backward peoples approach
nudity, if at all, with the attitude we have come to expect
from our painters and sculptors — the attitude made up of
all the senses at once. To a person using the whole sensorium, nudity is the richest possible expression of structural
form. But to the highly visual and lopsided sensibility of
industrial societies, the sudden confrontation with tactile
flesh is heady music, indeed.

There is a movement toward a new equilibrium today, as
we become aware of the preference for coarse, heavy textures
and sculptural shapes in dress. There is, also, the ritualistic
exposure of the body indoors and out-of-doors. Psychologists
have long taught us that much of our hearing takes place

through the skin itself. After centuries of being fully clad and of being contained in uniform visual space, the electric age ushers us into a world in which we live and breathe and listen with the entire epidermis. Of course, there is much zest of novelty in this cult, and the eventual equilibrium among the senses will slough off a good deal of the new ritual, both in clothing and in housing. Meantime, in both new attire and new dwellings, our unified sensibility cavorts amidst a wide range of awareness of materials and colors which makes ours one of the greatest ages of music, poetry, painting, and architecture alike.

13

Housing:
New Look and New Outlook

Here McLuhan offers a succinct comment on media as educators: "In making heat and energy accessible socially, to the family or the group, housing fosters new skills and new learning, performing the basic functions of all other media." This is a hint and a lead-in to another passage indicating clearly that it is irrelevant to McLuhan's purpose to make a distinction between a small group of media used for communication (press, radio, television) and the broad group of all other media he surveys: "Clothing and housing as extensions of skin and heat-control mechanisms, are media of communication … in the sense that they shape and rearrange the patterns of human association and community" (emphasis added). The principle whose assimilation could have saved Narcissus (extensions alter perceptions) is repeated with reference to electric lighting (first mentioned in the opening pages of Chapter One) and linked to the theme there: "In this domain, the medium is the message, and when the light is on there is a world of sense that disappears when the light is off."

— *(Editor)*

If clothing is an extension of our private skins to store and channel our own heat and energy, housing is a collective means of achieving the same end for the family or the group. Housing as shelter is an extension of our bodily heat-control mechanisms—a collective skin or garment. Cities are an even further extension of bodily organs to accommodate the needs of large groups. Many readers are familiar with the way in which James Joyce organized *Ulysses* by assigning the various city forms of walls, streets, civic buildings, and media to the various bodily organs. Such a parallel between the city and the human body enabled Joyce to establish a further parallel between ancient Ithaca and modern Dublin, creating a sense of human unity in depth, transcending history.

Baudelaire originally intended to call his *Fleurs du Mal, Les Limbes,* having in mind the city as corporate extensions of our physical organs. Our letting-go of ourselves, self-alienations, as it were, in order to amplify or increase the power of various functions, Baudelaire considered to be flowers of growths of evil. The city as amplification of human lusts and sensual striving had for him an entire organic and psychic unity.

Literate man, civilized man, tends to restrict and enclose space and to separate functions, whereas tribal man had freely extended the form of his body to include the universe. Acting as an organ of the cosmos, tribal man accepted his bodily functions as modes of participation in the divine energies. The human body in Indian religious thought was ritually related to the cosmic image, and this in turn was assimilated into the form of house. Housing was an image of both the body and the universe for tribal and nonliterate societies. The building of the house with its hearth as fire-altar was ritually associated with the act of creation. This same ritual was even more deeply embedded in the building of the ancient cities, their shape and process having been deliberately modeled as an act of divine praise. The city and the home in the tribal world (as in China and India today) can be accepted as iconic embodiments of the *word*, the divine *mythos*, the universal aspiration. Even in our present electric age, many people yearn for this inclusive strategy of acquiring significance for their own private and isolated beings.

Literate man, once having accepted an analytic technology of fragmentation, is not nearly so accessible to cosmic patterns as tribal man. He prefers separateness and compartmented spaces, rather than the open cosmos. He becomes less inclined to accept his body as a model of the universe, or to see his house—or any other of the media of communication, for that matter—as a ritual extension of his body. Once men have adopted the visual dynamic of the phonetic alphabet, they begin to lose the tribal man's obsession with cosmic order and ritual as recurrent in the physical organs and their social extension. Indifference to the cosmic, however, fosters intense concentration on minute segments and specialist tasks, which is the unique strength of Western man. For the specialist is one who never makes small mistakes while moving toward the grand fallacy.

Men live in round houses until they become sedentary and specialized in their work organization. Anthropologists have often noted this change from round to square without knowing its cause. The media analyst can help the anthropologist in this matter, although the explanation will not be obvious to people of visual culture. The visual man, likewise, cannot see much difference between the motion picture and TV, or between a Corvair and a Volkswagen, for this difference is not between two visual spaces, but between tactile and visual ones. A tent or a wigwam is not an enclosed or visual space. Neither is a cave nor a hole in the ground. These kinds of space—the tent, the wigwam, the igloo, the cave— are not "enclosed" in the visual sense because they follow dynamic lines of force, like a triangle. When enclosed, or translated into visual space, architecture tends to lose its tactile kinetic pressure. A square is the enclosure of a visual space; that is, it consists of space properties abstracted from manifest tensions. A triangle follows lines of force, this being the most economical way of anchoring a vertical object. A square moves beyond such kinetic pressures to enclose visual space relations, while depending upon diagonal anchors. This separation of the visual from direct tactile and kinetic pressure, and its translation into new dwelling spaces, occurs only when men have learned to practice specialization of their senses, and fragmentation of their work skills. The square room or house speaks the language of the sedentary specialist, while the round hut or igloo, like the conical wigwam, tells of the integral nomadic ways of food-gathering communities.

This entire discussion is offered at considerable risk of misapprehension because these are, spatially, highly technical matters. Nevertheless, when such spaces are understood, they offer the key to a great many enigmas, past and present. They explain the change from circular-dome architecture to gothic forms, a change occasioned by alteration in the ratio or proportion of the sense lives in the members of a society. Such a shift occurs with the extension of the body in new social technology and invention. A new extension sets up a new equilibrium among all of the senses and faculties leading, as we say, to a "new outlook"—new attitudes and preferences in many areas.

In the simplest terms, as already noted, housing is an effort to extend the body's heat-control mechanism. Clothing tackles the problem more directly but less fundamentally, and privately rather than socially. Both clothing and housing store warmth and energy and make these readily accessible for the execution of many tasks otherwise impossible. In making heat and energy accessible socially, to the family or the group, housing fosters new skills and new learning, performing the basic functions of all other media. Heat control is the key factor in housing, as well as in clothing. The Eskimo's dwelling is a good example. The Eskimo can go for days without food at 50 degrees below zero. The unclad native, deprived of nourishment, dies in a few hours.

It may surprise many to learn that the primitive shape of the igloo is, nonetheless, traceable to the primus stove. Eskimos have lived for ages in round stone houses, and, for the most part, still do. The igloo, made of snow blocks, is a fairly recent development in the life of this stone-age people. To live in such structures became possible with the coming of the white man and his portable stove. The igloo is an ephemeral shelter, devised for temporary use by trappers. The Eskimo became a trapper only after he had made contact with the white man; up until then he had been simply a food-gatherer. Let the igloo serve as an example of the way in which a new pattern is introduced into an ancient way of life by the intensification of a single factor — in this instance, artificial heat. In the same way, the intensification of a single factor in our complex lives leads naturally to a new balance among our technologically extended faculties, resulting in a new look and a new "outlook" with new motivations and inventions.

In the twentieth century we are familiar with the changes in housing and architecture that are the result of electric energy made available to elevators. The same energy devoted to lighting has altered our living and working spaces even more radically. Electric light abolished the divisions of night and day, of inner and outer, and of the subterranean and the terrestrial. It altered every consideration of space for work and production as much as the other electric media had altered the space-time experience of society. All this is reasonably familiar. Less familiar is the architectural revolution made possible by improvements in heating centuries ago.

With the mining of coal on a large scale in the Renaissance, inhabitants in the colder climates discovered great new resources of personal energy. New means of heating permitted the manufacture of glass and the enlargement of living quarters and the raising of ceilings. The Burgher house of the Renaissance became at once bedroom, kitchen, workshop, and sale outlet.

Once housing is seen as group (or corporate) clothing and heat control, the new means of heating can be understood as causing change in spatial form. Lighting, however, is almost as decisive as heating in causing these changes in architectural and city spaces. That is the reason why the story of glass is so closely related to the history of housing. The story of the mirror is a main chapter in the history of dress and manners and the sense of the self.

Recently an imaginative school principal in a slum area provided each student in the school with a photograph of himself. The classrooms of the school were abundantly supplied with large mirrors. The result was an astounding increase in the learning rate. The slum child has ordinarily very little visual orientation. He does not see himself as becoming something. He does not envisage distant goals and objectives. He is deeply involved in his own world from day to day, and can establish no beachhead in the highly specialized sense life of visual man. The plight of the slum child, via the TV image, is increasingly extended to the entire population.

Clothing and housing, as extensions of skin and heat-control mechanisms, are media of communication, first of all, in the sense that they shape and rearrange the patterns of human association and community. Varied techniques of lighting and heating would seem only to give new flexibility and scope to what is the basic principle of these media of clothing and housing; namely, their extension of our bodily heat-control mechanisms in a way that enables us to attain some degree of equilibrium in a changing environment.

Modern engineering provides means of housing that range from the space capsule to walls created by air jets. Some firms now specialize in providing large buildings with inside walls and floors that can be moved at will. Such flexibility naturally tends toward the organic. Human sensitivity seems once more to be attuned to the universal currents that made of tribal man a cosmic skin-diver.

It is not only the *Ulysses* of James Joyce that testifies to this trend. Recent studies of the Gothic churches have stressed the organic aims of their builders. The saints took the body seriously as the symbolic vesture of the spirit, and they regarded the Church as a second body, viewing its every detail with great completeness. Before James Joyce provided his detailed image of the metropolis as a second body, Baudelaire had provided a similar "dialogue" between the parts of the body extended to form the metropolis, in his *Fleurs du Mal*.

Electric lighting has brought into the cultural complex of the extensions of man in housing and city, an organic flexi-bility unknown to any other age. If color photography has created "museums without walls," how much more has electric lighting created space without walls, and day without night. Whether in the night city, the night highway, or the night ball game, sketching and writing with light have moved from the domain of the pictorial photograph to the live, dynamic spaces created by out-of-door lighting.

Not many ages ago, glass windows were unknown luxu-ries. With light control by glass came also a means of controlling the regularity of domestic routine, and steady application to crafts and trade without regard to cold or rain. The world was put in a frame. With electric light not only can we carry out the most precise operations with no regard for time or place or climate, but we can photograph the submicroscopic as easily as we can enter the subterranean world of the mine and of the cave-painters.

Lighting as an extension of our powers affords the clearest-cut example of how such extensions alter our perceptions. If people are inclined to doubt whether the wheel or typography or the plane could change our habits of sense perception, their doubts end with electric lighting. In this domain, the medium is the message, and when the light is on there is a world of sense that disappears when the light is off.

"Painting with light" is jargon from the world of stage-electricity. The uses of light in the world of motion, whether in the motorcar or the movie or the microscope, are as diverse as the uses of electricity in the world of power. Light is information without "content," much as the missile is a vehicle without the additions of wheel or highway. As the missile is a self-contained transportation system that consumes not only its fuel but its engine, so light is a self-contained communication system in which the medium is the message.

The recent development of the laser ray has introduced new possibilities for light. The laser ray is an amplification of light by intensified radiation. Concentration of radiant energy has made available some new properties in light. The laser ray—by thickening light, as it were—enables it to be modulated to carry information as do radio waves. But because of its greater intensity, a single laser beam can carry as much information as all the combined radio and TV channels in the United States. Such beams are not within the range of vision, and may well have a military future as lethal agents.

From the air at night, the seeming chaos of the urban area manifests itself as a delicate embroidery on a dark velvet ground. Gyorgy Kepes has developed these aerial effects of the city at night as a new art form of "landscape by light through" rather than "light on." His new electric landscapes have complete congruity with the TV image, which also exists

175

by light *through* rather than by light *on*.

The French painter André Girard began painting directly on film before the photographic movies became popular. In that early phase it was easy to speculate about "painting with light" and about introducing movement into the art of painting. Said Girard:

> I would not be surprised if, fifty years from now, almost no one would pay attention to paintings whose sub-jects remain *still* in their always too-narrow frames.

The coming of TV inspired him anew:

> Once I saw suddenly, in a control room, the sensi-tive eye of the camera presenting to me, one after another, the faces, the landscapes, the expressions of a big painting of mine in an order which I had never thought of. I had the feeling of a composer listening to one of his operas, all scenes mixed up in an order different from the one he wrote. It was like seeing a building from a fast elevator that showed you the roof before the basement, and made quick stops at some floors but not others.

Since that phase, Girard has worked out new techniques of control for painting with light in association with CBS and NBC technicians. The relevance of his work for housing is that it enables us to conceive of totally new possibilities for architectural and artistic modulation of space. Painting with light is a kind of housing-without-walls. The same electric technology, extended to the job of providing global thermo-static controls, points to the obsolescence of housing as an extension of the heat-control mechanisms of the body. It is equally conceivable that the electric extension of the process of collective consciousness, in making consciousness-without-walls, might render language walls obsolescent. Languages are stuttering extensions of our five senses, in varying ratios and wavelengths. An immediate simulation of consciousness would by-pass speech in a kind of massive extrasensory perception, just as global thermostats could by-pass those extensions of skin and body that we call houses. Such an extension of the process of consciousness by electric simulation may easily occur in the 1960s.

14

Money:
The Poor Man's Credit Card

For sheer memorable value, few passages of Understanding Media *surpass the reference to be found here to space ships designed to be edible. As for the principal topic, it is instructive to note the two passages that explicitly assimilate money and language, a comparison that allows McLuhan to add to his earlier commentaries on word magic (p. 120) and to declare that there is a magic aspect in all media. In this respect, there is a vein of semiotic reflection to be mined in the observation that "the alphabet is a one-way process of reduction of non-literate cultures into the specialist visual fragments of our Western world." This is the cultural corollary to a preceding passage characterizing language as a reducer and distorter of experience. The or that has proved valuable in earlier chapters occurs here to complete McLuhan's primary principle: "Since all media are extensions of ourselves,* or translations of some part of us into various materials, *any study of one medium helps us to understand all the others" (emphasis added). This leads to a further qualification: "The object, then, stores work and information or technical knowledge to the extent that something has been done to it. When the one object is exchanged for another, it is already assuming the function of money ..."*

— *(Editor)*

Central to modern psychoanalytical theory is the relation between the money complex and the human body. Some analysts derive money from the infantile impulse to play with feces. Ferenczi, in particular, calls money "nothing other than odorless dehydrated filth that has been made to shine." Ferenczi, in his concept of money, is elaborating Freud's concept of "Character and Anal Erotism." Although this idea of linking "filthy lucre" with the anal has continued in the main lines of psychoanalysis, it does not correspond sufficiently to the nature and function of money in society to provide a theme for the present chapter.

Money began in nonliterate cultures as a commodity, such as whales' teeth on Fiji; or rats on Easter Island, which later were considered a delicacy, were valued as a luxury, and thus became a means of mediation or barter. When the Spaniards were besieging Leyden in 1574, leather money was issued, but as hardship increased the population boiled and ate the new currency.

In literate cultures, circumstances may reintroduce commodity money. The Dutch, after the German occupation of World War II, were avid for tobacco. Since the supply was small, objects of high value such as jewels, precision instruments, and even houses were sold for small quantities of cigarettes. The *Reader's Digest* recorded an episode from the early occupation of Europe in 1945, describing how an unopened pack of cigarettes served as currency, passing from hand to hand, translating the skill of one worker into the skill of another as long as no one broke the seal.

Money always retains something of its commodity and community character. In the beginning, its function of extending the grasp of men from their nearest staples and commodities to more distant ones is very slight. Increased mobility of grasp and trading is small at first. So it is with the emergence of language in the child. In the first months grasping is reflexive, and the power to make voluntary release comes only toward the end of the first year. Speech comes with the development of the power to let go of objects. It gives the power of detachment from the environment that is also the power of great mobility in knowledge of the environment. So it is with the growth of the idea of money as currency rather than commodity. Currency is a way of letting go of the immediate staples and commodities that at first serve as money, in order to extend trading to the whole social complex. Trading by currency is based on the principle of grasping and letting go in an oscillating cycle. The one hand retains the article with which it tempts the other party. The other hand is extended in demand toward the object which is desired in exchange. The first hand lets go as soon as the second object is touched, somewhat in the manner of a trapeze artist exchanging one bar for another. In fact, Elias Canetti in *Crowds and Power* argues that the trader is involved in one of the most ancient of all pastimes, namely that of climbing trees and swinging from limb to limb. The primitive grasping, calculating, and timing of the greater arboreal apes he sees as a translation into financial terms of one of the oldest movement patterns. Just as the hand among the branches of the trees learned a pattern of grasping that was quite removed from the moving of food to mouth, so the trader and the financier have developed enthralling abstract activities that are extensions of the avid climbing and mobility of the greater apes.

Like any other medium, it is a staple, a natural resource. As an outward and visible form of the urge to change and to exchange, it is a corporate image, depending on society for its institutional status. Apart from communal participation, money is meaningless, as Robinson Crusoe discovered when he found the coins in the wrecked ship:

I smiled to myself at the sight of this money. "O drug!" said I aloud, "What are thou good for? Thou art not worth to me—no, not the taking off the ground: one of those knives is worth all this heap: I have no manner of use for thee; e'en remain where thou art and, go to the bottom, as a creature whose life is not worth saving."

However, upon second thoughts, I took it away; and wrapping it all in a piece of canvas, I began to think of making another raft ...

Primitive commodity money, like the magical words of nonliterate society, can be a storehouse of power, and has often become the occasion of feverish economic activity. The natives of the South Seas, when they are so engaged, seek no economic advantage. Furious application to production may be followed by deliberate destruction of the products in order to achieve moral prestige. Even in these "potlatch" cultures, however, the effect of the currencies was to expedite and to accelerate human energies in a way that had become universal in the ancient world with the technology of the phonetic alphabet. Money, like writing, has the power to specialize and to rechannel human energies and to separate functions, just as it translates and reduces one kind of work to another. Even in the electronic age it has lost none of this power.

Potlatch is very widespread, especially where there is ease of food-gathering or food-production. For example, among the Northwest coast fishermen, or rice-planters of Borneo, huge surpluses are produced that have to be destroyed or class differences would arise that would destroy the traditional social order. In Borneo the traveler may see tons of rice exposed to rains in rituals, and great art constructions, involving tremendous efforts, smashed.

At the same time, in these primitive societies, while money may release frantic energies in order to charge a bit of copper with magical prestige, it can buy very little. Rich and poor necessarily live in much the same manner. Today, in the electronic age, the richest man is reduced to having much the same entertainment, and even the same food and vehicles as the ordinary man.

The use of a commodity such as money naturally increases its production. The nonspecialist economy of Virginia in the seventeenth century made the elaborate European currencies quite dispensable. Having little capital, and wishing to put as little of this capital as possible into the shape of money, the Virginians turned to commodity money in some instances. When a commodity like tobacco was legislated into legal tender, it had the effect of stimulating the production of tobacco, just as the establishment of metallic currencies advanced the mining of metals.

Money, as a social means of extending and amplifying work and skill in an easily accessible and portable form, lost much of its magical power with the coming of representative money, or paper money. Just as speech lost its magic with writing, and further with printing, when printed money supplanted gold, the compelling aura of it disappeared. Samuel Butler in *Erewhon* (1872) gave clear indications in his treatment of the mysterious prestige conferred by precious metals. His ridicule of the money medium took the form of presenting the old reverent attitude to money in a new social context. This new kind of abstract, printed money of the high industrial age, however, simply would not sustain the old attitude:

> This is the true philanthropy. He who makes a colossal fortune in the hosiery trade, and by his energy has succeeded in reducing the price of woollen goods by the thousandth part of a penny in the pound—this man is worth ten professional philanthropists. So strongly are the Erewhonians impressed with this, that if a man has made a fortune of over £20,000 a

year they exempt him from all taxation, considering
him a work of art, and too precious to be meddled
with; they say, "How very much he must have done
for society before society could have been prevailed
upon to give him so much money"; so magnificent
an organization overawes them; they regard it as a
thing dropped from heaven.

   "Money," they say, "is the symbol of duty, it
is the sacrament of having done for mankind that
which mankind wanted. Mankind may not be a
very good judge, but there is no better." This used
to shock me at first, when I remembered that it had
been said on high authority that they who have riches
shall enter hardly into the kingdom of heaven; but
the influence of Erewhon had made me begin to
see things in a new light, and I could not help thinking that they who have not riches shall enter more
hardly still.

Earlier in the book, Butler had ridiculed the cash-register
morality and religion of an industrialized world, under
the guise of the "Musical Banks," with clergy in the role of
cashiers. In the present passage, he perceives money as "the
sacrament of having done for mankind that which mankind
wanted." Money, he is saying, is the "outward and visible sign
of an inward and invisible grace."

Money as a social medium or extension of an inner
wish and motive creates social and spiritual values, as
happens even in fashions in women's dress. A current ad
underlines this aspect of dress as currency (that is, as social
sacrament or outward and visible sign): "The important
thing in today's world of fashion is to appear to be wearing a popular fabric." Conformity to this fashion literally
gives *currency* to a style or fabric, creating a social medium
that increases wealth and expression thereby. Does not this
stress how money, or any medium whatever, is constituted
and made efficacious? When men become uneasy about

such social values achieved by uniformity and repetition, doing for mankind that which mankind wants, we can take it as a mark of the decline of mechanical technology.

"Money talks" because money is a metaphor, a transfer, and a bridge. Like words and language, money is a storehouse of communally achieved work, skill, and experience. Money, however, is also a specialist technology like writing; and as writing intensifies the visual aspect of speech and order, and as the clock visually separates time from space, so money separates work from the other social functions. Even today money is a language for translating the work of the farmer into the work of the barber, doctor, engineer, or plumber. As a vast social metaphor, bridge, or translator, money—like writing—speeds up exchange and tightens the bonds of interdependence in any community. It gives great spatial extension and control to political organizations, just as writing does, or the calendar. It is action at a distance, both in space and in time. In a highly literate, fragmented society, "Time is money," and money is the store of other people's time and effort.

During the Middle Ages the idea of the *fisc* or "the King's purse" kept the notion of money in relation to language ("the King's English") and to communication by travel ("the King's highway"). Before the advent of printing, it was quite natural for the means of communication to be regarded as extensions of a single body. In an increasingly literate society, money and the clock assumed a high degree of visual or fragmented stress. In practice, our Western use of money as store and translator of communal work and skill has depended upon long accustomation to the written word, and upon the power of the written word to specialize, to delegate, and to separate functions in an organization.

When we look at the nature and uses of money in nonliterate societies, we can better understand the ways in which writing helps to establish currencies. Uniformity of commodities, combined with a fixed-price system such as we now take for granted, does not become possible until printing prepares the ground. "Backward" countries take a long time to reach economic "take-off" because they do not undergo the extensive processing of print with its psychological conditioning in the ways of uniformity and repeatability. In general, the West is little aware of the way in which the world of prices and numbering is supported by the pervasive visual culture of literacy.

Nonliterate societies are quite lacking in the psychic resources to create and sustain the enormous structures of statistical information that we call markets and prices. Far easier is the organization of production than is the training of whole populations in the habits of translating their wishes and desires statistically, as it were, by means of market mechanisms of supply and demand, and the visual technology of prices. It was only in the eighteenth century that the West began to accept this form of extension of its inner life in the new statistical pattern of marketing. So bizarre did this new mechanism appear to thinkers of that time that they called it a "Hedonistic calculus." Prices then seemed to be comparable, in terms of feelings and desires, to the vast world of space that had yielded its inequities earlier to the translating power of the differential calculus. In a word, the fragmentation of the inner life by prices seemed as mysterious in the eighteenth century, as the minute fragmentation of space by means of calculus had seemed a century earlier.

The extreme abstraction and detachment represented by our pricing system is quite unthinkable and unusable amidst populations for whom the exciting drama of price haggling occurs with every transaction.

Today, as the new vortices of power are shaped by the instant electric interdependence of all men on this planet, the visual factor in social organization and in personal experience recedes and money begins to be less and less a means of storing or exchanging work and skill. Automation, which is electronic, does not represent physical work so much as programmed knowledge. As work is replaced by the sheer movement of information, money as a store of work merges with the informational forms of credit and credit card. From coin to paper currency, and from currency to credit card there is a steady progression toward commercial exchange as the movement of information itself. This trend toward an inclusive information is the kind of image represented by the credit card, and approaches once more the character of tribal money. For tribal society, not knowing the specialisms of job or of work, does not specialize money either. Its money can be eaten, drunk, or worn like the new space ships that are now designed to be edible.

"Work," however, does not exist in a nonliterate world. The primitive hunter or fisherman did no work, any more than does the poet, painter, or thinker of today. Where the whole man is involved there is no work. Work begins with the division of labor and the specialization of functions and tasks in sedentary, agricultural communities. In the computer age we are once more totally involved in our roles. In the electric age the "job of work" yields to dedication and commitment, as in the tribe.

In nonliterate societies money relates itself to the other organs of society quite simply. The role of money is enormously increased after money begins to foster specialism and separation of social functions. Money becomes, in fact, the principal means of interrelating the ever more specialist activities of literate society. The fragmenting power of the visual sense, as literacy separates it from the other senses, is a fact more easily identified now in the electronic age. Nowadays, with computers and electric programming, the means of storing and moving information become less and less visual and mechanical, while increasingly integral and organic. The total field created by the instantaneous electric forms cannot be visualized any more than the velocities of electronic particles can be visualized. The instantaneous creates an interplay among time and space and human occupations, for which the older forms of currency exchange become increasingly inadequate. A modern physicist who attempted to employ visual models of perception in organizing atomic data would not be able to get anywhere near the nature of his problems. Both time (as measured visually and segmentally) and space (as uniform, pictorial, and enclosed) disappear in the electronic age of instant information. In the age of instant information man ends his job of fragmented specializing and assumes the role of information-gathering. Today information-gathering resumes the inclusive concept of "culture" exactly as the primitive food-gatherer worked in complete equilibrium with his entire environment. Our quarry now, in this new nomadic and "workless" world, is knowledge and insight into the creative processes of life and society.

Men left the closed world of the tribe for the "open society," exchanging an ear for an eye by means of the technology of writing. The alphabet in particular enabled them to break out of the charmed circle and resonating magic of the tribal world. A similar process of economic change from the closed to the open society, from mercantilism and the economic protection of national trade to the open market ideal of the free-traders, was accomplished in more recent times by means of the printed word, and by moving from metallic to paper currencies. Today, electric technology puts the very concept of money in jeopardy, as the new dynamics of human interdependence shift from fragmenting media such as printing to inclusive or mass media like the telegraph.

Since all media are extensions of ourselves, or translations of some part of us into various materials, any study of one medium helps us to understand all the others. Money is no exception. The primitive or nonliterate use of money is especially enlightening, since it manifests an easy acceptance of staple products as media of communication. The nonliterate man can accept any staple as money, partly because the staples of a community are as much media of communication as they are commodities. Cotton, wheat, cattle, tobacco, timber, fish, fur, and many other products have acted as major shaping forces of community life in many cultures. When one of these staples becomes dominant as a social bond, it serves, also, as a store of value, and as a translator or exchanger of skills and tasks.

The classic curse of Midas, his power of translating all he touched into gold, is in some degree the character of any medium, including language. This myth draws attention to a magic aspect of all extensions of human sense and body; that is, to all technology whatever. All technology has the Midas touch. When a community develops some extension of itself, it tends to allow all other functions to be altered to accommodate that form.

Language, like currency, acts as a store of perception and as a transmitter of the perceptions and experience of one person or of one generation to another. As both a translator and storehouse of experience, language is, in addition, a reducer and a distorter of experience. The very great advantage of accelerating the learning process, and of making possible the transmission of knowledge and insight across time and space, easily overrides the disadvantages of linguistic codifications of experience. In modern mathematics and science there are increasingly more and more nonverbal ways of codifying experience.

Money, like language a store of work and experience, acts also as translator and transmitter. Especially since the written word has advanced the separation of social functions, money is able to move away from its role as store of work. This role is obvious when a staple or commodity like cattle or fur is used as money. As money separates itself from the commodity form and becomes a specialist agent of exchange (or translator of values), it moves with greater speed and in ever greater volume.

Even in recent times, the dramatic arrival of paper currency, or "representative money," as a substitute for commodity money caused confusions. Much in the same way, the Gutenberg technology created a vast new republic of letters, and stirred great confusion about the boundaries between the realms of literature and life. Representative money, based on print technology, created new speedy dimensions of credit that were quite inconsistent with the inert mass of bullion and of commodity money. Yet all efforts were bent to make the speedy new money behave like the slow bullion coach. J. M. Keynes stated this policy in *A Treatise on Money*:

> Thus the long age of Commodity Money has at last passed finally away before the age of Representative Money. Gold has ceased to be a coin, a hoard, a tangible claim to wealth, of which the value cannot slip away so long as the hand of the individual clutches the material stuff. It has become a much more abstract thing—just a standard of value; and it only keeps this nominal status by being handed round from time to time in quite small quantities amongst a group of Central Banks, on the occasions when one of them has been inflating or deflating its managed representative money in a different degree from what is appropriate to the behavior of its neighbors.

Paper, or representative money, has specialized itself away from the ancient role of money as a store of work into the equally ancient and basic function of money as transmitter and expediter of any kind of work into any other kind. Just as the alphabet was a drastic visual abstraction from the rich hieroglyphic culture of the Egyptians, so it also reduced and translated that culture into the great visual vortex of the

Graeco-Roman world. The alphabet is a one-way process of reduction of nonliterate cultures into the specialist visual fragments of our Western world. Money is an adjunct of that specialist alphabetic technology, raising even the Gutenberg form of mechanical repeatability to new intensity. As the alphabet neutralized the divergencies of primitive cultures by translation of their complexities into simple visual terms, so representative money reduced moral values in the nineteenth century. As paper expedited the power of the alphabet to reduce the oral barbarians to Roman uniformity of civilization, so paper money enabled Western industry to blanket the globe.

Shortly before the advent of paper money, the greatly increased volume of information movement in European newsletters and newspapers created the image and concept of National Credit. Such a corporate image of credit depended, then as now, on the fast and comprehensive information movement that we have taken for granted for two centuries and more. At that stage of the emergence of public credit, money assumed the further role of translating, not just local, but national stores of work from one culture to another.

One of the inevitable results of acceleration of information movement and of the translating power of money is the opportunity of enrichment for those who can anticipate this transformation by a few hours or years, as the case may be. We are particularly familiar today with examples of enrichment by means of advance information in stocks and bonds and real estate. In the past, when wealth was not so obviously related to information, an entire social class could monopolize the wealth resulting from a casual shift in technology. Keynes' report of just such an instance, in his study of "Shakespeare and the Profit Inflations," explains that since new wealth and bullion fall first to the governing classes, they experience a sudden buoyancy and euphoria, a glad release from the habitual stress and anxieties that fosters a prosperity, which in turn inspires the starving artist in his garret to invent new triumphant rhythms and exultant forms of painting and poetry. As long as profits leap well ahead of wages, the governing class cavorts in a style that inspires the greatest conceptions in the bosom of the impecunious artist. When, however, profits and wages keep in reasonable touch, this abounding joy of the governing class is correspondingly diminished, and art then cannot benefit from prosperity.

Keynes discovered the dynamics of money as a medium. The real task of a study of this one medium is identical with that of the study of all media; namely, as Keynes wrote, "to treat the problem dynamically, analyzing the different elements involved, in such a manner as to exhibit the causal process by which the price level is determined, and the method of transition from one position of equilibrium to another."

In a word, money is not a closed system, and does not have its meaning alone. As a translator and amplifier, money has exceptional powers of substituting one kind of thing for another. Information analysts have come to the conclusion that the degree to which one resource can be substituted for another increases when information increases. As we know more, we rely less on any one food or fuel or raw material. Clothes and furniture can now be made from many different materials. Money, which had been for many centuries the principal transmitter and exchanger of information, is now having its function increasingly transferred to science and automation.

Today, even natural resources have an informational aspect. They exist by virtue of the culture and skill of some community. The reverse, however, is true also. All media—or extension of man—are natural resources that exist by virtue of the shared knowledge and skill of a community. It was awareness of this aspect of money that hit Robinson Crusoe very hard when he visited the wreck, resulting in the meditation quoted at the beginning of this chapter.

When there are goods but no money, some sort of barter —or direct exchange of one product for another—has to occur. When, however, in nonliterate societies goods are used in direct exchange, then it is easiest to note their tendency to include the function of money. Some work has been done to some material, if only in bringing it from a distance. The object, then, stores work and information or technical knowledge to the extent that something has been done to it. When the one object is exchanged for another, it is already assuming the function of money, as translator or reducer of multiple things to some common denominator. The common denominator (or translator) is, however, also a time-saver and expediter. As such, money is time, and it would be hard to separate labor-saving from time-saving in this operation.

There is a mystery about the Phoenicians, who, although they were avid maritime traders, adopted coinage later than the landed Lydians. The reason assigned for this delay may not explain the Phoenician problem, but it draws sharp attention to a basic fact about money as a medium; namely, that those who traded by caravan required a light and portable medium of payment. This need was less for those who, like the Phoenicians, traded by sea. Portability, as a means of expediting and extending the effective distance of action, was also notably illustrated by papyrus. The alphabet was one thing when applied to clay or stone, and quite another when set down on light papyrus. The resulting leap in speed and space created the Roman Empire.

In the industrial age the increasingly exact measurement of work revealed time-saving as a major aspect of labor-saving. The media of money and writing and clock began to converge into an organic whole again that has brought us as close to the total involvement of man in his work, as of native in a primitive society, or of artist in his studio.

Money in one of its features provides a natural transition to number because the money hoard or collection has much

in common with the crowd. Moreover, the psychological patterns of the crowd and those associated with accumulations of
wealth are very close. Elias Canetti stresses that the dynamic
which is basic to crowds is the urge to rapid and unlimited
growth. The same power dynamic is characteristic of large
concentrations of wealth or treasure. In fact, the modern unit
of treasure in popular use is the *million*. It is a unit acceptable
to any type of currency. Always associated with the *million* is
the idea that it can be reached by a rapid speculative scramble.
In the same way, Canetti explains how the ambition to see
numbers mounting up was typical of Hitler's speeches.

Not only do crowds of people and piles of money strive
toward increase, but they also breed uneasiness about the
possibility of disintegration and deflation. This two-way
movement of expansion and deflation seems to be the cause of
the restlessness of crowds and the uneasiness that goes with
wealth. Canetti spends a good deal of analysis on the psychic
effects of the German inflation after the First World War. The
depreciation of the citizen went along with that of the German
mark. There was a loss of face and of worth in which the personal and monetary units became confused.

15

Clocks:
The Scent of Time

*McLuhan upbraids other thinkers here: Parkinson "squints,"
Eliade "is unaware," Mumford "takes no account of the phonetic alphabet" in treating the subject of time. McLuhan is in
overdrive in this chapter, moving within a single paragraph
through the entire evolution of technology, explaining that,
under electricity, synchronization is no longer simultaneity,
and returning to the subject of clocks. There is also commentary
on literacy as asceticism, the connections among the fifth century B.C., the Renaissance, and the twentieth century. Myth,
alluded to in the early pages of the book in relation to the electric
age, is connected here to iconicity and the multifaceted. The
chapter opens with the first of two references (a third will come
in Chapter 21, The Press) to alarm clocks as apparel, a passage
that inspired Umberto Eco to a monologue of the deaf (see Critical
Reception, p. 552-57).*

— (Editor)

Writing on *Communication in Africa*, Leonard Doob observes: "The turban, the sword and nowadays the alarm clock are worn or carried to signify high rank." Presumably it will be rather long before the African will watch the clock in order to be punctual.

Just as a great revolution in mathematics came when positional, tandem numbers were discovered (302 instead of 32, and so on), so great cultural changes occurred in the West when it was found possible to fix time as something that happens between two points. From this application of visual, abstract, and uniform units came our Western feeling for time as duration. From our division of time into uniform, visualizable units comes our sense of duration and our impatience when we cannot endure the delay between events. Such a sense of impatience, or of time as duration, is unknown among nonliterate cultures. Just as *work* began with the division of labor, duration begins with the division of time, and especially with those subdivisions by which mechanical clocks impose uniform succession on the time sense.

As a piece of technology, the clock is a machine that produces uniform seconds, minutes, and hours on an assembly-line pattern. Processed in this uniform way, time is separated from the rhythms of human experience. The mechanical clock, in short, helps to create the image of a numerically quantified and mechanically powered universe. It was in the world of the medieval monasteries, with their need for a rule and for synchronized order to guide communal life, that the clock got started on its modern developments. Time measured not by the uniqueness of private experience but by abstract uniform units gradually pervades all sense life, much as does the technology of writing and printing. Not only work, but also eating and sleeping, came to accommodate themselves to the clock rather than to organic needs. As the pattern of arbitrary and uniform measurement of time extended itself across society, even clothing began to undergo annual alteration in a way convenient for industry. At that point, of course, mechanical measurement of time as a principle of applied knowledge joined forces with printing and assembly line as means of uniform fragmentation of processes.

The most integral and involving time sense imaginable is that expressed in the Chinese and Japanese cultures. Until the coming of the missionaries in the seventeenth century, and the introduction of the mechanical clocks, the Chinese and Japanese had for thousands of years measured time by graduations of incense. Not only the hours and days, but the seasons and zodiacal signs were simultaneously indicated by a succession of carefully ordered scents. The sense of smell, long considered the root of memory and the unifying basis of individuality, has come to the fore again in the experiments of Wilder Penfield. During brain surgery, electric probing of brain tissue revived many memories of the patients. These evocations were dominated and unified by unique scents and odors that structured these past experiences. The sense of smell is not only the most subtle and delicate of the human senses; it is, also, the most iconic in that it involves the entire human sensorium more fully than any other sense. It is not surprising, therefore, that highly literate societies take steps to reduce or eliminate odors from the environment. B.O., the unique signature and declaration of human individuality, is a bad word in literate societies. It is far too involving for our habits of detachment and specialist attention. Societies that measured time scents would tend to be so cohesive and so profoundly unified as to resist every kind of change.

Lewis Mumford has suggested that the clock preceded the printing press in order of influence on the mechanization of society. But Mumford takes no account of the phonetic alphabet as the technology that had made possible the visual and uniform fragmentation of time. Mumford, in fact, is unaware of the alphabet as the source of Western mechanism, just as he is unaware of mechanization as the translation of society from audile-tactile modes into visual values. Our new electric technology is organic and non-mechanical in tendency because it extends, not our eyes, but our central nervous systems as a planetary vesture. In the space-time world of electric technology, the older mechanical time begins to feel unacceptable, if only because it is uniform.

Modern linguistic studies are structural rather than literary, and owe much to the new possibilities of computers for translation. As soon as an entire language is examined as a unified system, strange pockets appear. Looking at the usage scale of English, Martin Joos has wittily designated "five clocks of style," or five different zones and independent cultural climates. Only one of these zones is the area of *responsibility*. This is the zone of homogeneity and uniformity that ink-browed Gutenberg rules as his domain. It is the style-zone of Standard English pervaded by Central Standard Time, and within this zone the dwellers, as it were, may show varying degrees of punctuality.

Edward T. Hall in *The Silent Language* discusses how "Time Talks: American Accents," contrasting our time-sense with that of the Hopi Indians. Time for them is not a uniform succession or duration, but a pluralism of many kinds of things co-existing. "It is what happens when the corn matures or a sheep grows up.... It is the natural process that takes place while living substance acts out its life drama." Therefore, as many kinds of time exist for them as there are kinds of life. This, also, is the kind of time-sense held by the modern physicist and scientist. They no longer try to contain events in time, but think of each thing as making its own time and its own space. Moreover, now that we live electrically in an instantaneous world, space and time interpenetrate each other totally in a space-time world. In the same way, the painter, since Cézanne, has recovered the *plastic image* by which all of the senses coexist in a unified pattern. Each object and each set of objects engenders its own unique space by the relations it has among others visually or musically. When this awareness recurred in the Western world, it was denounced as the merging of all things in a flux. We now realize that this anxiety was a natural literary and visual response to the new nonvisual technology.

J. Z. Young, in *Doubt and Certainty in Science*, explains how electricity is not something that is conveyed by or contained in anything, but is something that occurs when two or more bodies are in special positions. Our language derived from phonetic technology cannot cope with this new view of knowledge. We still talk of electric current "flowing," or we speak of the "discharge" of electric energy like the lineal firing of guns. But quite as much as with the esthetic magic of painterly power, "electricity is the condition we observe when there are certain spatial relations between things." The painter learns how to adjust relations among things to release new perception, and the chemist and physicist learn how other relations release other kinds of power. Less and less, in the electric age, can we find any good reason for imposing the same set of relations on every kind of object or group of objects. Yet in the ancient world the only means of achieving power was getting a thousand slaves to act as one man. During the Middle Ages the communal clock extended by the bell permitted high coordination of the energies of small communities. In the Renaissance the clock combined with the uniform respectability of the new typography to extend the power of social organization almost to a national scale. By the nineteenth century it had provided a technology of cohesion that was inseparable from industry and transport, enabling an entire metropolis to act almost as an automaton. Now in the electric age of decentralized power and information we begin to chafe under the uniformity of clock-time. In this age of space-time we seek multiplicity, rather than repeatability, of rhythms. This is the difference between marching soldiers and ballet.

It is a necessary approach in understanding media and technology to realize that when the spell of the gimmick or an extension of our bodies is new, there comes *narcosis* or numbing to the newly amplified area. The complaints about clocks did not begin until the electric age had made their mechanical sort of time starkly incongruous. In our electric century the mechanical time-kept city looks like an aggregation of somnambulists and zombies, made familiar in the early part of T. S. Eliot's *The Waste Land*.

On a planet reduced to village size by new media, cities themselves appear quaint and odd, like archaic forms already overlaid with new patterns of culture. However, when mechanical clocks had been given great new force and practicality by mechanical writing, as printing was at first called, the response to the new time sense was very ambiguous and even mocking. Shakespeare's sonnets are full of the twin themes of immortality of fame conferred by the engine of print, as well as the petty futility of daily existence as measured by the clock:

> When I doe count the clock that tels the time,
> And see the brave day sunck in hidious night …
> Then of thy beauty do I question make
> That thou among the wastes of time must goe.
> ("Sonnet X")

In *Macbeth*, Shakespeare links the twin technologies of print and mechanical time in the familiar soliloquy, to manifest the disintegration of Macbeth's world:

> Tomorrow, and tomorrow, and tomorrow
> Creeps in this petty pace from day to day,
> To the last syllable of recorded time.

Time, as hacked into uniform successive bits by clock and print together, became a major theme of the Renaissance neurosis, inseparable from the new cult of precise measurement in the sciences. In "Sonnet LX," Shakespeare puts mechanical time at the beginning, and the new engine of immortality (print) at the end:

> Like as the waves make towards the pibled shore,
> So do our minuites hasten to their end,
> Each changing place with that which goes before,
> In sequent toil all forwards do contend …
> And yet to times in hope, my verse shall stand
> Praising thy worth, dispight his cruell hand.

John Donne's poem on "The Sun Rising" exploits the contrast between aristocratic and bourgeois time. The one trait that most damned the *bourgeoisie* of the nineteenth century was their punctuality, their pedantic devotion to mechanical time and sequential order. As space-time flooded through the gates of awareness from the new electric technology, all mechanical observance became distasteful and even ridiculous. Donne had the same ironic sense of the irrelevance of clock-time, but pretended that in the kingdom of love even the great cosmic cycles of time were also petty aspects of the clock:

> Busy old fool, unruly Sun,
> Why dost thou thus
> Through windows, and through curtains call on us?
> Must to thy motions lovers' seasons run?
> Saucy, pedantic wretch, go chide
> Late school-boys, and sour prentices,
> Go tell Court-huntsmen, that the King will ride,
> Call country ants to harvest offices,
> Love, all alike, no season knows nor clime,
> Nor hours, days, months, which are the rags of time.

Much of Donne's twentieth-century vogue was due to his challenging the authority of the new Gutenberg age to invest him with the stigmata of uniform repeatable typography and with the motives of precise visual measurement. In like manner, Andrew Marvell's "To his Coy Mistress" was full of contempt for the new spirit of measurement and calculation of time and virtue:

> Had we but world enough and time,
> This coyness, lady, were no crime.
> We would sit down and think which way
> To walk, and pass our long love's day ...
> An hundred years should go to praise
> Thine eyes, and on thy forehead gaze;
> Two hundred to adore each breast,
> But thirty thousand to the rest;
> An age at least to every part,
> And the last age should show your heart,
> For lady, you deserve this state,
> Nor would I love at lower rate.

Marvell merged the rates of exchange with the rates of praise suited to the conventional and fashionably fragmented outlook of his inamorata. For her box-office approach to reality, he substituted another time-structure, and a different model of perception. It is not unlike Hamlet's "Look on this picture and on that." Instead of a quiet bourgeois translation of the medieval love code into the language of the new middle-class tradesman, why not a Byronic caper to the farther shores of ideal love?

> But at my back I always hear
> Time's winged chariot hurrying near;
> And yonder all before us lie
> Deserts of vast eternity.

Here is the new lineal perspective that had come to painting with Gutenberg, but that had not entered the verbal universe until Milton's *Paradise Lost*. Even written language had resisted for two centuries the abstract visual order of lineal succession and vanishing point. The next age after Marvell, however, took to landscape poetry and the subordination of language to special visual effects.

But Marvell concluded his reverse strategy for the conquest of bourgeois clock-time with the observation:

> Thus, though we cannot make our sun
> Stand still, yet we will make him run.

He proposed that his beloved and he should transform themselves into a cannonball and fire themselves at the sun to make it run. Time can be defeated, as it were, by reversal of its characteristics if only it be speeded up enough. Experience of this fact awaited the electronic age, which found that instant speeds abolish time and space, and return man to an integral and primitive awareness.

Today not only clock-time, but the wheel itself, is obsolescent and is retracting into animal form under the impulse of greater and greater speeds. In the poem above, Andrew Marvell's intuition that clock-time could be defeated by speed was quite sound. At present the mechanical begins to yield to organic unity under conditions of electric speeds. Man now can look back at two or three thousand years of varying degrees of mechanization with full awareness of the mechanical as an interlude between two great organic periods of culture. In 1911 the Italian sculptor Boccioni said, "We are primitives of an unknown culture." Half a century later we know a bit more about the new culture of the electronic age, and that knowledge has lifted the mystery surrounding the machine.

As contrasted with the mere tool, the machine is an extension or outering of a process. The tool extends the fist, the nails, the teeth, the arm. The wheel extends the feet in rotation or sequential movement. Printing, the first complete mechanization of a handicraft, breaks up the movement of the hand into a series of discrete steps that are as repeatable as the wheel is rotary. From this analytic sequence came the assembly-line principle, but the assembly line is now obsolete in the electric age because synchronization is no longer sequential. By electric tapes, synchronization of any number of different acts can be simultaneous. Thus the mechanical principle of analysis in series has come to an end. Even the wheel has now come to an end in principle, although the mechanical stratum of our culture carries it still as part of an accumulated momentum, an archaic configuration.

The modern clock, mechanical in principle, embodied the wheel. The clock has ceased to have its older meanings and functions. Plurality-of-times succeeds uniformity-of-time. Today it is only too easy to have dinner in New York and indigestion in Paris. Travelers also have the daily experience of being at one hour in a culture that is still 3000 B.C., and at the next hour in a culture that is 1900 A.D. Most of North American life is, in its externals, conducted on nineteenth-century lines. Our inner experience, increasingly at variance with these mechanical patterns, is electric, inclusive, and mythic in mode. The mythic or iconic mode of awareness substitutes the multi-faceted for point-of-view.

Historians agree on the basic role of the clock in monastic life for the synchronization of human tasks. The acceptance of such fragmenting of life into minutes and hours was unthinkable, save in highly literate communities. Readiness to submit the human organism to the alien mode of mechanical time was as dependent upon literacy in the first Christian centuries as it is today. For the clock to dominate, there has to be the prior acceptance of the visual stress that is inseparable from phonetic literacy. Literacy is itself an abstract asceticism that prepares the way for endless patterns of privation in the human community. With universal literacy, time can take on the character of an enclosed or pictorial space that can be divided and subdivided. It can be filled in. "My schedule is filled up." It can be kept free: "I have a free week next month." And as Sebastian de Grazia has shown in *Of Time, Work and Leisure,* all the free time in the world is not leisure, because leisure accepts neither the division of labor that constitutes "work," nor the divisions of time that create "full time" and "free time." Leisure excludes time as a container. Once time is mechanically or visually enclosed, divided, and filled, it is possible to use it more and more efficiently. Time can be transformed into a labor-saving machine, as Parkinson reveals in his famous "Parkinson's Law."

The student of the history of the clock will find that a totally new principle entered with the invention of the mechanical clock. The earliest mechanical clocks had retained the old principle of the continuous action of the driving force, such as was used in the water clock and in the water wheel. It was about 1300 A.D. that the step was taken of momentarily interrupting rotary movement by a crown rod and balance wheel. This function was called "escapement" and was the means of literally translating the continuous force of the wheel into the visual principle of uniform but segmented succession. Escapement introduced the reciprocal reversing action of the hands in rotating a spindle forward and backward. The meeting in the mechanical clock of this ancient extension of hand movement with the forward rotary motion of the wheel was, in effect, the translation of hands into feet, and feet into hands. Perhaps no more difficult technological extension of interinvolved bodily appendages could be found. The source of the energy of the clock was thus separated from the hands, or the source of information, by technological translation. Escapement as a translation of one kind of wheel space into uniform and visual space is thus a direct anticipation of the infinitesimal calculus that translates any kind of space or movement into a uniform, continuous, and visual space.

Parkinson, sitting on the fence between the mechanical and the electric uses of work and time, is able to provide us with real entertainment by simply squinting, now with one eye, now with the other, at the time and work picture. Cultures like ours, poised at the point of transformation, engender both tragic and comic awareness in great abundance. It is the maximal interplay of diverse forms of perception and experience that makes great the cultures of the fifth century B.C., the sixteenth century, and the twentieth century. But few people have enjoyed living in these intense periods when all that ensures familiarity and security dissolves and is reconfigured in a few decades.

It was not the clock, but literacy reinforced by the clock, that created abstract time and led men to eat, not when they were hungry, but when it was "*time* to eat." Lewis Mumford makes a telling observation when he says that the abstract mechanical time-sense of the Renaissance enabled men to live in the classical past, and to tear themselves out of their own present. Here again, it was the printing press that made possible the recreation of the classic past by mass production of its literature and texts. The establishment of a mechanical and abstract time *pattern* soon extends itself to periodic alteration of clothing styles, much in the same way that mass production extends itself to periodic publication of newspapers and magazines. Today we take for granted that the job of *Vogue* magazine is to alter the dress styles as part of the process of its being printed at all. When a thing is current, it creates currency; fashion creates wealth by moving textiles and making them ever more current. This process we have seen at work in the section on "Money." Clocks are mechanical media that transform tasks and create new work and wealth by accelerating the pace of human association. By coordinating and accelerating human meetings and goings-on, clocks increase the sheer quantity of human exchange.

It is not really incongruous, therefore, when Mumford associates "the clock and the printing press and the blast furnace" as the giant innovations of the Renaissance. The clock, as much as the blast furnace, speeded the melting of materials and the rise of smooth conformity in the contours of social life. Long before the industrial revolution of the later

209

eighteenth century, people complained that society had become a "prose machine" that whisked them through life at a dizzy pace.

The clock dragged man out of the world of seasonal rhythms and recurrence, as effectively as the alphabet had released him from the magical resonance of the spoken word and the tribal trap. This dual translation of the individual out of the grip of Nature and out of the clutch of the tribe was not without its own penalties. But the return to Nature and the return to the tribe are, under electric conditions, fatally simple. We need beware of those who announce programs for restoring man to the original state and language of the race. These crusaders have never examined the role of media and technology in tossing man about from dimension to dimension. They are like the somnambulistic African chief with the alarm clock strapped to his back.

Mircea Eliade, professor of comparative religion, is unaware, in *The Sacred and the Profane*, that a "sacred" universe in his sense is one dominated by the spoken word and by auditory media. A "profane" universe, on the other hand, is one dominated by the visual sense. The clock and the alphabet, by hacking the universe into visual segments, ended the music of interrelation. The visual desacralizes the universe and produces the "nonreligious man of modern societies."

Historically, however, Eliade is useful in recounting how, before the age of the clock and the time-kept city, there was for tribal man a cosmic clock and a sacred time of the cosmogony itself. When tribal man wanted to build a city or a house, or cure an illness, he wound up the cosmic clock by an elaborate ritual reenactment or recitation of the original process of creation. Eliade mentions that in Fiji "the ceremony for installing a new ruler is called 'creation of the world.' " The same drama is enacted to help the growth of crops. Whereas modern man feels obligated to be punctual and conservative of time, tribal man bore the responsibility for keeping the cosmic clock supplied with energy. But electric or ecological man (man of the total field) can be expected to surpass the old tribal cosmic concern with the Africa within.

Primitive man lived in a much more tyrannical cosmic machine than Western literate man has ever invented. The world of the ear is more embracing and inclusive than that of the eye can ever be. The ear is hypersensitive. The eye is cool and detached. The ear turns man over to universal panic while the eye, extended by literacy and mechanical time, leaves some gaps and some islands free from the unremitting acoustic pressure and reverberation.

16

The Print:
How to Dig It

It is surprising that this rich chapter is so seldom referred to in discussions of McLuhan's writings. He works on several fronts at once, giving pre-Gutenberg printing of text and image on paper by woodcut its due in the development of modern science and technology, discussing the print's advantage over words as a store of information, situating it in the progression from medieval manuscripts to modern newspapers, and calling attention to the feature of tactility that unites the beginning and the end of this evolution of media. McLuhan's perspective explains the enthusiastic reception accorded to woodcuts and photography alike in print-dominated cultures. This chapter's subtitle, text, and conclusion anticipate the link to the cool medium of comics to be discussed in Chapter 17. Just how vast the implications of McLuhan's approach to media studies are for the intellectual enterprise emerges clearly as the chapter draws to a close with a comment on the reductive impulse that led Descartes to import mathematical rigor as a requisite for philosophical investigation: ("This striving for an irrelevant precision served only to exclude from philosophy most of the questions of philosophy; and that great kingdom of philosophy was soon parceled out into the wide range of uncommunicating sciences and specialties we know today.")

— (Editor)

The art of making pictorial statements in a precise and re-peatable form is one that we have long taken for granted in the West. But it is usually forgotten that without prints and blueprints, without maps and geometry, the world of modern sciences and technologies would hardly exist.

In the time of Ferdinand and Isabella and other maritime monarchs, maps were top-secret, like new electronic discoveries today. When the captains returned from their voyages, every effort was made by the officers of the crown to obtain both originals and copies of the maps made during the voyage. The result was a lucrative black-market trade, and secret maps were widely sold. The sort of maps in question had nothing in common with those of later design, being in fact more like diaries of different adventures and experiences. For the later perception of space as uniform and continuous was unknown to the medieval cartographer, whose efforts resembled modern nonobjective art. The shock of the new Renaissance space is still felt by natives who encounter it today for the first time. Prince Modupe tells in his autobiography, *I Was a Savage*, how he had learned to read maps at school, and how he had taken back home to his village a map of a river his father had traveled for years as a trader.

> … my father thought the whole idea was absurd. He refused to identify the stream he had crossed at Bomako, where it is no deeper, he said, than a man is high, with the great widespread waters of the vast Niger delta. Distances as measured in miles had no meaning for him … Maps are liars, he told me briefly. From his tone of voice I could tell that I had offended him in some way not known to me at the time. The things that hurt one do not show on a map The truth of a place is in the joy and the hurt that come from it. I had best not put my trust in anything as inadequate

as a map, he counseled ... I understand now, although
I did not at the time, that my airy and easy sweep of
map-traced staggering distances belittled the journeys
he had measured on tired feet. With my big map-talk,
I had effaced the magnitude of his cargo-laden, heat-
weighted treks.

All the words in the world cannot describe an object like
a bucket, although it is possible to tell in a few words how
to *make* a bucket. This inadequacy of words to convey visual
information about objects was an effectual block to the
development of the Greek and Roman sciences. Pliny the
Elder reported the inability of the Greek and Latin botanists
to devise a means of transmitting information about plants
and flowers:

Hence it is that other writers have confined themselves to a verbal description of the plants; indeed
some of them have not so much as described them
even, but have contented themselves for the most
part with a bare recital of their names ...

We are confronted here once more with that basic function of media—to store and to expedite information. Plainly,
to store is to expedite, since what is stored is also more
accessible than what has to be gathered. The fact that visual
information about flowers and plants cannot be stored verb-
ally also points to the fact that science in the Western world
has long been dependent on the visual factor. Nor is this
surprising in a literate culture based on the technology of the
alphabet, one that reduces even spoken language to a visual
mode. As electricity has created multiple non-visual means
of storing and retrieving information, not only culture but
science also has shifted its entire base and character. For the
educator, as well as the philosopher, exact knowledge of
what this shift means for learning and the mental process
is not necessary.

Well before Gutenberg's development of printing from movable types, a great deal of printing on paper by woodcut had been done. Perhaps the most popular form of this kind of block printing of text and image had been in the form of the *Biblia Pauperum*, or Bibles of the Poor. Printers in this woodcut sense preceded typographic printers, though by just how long a period it is not easy to establish, because these cheap and popular prints, despised by the learned, were not preserved any more than are the comic books of today. The great law of bibliography comes into play in this matter of the printing that precedes Gutenberg: "The more there were, the fewer there are." It applies to many items besides printed matter—to the postage stamp and to the early forms of radio receiving sets.

Medieval and Renaissance man experienced little of the separation and specialty among the arts that developed later. The manuscript and the earlier printed books were read aloud, and poetry was sung or intoned. Oratory, music, literature, and drawing were closely related. Above all, the world of the illuminated manuscript was one in which lettering itself was given plastic stress to an almost sculptural degree. In a study of the art of Andrea Mantegna, the illuminator of manuscripts, Millard Meiss mentions that, amidst the flowery and leafy margins of the page, Mantegna's letters "rise like monuments, stony, stable and finely cut.... Palpably soled and weighty, they stand boldly before the colored ground, upon which they often throw a shadow...."

The same feeling for the letters of the alphabet as engraved icons has returned in our own day in the graphic arts and in advertising display. Perhaps the reader will have encountered the sense of this coming change in Rimbaud's sonnet on the vowels, or in some of Braque's paintings. But ordinary newspaper headline style tends to push letters toward the iconic form, a form that is very near to auditory resonance, as it is also to tactile and sculptural quality.

Perhaps the supreme quality of the print is one that is lost on us, since it has so casual and obvious an existence. It is simply that it is a pictorial statement that can be repeated precisely and indefinitely—at least as long as the printing surface lasts. Repeatability is the core of the mechanical principle that has dominated our world, especially since the Gutenberg technology. The message of the print and of typography is primarily that of repeatability. With typography, the principle of movable type introduced the means of mechanizing any handicraft by the process of segmenting and fragmenting an integral action. What had begun with the alphabet as a separation of the multiple gestures and sights and sounds in the spoken word, reached a new level of intensity, first with the woodcut and then with typography. The alphabet left the visual component as supreme in the word, reducing all other sensuous facts of the spoken word to this form. This helps to explain why the woodcut, and even the photograph, were so eagerly welcomed in a literate world. These forms provide a world of inclusive gesture and dramatic posture that necessarily is omitted in the written word.

The print was eagerly seized upon as a means of imparting information, as well as an incentive to piety and meditation. In 1472 the *Art of War* by Volturius was printed at Verona, with many woodcuts to explain the machinery of war. But the uses of the woodcut as an aid to contemplation in Books of Hours, Emblems, and Shepherds' Calendars continued for two hundred years on a large scale.

It is relevant to consider that the old prints and woodcuts, like the modern comic strip and comic book, provide very little data about any particular moment in time, or aspect in space, of an object. The viewer, or reader, is compelled to participate in completing and interpreting the few hints provided by the bounding lines. Not unlike the character of the woodcut and the cartoon is the TV image, with its very low degree of data about objects, and the resulting high degree of participation by the viewer in order to complete what is only hinted at in the mosaic mesh of dots. Since the advent of TV, the comic book has gone into a decline.

It is, perhaps, obvious enough that if a cool medium involves the viewer a great deal, a hot medium will not. It may contradict popular ideas to say that typography as a hot medium involves the reader much less than did manuscript, or to point out that the comic book and TV as cool media involve the user, as maker and participant, a great deal.

After the exhaustion of the Graeco-Roman pools of slave labor, the West had to technologize more intensively than the ancient world had done. In the same way the American farmer, confronted with new tasks and opportunities, and at the same time with a great shortage of human assistance, was goaded into a frenzy of creation of labor-saving devices. It would seem that the logic of success in this matter is the ultimate retirement of the work force from the scene of toil. In a word, automation. If this, however, has been the motive behind all of our human technologies, it does not follow that we are prepared to accept the consequences. It helps to get one's bearings to see the process at work in remote times when work meant specialist servitude, and leisure alone meant a life of human dignity and involvement of the whole man.

The print in its clumsy woodcut-phase reveals a major aspect of language; namely, that words cannot bear sharp definition in daily use. When Descartes surveyed the philosophical scene at the beginning of the seventeenth century, he was appalled at the confusion of tongues and began to strive toward a reduction of philosophy to precise mathematical form. This striving for an irrelevant precision served only to exclude from philosophy most of the questions of philosophy; and that great kingdom of philosophy was soon parceled out into the wide range of uncommunicating sciences and specialties we know today. Intensity of stress on visual blueprinting and precision is an explosive force that fragments the world of power and knowledge alike. The increasing precision and quantity of visual information transformed the print into a three-dimensional world of perspective and fixed point of view. Hieronymus Bosch, by means of paintings that interfused medieval forms in Renaissance space, told what it felt like to live straddled between the two worlds of the old and the new during this revolution. Simultaneously, Bosch provided the older kind of plastic, tactile image but placed it in the intense new visual perspective. He gave at once the older medieval idea of unique, discontinuous space, superimposed on the new idea of uniform, connected space. This he did with earnest nightmare intensity.

Lewis Carroll took the nineteenth century into a dream world that was as startling as that of Bosch, but built on reverse principles. *Alice in Wonderland* offers as norm that continuous time and space that had created consternation in the Renaissance. Pervading this uniform Euclidean world of familiar space-and-time, Carroll drove a fantasia of discontinuous space-and-time that anticipated Kafka, Joyce, and Eliot. Carroll, the mathematical contemporary of Clerk Maxwell, was quite *avant-garde* enough to know about the non-Euclidean geometries coming into vogue in his time. He gave the confident Victorians a playful foretaste of Einsteinian time-and-space in *Alice in Wonderland*. Bosch had provided his era a foretaste of the new continuous time-and-space of uniform perspective. Bosch looked ahead to the modern world with horror, as Shakespeare did in *King Lear*, and as Pope did in *The Dunciad*. But Lewis Carroll greeted the electronic age of space-time with a cheer.

Nigerians studying at American universities are sometimes asked to identify spatial relations. Confronted with objects in sunshine, they are often unable to indicate in which direction shadows will fall, for this involves casting into three-dimensional perspective. Thus sun, objects, and observer are experienced separately and regarded as independent of one another. For medieval man, as for the native, space was not homogeneous and did not *contain* objects. Each thing made its own space, as it still does for the native (and equally for the modern physicist). Of course this does not mean that native artists do not relate things. They often contrive the most complicated, sophisticated configurations. Neither artist nor observer has the slightest trouble recognizing and interpreting the pattern, but only when it is a traditional one. If you begin to modify it, or translate it into another medium (three dimensions, for instance), the native fails to recognize it.

An anthropological film showed a Melanesian carver cutting out a decorated drum with such skill, coordination, and ease that the audience several times broke into applause — it became a song, a ballet. But when the anthropologist asked the tribe to build crates to ship these carvings in, they struggled unsuccessfully for three days to make two planks intersect at a 90-degree angle, then gave up in frustration. They couldn't crate what they had created.

In the low definition world of the medieval woodcut, each object created its own space, and there was no rational connected space into which it must fit. As the retinal impression is intensified, objects cease to cohere in a space of their own making, and, instead, become "contained" in a uniform, continuous, and "rational" space. Relativity theory in 1905 announced the dissolution of uniform Newtonian space as an illusion or fiction, however useful. Einstein pronounced the doom of continuous or "rational" space, and the way was made clear for Picasso and the Marx brothers and *MAD*.

17

Comics:
MAD Vestibule to TV

A topic that McLuhan had explored extensively in The Mechanical Bride *is admirably compressed into a few pages. Popular usage, once again, provides McLuhan with a clue to media mediating culture and technology—here* square *is an electric Everyman's judgment on all* forms *of high definition. It is also a rich echo of the anecdote at the end of Chapter 16, The Print, about the Melanesian carvers who could not crate what they created. McLuhan may shock here with the observation that "the TV image has ended the consumer phase of American culture." If it strikes the reader as counter-factual, it should provoke a reflection on the full meaning of his phrase "trained incapacity."*

— (Editor)

It was thanks to the print that Dickens became a comic writer. He began as a provider of copy for a popular cartoonist. To consider the comics here, after "The Print," is to fix attention upon the persistent print-like, and even crude woodcut, characteristics of our twentieth-century comics. It is by no means easy to perceive how the same qualities of print and woodcut could reappear in the mosaic mesh of the TV image. TV is so difficult a subject for literary people that it has to be approached obliquely. From the three million dots per second on TV, the viewer is able to accept, in an iconic grasp, only a few dozen, seventy or so, from which to shape an image. The image thus made is as crude as that of the comics. It is for this reason that the print and the comics provide a useful approach to understanding the TV image, for they offer very little visual information or connected detail. Painters and sculptors, however, can easily understand TV, because they sense how very much tactile involvement is needed for the appreciation of plastic art.

The structural qualities of the print and woodcut obtain, also, in the cartoon, all of which share a participational and do-it-yourself character that pervades a wide variety of media experiences today. The print is clue to the comic cartoon, just as the cartoon is clue to understanding the TV image.

Many a wrinkled teenager recalls his fascination with that pride of the comics, the "Yellow Kid" of Richard F. Outcault. On first appearance, it was called "Hogan's Alley" in the New York *Sunday World*. It featured a variety of scenes of kids from the tenements, Maggie and Jiggs as children, as it were. This feature sold many papers in 1898 and thereafter. Hearst soon bought it, and began large-scale comic supplements. Comics (as already explained in the chapter on The Print), being low in definition, are a highly participational form of expression, perfectly adapted to the mosaic form of the newspaper. They provide, also, a sense of continuity from one day to the next. The individual news item is very low in information, and requires completion or fill-in by the reader, exactly as does the TV image, or the wirephoto. That is the reason why TV hit the comic-book world so hard. It was a real rival, rather than a complement. But TV hit the pictorial ad world even harder, dislodging the sharp and glossy, in favor of the shaggy, the sculptural, and the tactual. Hence the sudden eminence of *MAD* magazine which offers, merely, a ludicrous and cool replay of the forms of the hot media of photo, radio, and film. *MAD* is the old print and woodcut image that recurs in various media today. Its type of configuration will come to shape all of the acceptable TV offerings.

The biggest casualty of the TV impact was Al Capp's "Li'l Abner." For eighteen years Al Capp had kept Li'l Abner on the verge of matrimony. The sophisticated formula used with his characters was the reverse of that employed by the French novelist Stendhal, who said, "I simply involve my people in the consequences of their own stupidity and then give them brains so they can suffer." Al Capp, in effect, said, "I simply involve my people in the consequences of their own stupidity and then *take away* their brains so that they can do nothing about it." Their inability to help themselves created a sort of parody of all the other suspense comics. Al Capp pushed suspense into absurdity. But readers have long enjoyed the fact that the Dogpatch predicament of helpless ineptitude was a paradigm of the human situation, in general.

With the arrival of TV and its iconic mosaic image, the everyday life situations began to seem very square, indeed. Al Capp suddenly found that his kind of distortion no longer worked. He felt that Americans had lost their power to laugh at themselves. He was wrong. TV simply involved everybody in everybody more deeply than before. This cool medium, with its mandate of participation in depth, required Capp to refocus the Li'l Abner image. His confusion and dismay were a perfect match for the feelings of those in every major American enterprise. From *Life* to General Motors, and from the classroom to the Executive Suite, a refocusing of aims and images to permit ever more audience involvement and participation has been inevitable. Capp said: "But now America has changed. The humorist feels the change more, perhaps, than anyone. Now there are things about America we can't kid."

Depth involvement encourages everyone to take himself much more seriously than before. As TV cooled off the American audience, giving it new preferences and new orientation of sight and sound and touch and taste, Al Capp's wonderful brew also had to be toned down. There was no more need to kid Dick Tracy or the suspense routines. As *MAD* magazine discovered, the new audience found the scenes and themes of ordinary life as funny as anything in remote Dogpatch. *MAD* magazine simply transferred the world of ads into the world of the comic book, and it did this just when the TV image was beginning to eliminate the comic book by direct rivalry. At the same time, the TV image rendered the sharp and clear photographic image as blur and blear. TV cooled off the ad audience until the continuing vehemence of the ads and entertainment suited the program of the *MAD* magazine world very well. TV, in fact, turned the previous hot media of photo, film, and radio into a comic-strip world by simply featuring them as overheated packages. Today the ten-year-old clutches his or her *MAD* ("Build up your Ego with *MAD*") in the same way that the Russian beatnik treasures an old Presley tape obtained from a G.I. broadcast. If the "Voice of America" suddenly switched to jazz, the Kremlin would have reason to crumble. It would be almost as effective as if the Russian citizens had copies of Sears Roebuck catalogues to goggle at, instead of our dreary propaganda for the American way of life.

Picasso has long been a fan of American comics. The highbrow, from Joyce to Picasso, has long been devoted to American popular art because he finds in it an authentic imaginative reaction to official action. Genteel art, on the other hand, tends merely to evade and disapprove of the blatant modes of action in a powerful high definition, or "square," society. Genteel art is a kind of repeat of the specialized acrobatic feats of an industrialized world. Popular art is the clown reminding us of all the life and faculty that we have omitted from our daily routines. He ventures to perform the specialized routines of the society, acting as integral man. But integral man is quite inept in a specialist situation. This, at least, is one way to get at the art of the comics, and the art of the clown.

Today our ten-year-olds, in voting for *MAD*, are telling us in their own way that the TV image has ended the consumer phase of American culture. They are now telling us what the eighteen-year-old beatniks were first trying to say ten years ago. The pictorial consumer age is dead. The iconic age is upon us. We now toss to the Europeans the package that concerned us from 1922 to 1952. They, in turn, enter their first consumer age of standardized goods. We move into our first depth-age of art-and-producer orientation. America is Europeanizing on as extensive a pattern as Europe is Americanizing.

Where does this leave the older popular comics? What about "Blondie" and "Bringing Up Father"? Theirs was a pastoral world of primal innocence from which young America has clearly graduated. There was still adolescence in those days, and there were still remote ideals and private dreams, and visualizable goals, rather than vigorous and ever-present corporate postures for group participation.

The chapter on The Print indicated how the cartoon is a do-it-yourself form of experience that has developed an ever more vigorous life as the electric age advanced. Thus, all electric appliances, far from being labor-saving devices, are new forms of work, decentralized and made available to everybody. Such is, also, the world of the telephone and the TV image that demands so much more of its users than does radio or movie. As a simple consequence of this participational and do-it-yourself aspect of the electric technology, every kind of entertainment in the TV age favors the same kind of personal involvement. Hence the paradox that, in the TV age, Johnny can't read because reading, as customarily taught, is too superficial and consumerlike an activity. Therefore the highbrow paperback, because of its depth character, may appeal to youngsters who spurn ordinary narrative offerings. Teachers today frequently find that students who can't read a page of history are becoming experts in code and linguistic analysis. The problem, therefore, is not that Johnny can't read, but that, in an age of depth involvement, Johnny can't visualize distant goals.

The first comic books appeared in 1935. Not having anything connected or literary about them, and being as difficult to decipher as the *Book of Kells*, they caught on with the young. The elders of the tribe, who had never noticed that the ordinary newspaper was as frantic as a surrealist art exhibition, could hardly be expected to notice that the comic books were as exotic as eighth-century illuminations. So, having noticed nothing about the *form*, they could discern nothing of the *contents*, either. The mayhem and violence were all they noted. Therefore, with naïve literary logic, they waited for violence to flood the world. Or, alternatively, they attributed existing crime to the comics. The dimmest-witted convict learned to moan, "It wuz comic books done this to me."

Meantime, the violence of an industrial and mechanical environment had to be lived and given meaning and motive in the nerves and viscera of the young. To live and experience anything is to translate its direct impact into many indirect forms of awareness. We provided the young with a shrill and raucous asphalt jungle, beside which any tropical animal jungle was as quiet and tame as a rabbit hutch. We called this normal. We paid people to keep it at the highest pitch of intensity because it paid well. When the entertainment industries tried to provide a reasonable facsimile of the ordinary city vehemence, eyebrows were raised.

It was Al Capp who discovered that until TV, at least, any degree of Scragg mayhem or Phogbound morality was accepted as funny. *He* didn't think it was funny. He put in his strip just exactly what he saw around him. But our trained incapacity to relate one situation to another enabled his sardonic realism to be mistaken for humor. The more he showed the capacity of people to involve themselves in hideous difficulties, along with their entire inability to turn a hand to help themselves, the more they giggled. "Satire," said Swift, "is a glass in which we see every countenance but our own."

The comic strip and the ad, then, both belong to the world of games, to the world of models and extensions of situations elsewhere. *MAD* magazine, world of the woodcut, the print, and the cartoon, brought them together with other games and models from the world of entertainment. *MAD* is a kind of newspaper mosaic of the ad as entertainment, and entertainment as a form of madness. Above all, it is a print- and woodcut-form of expression and experience whose sudden appeal is a sure index of deep changes in our culture. Our need now is to understand the formal character of print, comic and cartoon, both as challenging and changing the consumer-culture of film, photo, and press. There is no single approach to this task, and no single observation or idea that can solve so complex a problem in changing human perception.

18

The Printed Word:
Architect of Nationalism

Print fused the cultural worlds of ancient and medieval times into a third world, intensifying the explosion set off originally by the phonetic alphabet, but under electric technology, explosion has turned to implosion. This state of affairs creates a somnambulism even more powerful than the one that gripped Narcissus, whose somnambulism McLuhan ties to his fundamental reference point for the definition of media: "An extension appears to be an amplification of an organ, a sense or a function, that inspires the central nervous system to a self-protective gesture of numbing of the extended area, at least so far as direct inspection and awareness are concerned." And a new dimension is added to the theme of an earlier chapter on the transforming power of media: "A new medium is never an addition to an old one, nor does it leave the old one in peace."

— *(Editor)*

"You may perceive, Madam," said Dr. Johnson with a pugilistic smile, "that I am well-bred to a degree of needless scrupulosity." Whatever the degree of conformity the Doctor had achieved with the new stress of his time on white-shirted tidiness, he was quite aware of the growing social demand for visual presentability.

Printing from movable types was the first mechanization of a complex handicraft, and became the archetype of all subsequent mechanization. From Rabelais and More to Mill and Morris, the typographic explosion extended the minds and voices of men to reconstitute the human dialogue on a world scale that has bridged the ages. For if seen merely as a store of information, or as a new means of speedy retrieval of knowledge, typography ended parochialism and tribalism, psychically and socially, both in space and in time. Indeed the first two centuries of printing from movable types were motivated much more by the desire to see ancient and medieval books than by the need to read and write new ones. Until 1700 much more than 50 per cent of all printed books were ancient or medieval. Not only antiquity but also the Middle Ages were given to the first reading public of the printed word. And the medieval texts were by far the most popular.

Like any other extension of man, typography had psychic and social consequences that suddenly shifted previous boundaries and patterns of culture. In bringing the ancient and medieval worlds into fusion—or, as some would say, confusion—the printed book created a third world, the modern world, which now encounters a new electric technology or a new extension of man. Electric means of moving of information are altering our typographic culture as sharply as print modified medieval manuscript and scholastic culture.

Beatrice Warde has recently described in *Alphabet* an

electric display of letters painted by light. It was a Norman McLaren movie advertisement of which she asks:

> Do you wonder that I was late for the theatre that night, when I tell you that I saw two club-footed Egyptian A's ... walking off arm-in-arm with the unmistakable swagger of a music-hall comedy-team? I saw base-serifs pulled together as if by ballet shoes, so that the letters tripped off literally *sur les pointes* ... after forty centuries of the *necessarily static* Alphabet, I saw what its members could do in the fourth dimension of Time, "flux," movement. You may well say that I was electrified.

Nothing could be farther from typographic culture with its "place for everything and everything in its place."

Mrs. Warde has spent her life in the study of typography and she shows sure tact in her startled response to letters that are not printed by types but painted by light. It may be that the explosion that began with phonetic letters (the "dragon's teeth" sowed by King Cadmus) will reverse into "implosion" under the impulse of the instant speed of electricity. The alphabet (and its extension into typography) made possible the spread of the power that is knowledge, and shattered the bonds of tribal man, thus exploding him into an agglomeration of individuals. Electric writing and speed pour upon him, instantaneously and continuously, the concerns of all other men. He becomes tribal once more. The human family becomes one tribe again.

Any student of the social history of the printed book is likely to be puzzled by the lack of understanding of the psychic and social effects of printing. In five centuries explicit comment and awareness of the effects of print on human sensibility are very scarce. But the same observation can be made about all the extensions of man, whether it be clothing or the computer. An extension appears to be an amplification of an organ, a sense or a function, that inspires the central nervous system to a self-protective gesture of numbing of the extended area, at least so far as direct inspection and awareness are concerned. Indirect comment on the effects of the printed book is available in abundance in the work of Rabelais, Cervantes, Montaigne, Swift, Pope, and Joyce. They used typography to create new art forms.

Psychically the printed book, an extension of the visual faculty, intensified perspective and the fixed point of view. Associated with the visual stress on point of view and the vanishing point that provides the illusion of perspective there comes another illusion that space is visual, uniform and continuous. The linearity, precision, and uniformity of the arrangement of movable types are inseparable from these great cultural forms and innovations of Renaissance experience. The new intensity of visual stress and private point of view in the first century of printing were united to the means of self-expression made possible by the typographic extension of man.

Socially, the typographic extension of man brought in nationalism, industrialism, mass markets, and universal literacy and education. For print presented an image of repeatable precision that inspired totally new forms of extending social energies. Print released great psychic and social energies in the Renaissance, as today in Japan or Russia, by breaking the individual out of the traditional group while providing a model of how to add individual to individual in massive agglomeration of power. The same spirit of private enterprise that emboldened authors and artists to cultivate self-expression led other men to create giant corporations, both military and commercial.

Perhaps the most significant of the gifts of typography to man is that of detachment and noninvolvement—the power to act without reacting. Science since the Renaissance has exalted this gift which has become an embarrassment in the electric age, in which all people are involved in all others at all times. The very word "disinterested," expressing the loftiest detachment and ethical integrity of typographic man, has in the past decade been increasingly used to mean: "He couldn't care less." The same integrity indicated by the term "disinterested" as a mark of the scientific and scholarly temper of a literate and enlightened society is now increasingly repudiated as "specialization" and fragmentation of knowledge and sensibility. The fragmenting and analytic power of the printed word in our psychic lives gave us that "dissociation of sensibility" which in the arts and literature since Cézanne and since Baudelaire has been a top priority for elimination in every program of reform in taste and knowledge. In the "implosion" of the electric age the separation of thought and feeling has come to seem as strange as the departmentalization of knowledge in schools and universities. Yet is was precisely the power to separate thought and feeling, to be able to act without reacting, that split literate man out of the tribal world of close family bonds in private and social life.

Typography was no more an addition to the scribal art than the motorcar was an addition to the horse. Printing had its "horseless carriage" phase of being misconceived and misapplied during its first decades, when it was not uncommon for the purchaser of a printed book to take it to a scribe to have it copied and illustrated. Even in the early eighteenth century a "textbook" was still defined as a "Classick Author written very wide by the Students, to give room for an Interpretation dictated by the Master, &c., to be inserted in the Interlines" (O.E.D.). Before printing, much of the time in school and college classrooms was spent in making such texts. The classroom tended to be a *scriptorium* with a commentary. The student was an editor-publisher. By the same token the book market was a secondhand market of relatively scarce items. Printing changed learning and marketing processes alike. The book was the first teaching machine and also the first mass-produced commodity. In amplifying and extending the written word, typography revealed and greatly extended the structure of writing. Today, with the cinema and the electric speedup of information movement, the formal structure of the printed word, as of mechanism in general, stands forth like a branch washed up on the beach. A new medium is never an addition to an old one, nor does it leave the old one in peace. It never ceases to oppress the older media until it finds new shapes and positions for them. Manuscript culture had sustained an oral procedure in education that was called "scholasticism" at its higher levels; but by putting the same text in front of any given number of students or readers print ended the scholastic regime of oral disputation very quickly. Print provided a vast new memory for past writings that made a personal memory inadequate.

Margaret Mead has reported that when she brought several copies of the same book to a Pacific island there was great excitement. The natives had seen books, but only one copy of each, which they had assumed to be unique. Their astonishment at the identical character of several books was a natural response to what is after all the most magical and potent aspect of print and mass production. It involves a principle of extension by homogenization that is the key to understanding Western power. The open society is open by virtue of a uniform typographic educational processing that permits indefinite expansion of any group by additive means. The printed book based on typographic uniformity and repeatability in the visual order was the first teaching machine, just as typography was the first mechanization of a handicraft. Yet in spite of the extreme fragmentation or specialization of human action necessary to achieve the printed word, the printed book represents a rich composite of previous cultural inventions. The total effort embodied in the illustrated book in print offers a striking example of the variety of separate acts of invention that are requisite to bring about a new technological result.

The psychic and social consequences of print included an extension of its fissile and uniform character to the gradual homogenization of diverse regions with the resulting amplification of power, energy, and aggression that we associate with new nationalisms. Psychically, the visual extension and amplification of the individual by print had many effects. Perhaps as striking as any other is the one mentioned by Mr. E. M. Forster, who, when discussing some Renaissance types, suggested that "the printing press, then only a century old, had been mistaken for an engine of immortality, and men had hastened to commit to it deeds and passions for the benefit of future ages." People began to act as though immortality were inherent in the magic repeatability and extensions of print.

Another significant aspect of the uniformity and repeatability of the printed page was the pressure it exerted toward "correct" spelling, syntax, and pronunciation. Even more notable were the effects of print in separating poetry from song, and prose from oratory, and popular from educated speech. In the matter of poetry it turned out that, as poetry could be read without being heard, musical instruments could also be played without accompanying any verses. Music veered from the spoken word, to converge again with Bartók and Schoenberg.

With typography the process of separation (or explosion) of functions went on swiftly at all levels and in all spheres; nowhere was this matter observed and commented on with more bitterness than in the plays of Shakespeare. Especially in *King Lear*, Shakespeare provided an image or model of the process of quantification and fragmentation as it entered the world of politics and of family life. Lear at the very opening of the play presents "our darker purpose" as a plan of delegation of powers and duties:

> Only we shall retain
> The name, an all th'addition to a King;
> The sway, revenue, execution of the rest,
> Beloved sons, be yours: which to confirm,
> This coronet part between you.

This act of fragmentation and delegation blasts Lear, his kingdom, and his family. Yet to divide and rule was the dominant new idea of the organization of power in the Renaissance. "Our darker purpose" refers to Machiavelli himself, who had developed an individualist and quantitative idea of power that struck more fear in that time than Marx in ours. Print, then, challenged the corporate patterns of medieval organization as much as electricity now challenges our fragmented individualism.

The uniformity and repeatability of print permeated the Renaissance with the idea of time and space as continuous measurable quantities. The immediate effect of this idea was to desacralize the world of nature and the world of power alike. The new technique of control of physical processes by segmentation and fragmentation separated God and Nature as much as Man and Nature, or man and man. Shock at this departure from traditional vision and inclusive awareness was often directed toward the figure of Machiavelli, who had merely spelled out the new quantitative and neutral or scientific ideas of force as applied to the manipulation of kingdoms.

Shakespeare's entire work is taken up with the themes of the new delimitations of power, both kingly and private. No greater horror could be imagined in his time than the spectacle of Richard II, the sacral king, undergoing the indignities of imprisonment and denudation of his sacred prerogatives. It is in *Troilus and Cressida*, however, that the new cults of fissile, irresponsible power, public and private, are paraded as a cynical charade of atomistic competition:

239

Take the instant way;
For honour travels in a strait so narrow
Where one but goes abreast: keep, then, the path;
For emulation hath a thousand sons
That one by one pursue: if you give way,
Or hedge aside from the direct forthright,
Like to an enter'd tide they all rush by
And leave you hindmost …

(III, iii)

The image of society as segmented into a homogeneous mass of quantified appetites shadows Shakespeare's vision in the later plays.

Of the many unforeseen consequences of typography, the emergence of nationalism is, perhaps, the most familiar. Political unification of populations by means of vernacular and language groupings was unthinkable before printing turned each vernacular into an extensive mass medium. The tribe, an extended form of a family of blood relatives, is exploded by print, and is replaced by an association of men homogeneously trained to be individuals. Nationalism itself came as an intense new visual image of group destiny and status, and depended on a speed of information movement unknown before printing. Today nationalism as an image still depends on the press but has all the electric media against it. In business, as in politics, the effect of even jet-plane speeds is to render the older national groupings of social organization quite unworkable. In the Renaissance it was the speed of print and the ensuing market and commercial developments that made nationalism (which is continuity and competition in homogeneous space) as natural as it was new. By the same token, the heterogeneities and noncompetitive discontinuities of medieval guilds and family organization had become a great nuisance as speedup of information by print called for more fragmentation and uniformity of function. The Benvenuto Cellinis, the goldsmith-cum-painter-cum-sculptor-cum-writer-cum-condottiere, became obsolete.

Once a new technology comes into a social milieu it cannot cease to permeate that milieu until every institution is saturated. Typography has permeated every phase of the arts and sciences in the past five hundred years. It would be easy to document the processes by which the principles of continuity, uniformity, and repeatability have become the basis of calculus and of marketing, as of industrial production, entertainment, and science. It will be enough to point out that repeatability conferred on the printed book the strangely novel character of a uniformly priced commodity opening the door to price systems. The printed book had in addition the quality of portability and accessibility that had been lacking in the manuscript.

Directly associated with these expansive qualities was the revolution in expression. Under manuscript conditions the role of being an author was a vague and uncertain one, like that of a minstrel. Hence, self-expression was of little interest. Typography, however, created a medium in which it was possible to speak out loud and bold to the world itself, just as it was possible to circumnavigate the world of books previously locked up in a pluralistic world of monastic cells. Boldness of type created boldness of expression.

Uniformity reached also into areas of speech and writing, leading to a single tone and attitude to reader and subject spread throughout an entire composition. The "man of letters" was born. Extended to the spoken word, this literate *equitone* enabled literate people to maintain a single "high tone" in discourse that was quite devastating, and enabled nineteenth-century prose writers to assume moral qualities that few would now care to simulate. Permeation of the colloquial language with literate uniform qualities has flattened out educated speech till it is a very reasonable acoustic facsimile of the uniform and continuous visual effects of typography. From this technological effect follows the further fact that the humor, slang, and dramatic vigor of American-English speech are monopolies of the semi-literate.

These typographical matters for many people are charged with controversial values. Yet in any approach to understanding print it is necessary to stand aside from the form in question if its typical pressure and life are to be observed. Those who panic now about the threat of the newer media and about the revolution we are forging, vaster in scope than that of Gutenberg, are obviously lacking in cool visual detachment and gratitude for that most potent gift bestowed on Western man by literacy and typography: his power to act without reaction or involvement. It is this kind of specialization by dissociation that has created Western power and efficiency. Without this dissociation of action from feeling and emotion people are hampered and hesitant. Print taught men to say, "Damn the torpedoes. Full steam ahead!"

19

Wheel, Bicycle, and Airplane

James Joyce, the author most frequently quoted in Understanding Media, *lends McLuhan his* a-stone-aging *("astonishing") pun here. The same piece of Joycean whimsical wisdom contains the seed of a principle that McLuhan would make explicit in his later work, that of* retrieval. *This inevitable effect of the introduction of a new medium brings back older structures and environments, or older forms of action, human organization, and thought. McLuhan upstages Joyce's linguistic acrobatics by retrieving and reworking the pun on the ablative absolute from his introductory chapter, following it with the Latin phrase* ceteris paribus *("other things being equal"), itself an ablative absolute, creating a fresh sense in which the medium is the message, but not before he has created a linguistic metaphor for the operation of media: "When such ablatives intrude, they alter the syntax of society." And though the movies will get a later chapter all to themselves, it is here that McLuhan opens the subject, tying the flip from mechanical to organic form in the technology of the film medium to the often-mentioned reversal principle (see Subject Index), calling it "a pattern that appears in all human extensions."*

— (Editor)

The kinds of interplay between wheel, bicycle, and airplane are startling to those who have never thought about them. Scholars tend to work on the archeological assumption that things need to be studied in isolation. This is the habit of specialism that quite naturally derives from typographic culture. When a scholar like Lynn White ventures to make some interrelations, even in his own area of special historical study, he causes a good deal of unhappiness among his merely specialist colleagues. In his *Medieval Technology and Social Change* he explains how the feudal system was a social extension of the stirrup. The stirrup first appeared in the West in the eighth century A.D., having been introduced from the East. With the stirrup came mounted shock combat that called into existence a new social class. The European cavalier class had already existed to be armed, but to mount a knight in full armor required the combined resources of ten or more peasant holdings. Charlemagne demanded that less prosperous freemen merge their private farms to equip a single knight for the wars. The pressure of the new war technology gradually developed classes and an economic system that could provide numerous cavaliers in heavy armor. By about the year 1000 A.D. the old word *miles* had changed from "soldier" to "knight."

Lynn White has much to say, also, about horseshoes and horsecollars as revolutionary technology that increased the power and extended the range and speed of human action in the early Middle Ages. He is sensitive to the psychic and social implications of each technological extension of man, showing how the heavy wheel-plow brought about a new order in the field system, as well as in the diet of that age. "The Middle Ages were literally full of beans."

To come more directly to our subject of the wheel, Lynn White explains how the evolution of the wheel in the Middle

Ages was related to the development of the horsecollar and
the harness. The greater speed and endurance of the horse
was not available for cartage until the discovery of the collar.
But once evolved, this horse-harness led to the development
of wagons with pivoted front axles and brakes. The four-wheel
wagon capable of hauling heavy loads was a common feature
by the middle of the thirteenth century. The effects on town
life were extraordinary. Peasants began to live in cities while
going each day to their fields, almost in the manner of motorized Saskatchewan farmers. These latter live mainly in the
city, having no housing in the country beyond sheds for their
tractors and equipment.

With the coming of the horse-drawn bus and streetcar,
American towns developed housing that was no longer within
sight of shop or factory. The railroad next took over the development of the suburbs, with housing kept within walking
distance of the railroad stop. Shops and hotels around the
railroad gave some concentration and form to the suburb.
The automobile, followed by the airplane, dissolved this
grouping and ended the pedestrian, or human, scale of the
suburb. Lewis Mumford contends that the car turned the
suburban housewife into a full-time chauffeur. Certainly the
transformations of the wheel as expediter of tasks, and architect of ever-new human relations, is far from finished, but its
shaping power is waning in the electric age of information,
and that fact makes us much more aware of its characteristic
form as now tending toward the archaic.

Before the emergence of the wheeled vehicle, there was
merely the abrasive traction principle — runners, skids, and
skis preceded wheels for vehicles, just as the abrasive, semi-
rotary motion of the hand-operated spindle and drill preceded
the full, free rotary motion of the potter's wheel. There is a
moment of translation or "abstraction" needed to separate
the reciprocating movement of hand from the free movement
of wheel. "Doubtless the notion of the wheel came originally

from observing that rolling a log was easier than shoving it," writes Lewis Mumford in *Technics and Civilization*. Some might object that log-rolling is closer to the spindle operation of the hands than to the rotary movement of feet, and need never have got translated into the technology of wheel. Under stress, it is more natural to fragment our own bodily form, and to let part of it go into another material, than it is to transfer any of the motions of external objects into another material. To extend our bodily postures and motions into new materials, by way of amplification, is a constant drive for more power. Most of our bodily stresses are interpreted as needs for extending storage and mobility functions, such as occur, also, in speech, money and writing. All manner of utensils are a yielding to this bodily stress by means of extensions of the body. The need for storage and portability can readily be noted in vases, jars, and "slow matches" (stored fire).

Perhaps the main feature of all tools and machines— economy of gesture—is the immediate expression of any physical pressure which impels us to outer or to extend ourselves, whether in words or in wheels. Man can say it with flowers or plows or locomotives. In "Krazy Kat," Ignatz said it with bricks.

One of the most advanced and complicated uses of the wheel occurs in the movie camera and in the movie projector. It is significant that this most subtle and complex grouping of wheels should have been invented in order to win a bet that all four feet of a running horse were sometimes off the ground simultaneously. This bet was made between the pioneer photographer Eadweard Muybridge and horse-owner Leland Stanford, in 1889. At first, a series of cameras were set up side by side, each to snap an arrested moment of the horse's hooves in action. The movie camera and the projector were evolved from the idea of reconstructing mechanically the movement of feet. The wheel, that began as extended feet, took a great evolutionary step into the movie theater.

By an enormous speedup of assembly-line segments, the movie camera rolls up the real world on a spool, to be unrolled and translated later onto the screen. That the movie re-creates organic process and movement by pushing the mechanical principle to the point of reversal is a pattern that appears in all human extensions, whatever, as they reach a peak of performance. By speedup, the airplane rolls up the highway into itself. The road disappears into the plane at take-off, and the plane becomes a missile, a self-contained transportation system. At this point the wheel is reabsorbed into the form of bird or fish that the plane becomes as it takes to the air. Skin-divers need no path or road, and claim that their motion is like that of bird flight; their feet cease to exist as the progressive, sequential movement that is the origin of rotary action of the wheel. Unlike wing, or fin, the wheel is lineal and requires road for its completion.

It was the tandem alignment of wheels that created the velocipede and then the bicycle, for with the acceleration of wheel by linkage to the visual principle of mobile lineality, the wheel acquired a new degree of intensity. The bicycle lifted the wheel onto the plane of aerodynamic balance, and not too indirectly created the airplane. It was no accident that the Wright brothers were bicycle mechanics, or that early airplanes seemed in some ways like bicycles. The transformations of technology have the character of organic evolution because all technologies are extensions of our physical being. Samuel Butler raised great admiration in Bernard Shaw by his insight that the evolutionary process had been fantastically accelerated by transference to the machine mode. Shaw, however, was happy to leave the matter in this delightfully opaque state. Butler, himself, had at least indicated that machines were given vicarious powers of reproduction by their subsequent impact upon the very bodies that had brought them into being by extension. Response to the increased power and speed of our own extended bodies is one which engenders new extensions. Every technology creates new stresses and needs in the human beings who have engendered it. The new need and the new technological response are born of our embrace of the already existing technology—a ceaseless process.

Those familiar with the novels and plays of Samuel Beckett need not be reminded of the rich clowning he engenders by means of the bicycle. It is for him the prime symbol of the Cartesian mind in its acrobatic relation of mind and body in precarious imbalance. This plight goes with a lineal progression that mimics the very form of purposeful and resourceful independence of action. For Beckett, the integral being is not the acrobat but the clown. The acrobat acts as a specialist, using only a limited segment of his faculties. The clown is the integral man who mimes the acrobat in an elaborate drama of incompetence. Beckett sees the bicycle as the sign and symbol of specialist futility in the present electric age, when we must all interact and react, using all our faculties at once.

Humpty-Dumpty is the familiar example of the clown unsuccessfully imitating the acrobat. Just because all the King's horses and all the King's men couldn't put Humpty-Dumpty together again, it doesn't follow that electromagnetic automation couldn't have put Humpty-Dumpty back together. The integral and unified egg had no business sitting on a wall anyway. Walls are made of uniformly fragmented bricks that arise with specialisms and bureaucracies. They are the deadly enemies of integral beings like eggs. Humpty-Dumpty met the challenge of the wall with a spectacular collapse.

The same nursery rhyme comments on the consequences of the fall of Humpty-Dumpty. That is the point about the King's horses and men. They, too, are fragmented and specialized. Having no unified vision of the whole, they are helpless.

Humpty-Dumpty is an obvious example of integral wholeness. The mere existence of the wall already spelt his fall. James Joyce in *Finnegans Wake* never ceases to interlace these themes, and the title of the work indicates his awareness that "a-stone-aging" as it may be, the electric age is recovering the unity of plastic and iconic space, and *is* putting Humpty-Dumpty back together again.

The potter's wheel, like all other technologies, was the acceleration of an existing process. After nomad food-gathering had shifted to sedentary plowing and seeding, the need for storage increased. Pots were needed for more and more purposes. Men turned their powers to changing the forms of things by cultivation. Change to special production in local areas created the need for exchange and for transport. For this purpose sledges were used in Northern Europe before 5000 B.C., and human porters and pack-bearing animals preceded sledges naturally. The wheel under the sledge was an accelerator of feet, not of hand. With this acceleration of the feet came the need for road, just as with the extension of our backsides in the form of chair, came the need for table. The wheel is an ablative absolute of feet, as chair is the ablative absolute of backside. But when such ablatives intrude, they alter the syntax of society. There is no *ceteris paribus* in the world of media and technology. Every extension or acceleration effects new configurations in the over-all situation at once.

The wheel made the road, and moved produce faster from fields to settlements. Acceleration created larger and larger centers, more and more specialism, and more intense incentives, aggregates, and aggressions. So it is that the wheeled vehicle makes its appearance at once as a war chariot, just as the urban center, created by the wheel, makes its appearance as an aggressive stronghold. No further motivation than the compounding and consolidating of specialist skills by acceleration of the wheel is needed to explain the mounting degree of human creativity and destructiveness.

Lewis Mumford calls this urbanization "implosion," but it was really an explosion. Cities were made by the fragmenting of pastoral modes. The wheel and the road expressed and advanced this explosion by a radiational or center-margin pattern. Centralism depends on margins that are accessible by road and wheel. Maritime power does not assume this center-margin structure, and neither do desert and steppe cultures. Today with jet and electricity, urban centralism and specialism reverse into decentralism and interplay of social functions in ever more nonspecialist forms.

The wheel and the road are centalizers because they accelerate up to a point that ships cannot. But acceleration beyond a certain point, when it occurs by means of the automobile and the plane, creates decentralism in the midst of the older centralism. This is the origin of the urban chaos of our time. The wheel, pushed beyond a certain intensity of movement, no longer centralizes. All electric forms whatsoever have a decentralizing effect, cutting across the older mechanical patterns like a bagpipe in a symphony. It is too bad that Mr. Mumford has chosen the term "implosion" for the urban specialist explosion. "Implosion" belongs to the electronic age, as it belonged to the prehistoric cultures. All primitive societies are implosive, like the spoken word. But "technology is explicitness," as Lyman Bryson said; and explicitness or specialist extension of functions, is centralism and explosion of functions, and not implosion, contraction or simultaneity.

An airline executive who is much aware of the implosive character of world aviation asked a corresponding executive of each airline in the world to send him a pebble from outside his office. His idea was to build a little cairn of pebbles from all parts of the world. When asked, "So what?" he said that in one spot one could touch every part of the world because of aviation. In effect, he had hit upon the mosaic or iconic principle of simultaneous touch and interplay that is inherent in the implosive speed of the airplane. The same principle of implosive mosaic is even more characteristic of electric information movement of all kinds.

Centralism and extension of power by wheel and written word to the margins of empire are creative of the direct force, outside and external, to which men do not necessarily submit their minds. But implosion is the spell and incantation of the tribe and the family, to which men readily submit. Under technological explicitness, even of the urban centralist structure, some men managed to break out of the charmed circle of tribal magic. Mumford cites the words of the Chinese philosopher Mencius as a comment on this situation:

> When men are subdued by force they do not submit in their minds, but only because their strength is inadequate. When men are subdued by power in personality they are pleased to their very heart's core and do really submit.

As the expression of new specialist extensions of our bodies the congregating of people and supplies in centers by wheel and road called for endless reciprocal expansion in a spongelike action of intake and output, which has entrapped all urban structures everywhere in place and time. Mumford observes: "If I interpret the evidence correctly, the cooperative forms of urban polity were undermined and vitiated from the outset by the destructive death-oriented myths which attended … the exorbitant expansion of physical power and technological adroitness." To have such power by extension of their own bodies, men must explode the inner unity of their beings into explicit fragments. Today, in an age of implosion, we are playing the ancient explosion backward, as on a film. We can watch the pieces of man's being coming together again in an age that has so much power that the all-destructive use of it appears meaningless, even to the dim and skew of wit.

Historians see the forms of the great cities in the ancient world as manifesting all facets of human personality. Institutions, architectural and administrative, as extensions of our physical beings necessarily tend toward worldwide similarities. The central nervous system of the city was the citadel which included the great temple and palace of the king, invested with the dimensions and iconography of power and prestige. The extent to which this central core could extend its power safely depended on its power of acting at a distance. Not until the alphabet appeared, together with papyrus, could the citadel extend itself very far in space. (See the chapter on Roads and Paper Routes.) The ancient city, however, could appear as quickly as specialist man could separate his inner functions in space and architecture. To say that the cities of the Aztecs and the Peruvians resembled European cities is only to say that they shared and extended the same faculties in both regions. The question of direct physical influence and imitation as if by diffusion becomes irrelevant.

20

The Photograph:
The Brothel-without-Walls

The unifying thread of this chapter is syntax, *defined by McLuhan as "the net of rationality." He chronicles its disappearance from late prints, telegraph messages, and impressionist painting, reinforcing his argument for media effects by this cumulative evidence. The logic of the photograph, he declares, is "neither verbal nor syntactical," as a result of which print-dominated Western culture could only succumb to the photograph's trance-inducing power. Photography provided a means of autonomous representation of objects, pushing pictorial representation beyond the power of the painter's palette, thus ensuring another instance of the reversal principle coming into operation, in this case through "statement without syntax." Such statement, McLuhan says, is effectively equivalent to gesture, mime, and above all* gestalt *(see Glossary). Its development paved the way for poets and painters to explore the landscape of the human psyche. McLuhan punctuates this mid-point in his inventory of media by reminding the reader of the criteria for defining and studying them, closing the chapter with: "To understand the medium of the photograph is quite impossible, then, without grasping its relation to other media, both old and new. For media, as extensions of our physical and nervous systems, constitute a world of biochemical interactions that must ever seek new equilibrium as new extensions occur."*

— *(Editor)*

A photograph of "St. Peter's at a Moment of History" was the cover feature of *Life* magazine for June 14, 1963. It is one of the peculiar characteristics of the photo that it isolates single moments in time. The TV camera does not. The continuous scanning action of the TV camera provides, not the isolated moment or aspect, but the contour, the iconic profile and the transparency. Egyptian art, like primitive sculpture today, provided the significant outline that had nothing to do with a moment in time. Sculpture tends toward the timeless.

Awareness of the transforming power of the photo is often embodied in popular stories like the one about the admiring friend who said, "My, that's a fine child you have there!" Mother: "Oh, that's nothing. You should see his photograph." The power of the camera to be everywhere and to interrelate things is well indicated in the *Vogue* magazine boast (March 15, 1953): "A woman now, and without having to leave the country, can have the best of five (or more) nations hanging in her closet—beautiful and compatible as a statesman's dream." That is why, in the photographic age, fashions have come to be like the collage style in painting.

A century ago the British craze for the monocle gave to the wearer the power of the camera to fix people in a superior stare, as if they were objects. Erich von Stroheim did a great job with the monocle in creating the haughty Prussian officer. Both monocle and camera tend to turn people into things, and the photograph extends and multiplies the human image to the proportions of mass-produced merchandise. The movie stars and matinee idols are put in the public domain by photography. They become dreams that money can buy. They can be bought and hugged and thumbed more easily than public prostitutes. Mass-produced merchandise has always made some people uneasy in its prostitute aspect. Jean Genet's *The Balcony* is a play on this theme of society as a brothel environed by violence and horror. The avid desire of mankind to prostitute itself stands up against the chaos of revolution. The brothel remains firm and permanent amidst the most furious changes. In a word, photography has inspired Genet with the theme of the world since photography as a Brothel-without-Walls.

Nobody can commit photography alone. It is possible to have at least the illusion of reading and writing in isolation, but photography does not foster such attitudes. If there is any sense in deploring the growth of corporate and collective art forms such as the film and the press, it is surely in relation to the previous individualist technologies that these new forms corrode. Yet if there had been no prints or woodcuts and engravings, there would never have come the photograph. For centuries, the woodcut and the engraving had delineated the world by an arrangement of lines and points that had syntax of a very elaborate kind. Many historians of this visual syntax, like E. H. Gombrich and William M. Ivins, have been at great pains to explain how the art of the hand-written manuscript had permeated the art of the woodcut and the engraving until, with the halftone process, the dots and lines suddenly fell below the threshold of normal vision. Syntax, the net of rationality, disappeared from the later prints, just as it tended to disappear from the telegraph message and from the impressionist painting. Finally, in the *pointillisme* of Seurat, the world suddenly appeared *through* the painting. The direction of a syntactical point of view from outside *onto* the painting ended as literary form dwindled into headlines with the telegraph. With the photograph, in the same way, men had discovered how to make visual reports without syntax.

It was in 1839 that William Henry Fox Talbot read a paper to the Royal Society which had as title: "Some account of the Art of Photogenic Drawing, or the process by which Natural Objects may be made to delineate themselves without the aid of the artist's pencil." He was quite aware of photography as a kind of automation that eliminated the syntactical procedures of pen and pencil. He was probably less aware that he had brought the pictorial world into line with the new industrial procedures. For photography mirrored the external world automatically, yielding an exactly repeatable visual image. It was this all-important quality of uniformity and repeatability that had made the Gutenberg break between the Middle Ages and the Renaissance. Photography was almost as decisive in making the break between mere mechanical industrialism and the graphic age of electronic man. The step from the age of Typographic Man to the age of Graphic Man was taken with the invention of photography. Both daguerreotypes and photographs introduced light and chemistry into the making process. Natural objects delineated themselves by an exposure intensified by lens and fixed by chemicals. In the daguerreotype process there was the same stippling or pitting with minute dots that was echoed later in Seurat's *pointillisme*, and is still continued in the newspaper mesh of dots that is called "wire-photo." Within a year of Daguerre's discovery, Samuel F. B. Morse was taking photographs of his wife and daughter in New York City. Dots for the eye (photograph) and dots for the ear (telegraph) thus met on top of a skyscraper.

A further cross-fertilization occurred in Talbot's invention of the photo, which he imagined as an extension of the *camera obscura*, or pictures in "the little dark room," as the Italians had named the picture play-box of the sixteenth century. Just at the time when mechanical writing had been achieved by movable types, there grew up the pastime of looking at moving images on the wall of a dark room. If there is sunshine outside and a pinhole in one wall, then the images of the outer world will appear on the wall opposite. This new discovery was very exciting to painters, since it intensified the new illusion of perspective and of the third

dimension that is so closely related to the printed word. But
the early spectators of the moving image in the sixteenth
century saw those images upside down. For this reason the
lens was introduced — in order to turn the picture right side
up. Our normal vision is also upside down. Psychically, we
learn to turn our visual world right side up by translating
the retinal impression from visual into tactile and kinetic
terms. Right side up is apparently something we feel but
cannot see directly.

To the student of media, the fact that "normal" right-
side-up vision is a translation from one sense into another is
a helpful hint about the kinds of activity of distortion and
translation that any language or culture induces in all of us.
Nothing amuses the Eskimo more than for the white man to
crane his neck to see the magazine pictures stuck on the igloo
walls. For the Eskimo no more needs to look at a picture right
side up than does a child before he has learned his letters *on a
line*. Just why Westerners should be disturbed to find that
natives have to learn to read pictures, as we learn to read letters, is worth consideration. The extreme bias and distortion
of our sense-lives by our technology would seem to be a fact
that we prefer to ignore in our daily lives. Evidence that
natives do not perceive in perspective or sense the third dimension seems to threaten the Western ego-image and structure,
as many have found after a trip through the Ames Perception
Laboratory at Ohio State University. This lab is arranged to
reveal the various illusions we create for ourselves in what
we consider to be "normal" visual perception.

That we have accepted such bias and obliquity in a subliminal way through most of human history is clear enough.
Just why we are no longer content to leave our experience in
this subliminal state, and why many people have begun to
get very conscious about the unconscious, is a question well
worth investigation. People are nowadays much concerned
to set their houses in order, a process of self-consciousness

that has received large impetus from photography.

William Henry Fox Talbot, delighting in Swiss scenery, began to reflect on the *camera obscura* and wrote that "it was during these thoughts that the idea occurred to me ... how charming it would be if it were possible to cause these natural images to imprint themselves durably, and remain fixed on paper!" The printing press had, in the Renaissance, inspired a similar desire to give permanence to daily feelings and experience.

The method Talbot devised was that of printing positives chemically from negatives, to yield an exactly repeatable image. Thus the roadblock that had impeded the Greek botanists and had defeated their successors was removed. Most of the sciences had been, from their origins, utterly handicapped by the lack of adequate nonverbal means of transmitting information. Today, even subatomic physics would be unable to develop without the photograph.

The Sunday *New York Times* for June 15, 1958 reported:

Tiny Cells "Seen" By New Technique

Microphoretic Method Spots Million-Billionth Of Gram, London Designer Says

Samples of substances weighing less than a million-billionth of a gram can be analysed by a new British microscopic technique. This is the "microphoretic method" by Bernard M. Turner, a London bio-chemical analyst and instruments designer. It can be applied to the study of the cells of the brain and nervous system, cell duplication including that in cancerous tissue, and it will assist, it is believed, in the analyses of atmospheric pollution by dust .

In effect, an electric current pulls or pushes the different constituents of the sample into zones where they would normally be invisible.

However, to say that "the camera cannot lie" is merely to underline the multiple deceits that are now practiced in its name. Indeed, the world of the movie that was prepared by the photograph has become synonymous with illusion and fantasy, turning society into what Joyce called an "all-nights newsery reel," that substitutes a "reel" world for reality. Joyce knew more about the effects of the photograph on our senses, our language, and our thought processes than anybody else. His verdict on the "automatic writing" that is photography was the *abnihilization of the etym*. He saw the photo as at least a rival, and perhaps a usurper, of the word, whether written or spoken. But if *etym* (etymology) means the heart and core and moist substance of those beings that we grasp in words, then Joyce may well have meant that the photo was a new creation from nothing (*ab-nihil*), or even a reduction of creation to a photographic negative. If there is, indeed, a terrible nihilism in the photo and a substitution of shadows for substance, then we are surely not the worse for knowing it. The technology of the photo is an extension of our own being and can be withdrawn from circulation like any other technology if we decide that it is virulent. But amputation of such *extensions* of our physical being calls for as much knowledge and skill as are prerequisite to any other physical amputation.

If the phonetic alphabet was a technical means of *severing* the spoken word from its aspects of sound and gesture, the photograph and its development in the movie *restored* gesture to the human technology of recording experience. In fact, the snapshot of arrested human postures by photography directed more attention to physical and psychic posture than ever before. The age of the photograph has become the age of gesture and mime and dance, as no other age has ever been. Freud and Jung built their observations on the interpretation of the languages of both individual and collective postures and gestures with respect to dreams and to the ordinary acts of everyday life. The physical and psychic *gestalts*, or "still" shots, with which they worked were much owing to the posture world revealed by the photograph. The photograph is just as useful for collective, as for individual, postures and gestures, whereas written and printed language is biased toward the private and individual posture. Thus, the traditional figures of rhetoric were individual postures of mind of the private speaker in relation to an audience, whereas myth and Jungian archetypes are collective postures of the mind with which the written form could not cope, any more than it could command mime and gesture. Moreover, that the photograph is quite versatile in revealing and arresting posture and structure wherever it is used, occurs in countless examples, such as the analysis of bird-flight. It was the photograph that revealed the secret of bird-flight and enabled man to take off. The photo, in arresting bird-flight, showed that it was based on a principle of wing *fixity*. Wing movement was seen to be for propulsion, not for flight.

Perhaps the great revolution produced by photograph was in the traditional arts. The painter could no longer depict a world that had been much photographed. He turned, instead, to reveal the inner process of creativity in expressionism and in abstract art. Likewise, the novelist could no longer describe objects or happenings for readers who already knew what was happening by photo, press, film, and radio. The poet and novelist turned to those inward gestures of the mind by which we achieve insight and by which we make ourselves and our world. Thus art moved from outer matching to inner making. Instead of depicting a world that matched the world we already knew, the artists turned to presenting the creative process for public participation. He has given to us now the means of becoming involved in the making-process. Each development of the electric age attracts, and demands, a high degree of producer-orientation. The age of the consumer of processed and packaged goods is, therefore, not the present electric age, but the mechanical age that preceded it. Yet, inevitably, the age of the mechanical has had to overlap with the electric, as in such obvious instances as the internal combustion engine that requires the electric spark to ignite the explosion that moves its cylinders. The telegraph is an electric form that, when crossed with print and rotary presses, yields the modern newspaper. And the photograph is not a machine, but a chemical and light process that, crossed with the machine, yields the movie. Yet there is a vigor and violence in these hybrid forms that is self-liquidating, as it were. For in radio and TV—purely electric forms from which the mechanical principle has been excluded—there is an altogether new relation of the medium to its users. This is a relation of high participation and involvement that, for good or ill, no mechanism had ever evoked.

Education is ideally civil defense against media fall-out. Yet Western man has had, so far, no education or equipment for meeting any of the new media on their own terms. Literate man is not only numb and vague in the presence of film or photo, but he intensifies his ineptness by a defensive arrogance and condescension to "pop kulch" and "mass entertainment." It was in this spirit of bulldog opacity that the scholastic philosophers failed to meet the challenge of the printed book in the sixteenth century. The vested interests of acquired knowledge and conventional wisdom have always been by-passed and engulfed by new media. The study of this process, however, whether for the purpose of fixity or of change, has scarcely begun. The notion that self-interest confers a keener eye for recognizing and controlling the processes of change is quite without foundation, as witness the motorcar industry. Here is a world of obsolescence as surely doomed to swift erosion as was the enterprise of the buggy- and wagon-makers in 1915. Yet does General Motors, for example, know, or even suspect, anything about the effect of the TV image on the users of motorcars? The magazine enterprises are similarly undermined by the TV image and its effect on the advertising icon. The meaning of the new ad icon has not been grasped by those who stand to lose all. The same is true of the movie industry in general. Each of these enterprises lacks any "literacy" in any medium but its own, and thus the startling changes resulting from new hybrids and crossings of media catch them unawares.

To the student of media structures, every detail of the total mosaic of the contemporary world is vivid with meaningful life. As early as March 15, 1953, *Vogue* magazine announced a new hybrid, resulting from a cross between photograph and air travel:

> This first International Fashion Issue of *Vogue* is to mark a new point. We couldn't have done such an issue before. Fashion only got its internationalization papers a short time ago, and for the first time in one issue we can report on couture collections in five countries.

The advantages of such ad copy as high-grade ore in the lab of the media analyst can be recognized only by those trained in the language of vision and of the plastic arts in general. The copy writer has to be a strip-tease artist who has entire empathy with the immediate state of mind of the audience. Such, indeed, is also the aptitude of the popular novelist or song writer. It follows that any widely accepted writer or entertainer embodies and reveals a current set of attitudes that can be verbalized by the analyst. "Do you read me, Mac?" But were the words of the *Vogue* writer to be considered merely on literary or editorial grounds, their meaning would be missed, just as the copy in a pictorial ad is not to be considered as literary statement but as mime of the psychopathology of everyday life. In the age of the photograph, language takes on a graphic or iconic character, whose "meaning" belongs very little to the semantic universe, and not at all to the republic of letters.

If we open a 1938 copy of *Life*, the pictures or postures then seen as normal now give a sharper sense of remote time than do objects of real antiquity. Small children now attach the phrase "the olden days" to yesterday's hats and overshoes, so keenly are they attuned to the abrupt seasonal changes of visual posture in the world of fashions. But the basic experience here is one that most people feel for yesterday's newspaper, than which nothing could be more drastically out of fashion. Jazz musicians express their distaste for recorded jazz by saying, "it is as stale as yesterday's newspaper."

Perhaps that is the readiest way to grasp the meaning of the photograph in creating a world of accelerated transience. For the relation we have to "today's newspaper," or verbal jazz, is the same that people feel for fashions. Fashion is not a way of being informed or aware, but a way of being *with it*. That, however, is merely to draw attention to a negative aspect of the photograph. Positively, the effect of speeding up temporal sequence is to abolish time, much as the telegraph and cable abolished space. Of course the photograph does both. It wipes out our national frontiers and cultural barriers, and involves us in *The Family of Man*, regardless of any particular point of view. A picture of a group of persons of any hue whatever is a picture of people, not of "colored people." That is the logic of the photograph, politically speaking. But the logic of the photograph is neither verbal nor syntactical, a condition which renders literary culture quite helpless to cope with the photograph. By the same token, the complete transformation of human sense-awareness by this form involves a development of self-consciousness that alters facial expression and cosmetic makeup as immediately as it does our bodily stance, in public or in private. This fact can be gleaned from any magazine or movie of fifteen years back. It is not too much to say, therefore, that if outer posture is affected by the photograph, so with our inner postures and the dialogue with ourselves. The age of Jung and Freud is, above all, the age of the photograph, the age of the full gamut of self-critical attitudes.

This immense tidying-up of our inner lives, motivated by the new picture *gestalt* culture, has had its obvious parallels in our attempts to rearrange our homes and gardens and our cities. To see a photograph of the local slum makes the condition unbearable. The mere matching of the picture with reality provides a new motive for change, as it does a new motive for travel.

Daniel Boorstin in *The Image: or What Happened to the American Dream* offers a conducted *literary* tour of the new photographic world of travel. One has merely to look at the new tourism in a literary perspective to discover that it makes no sense at all. To the literary man who has read about Europe, in leisurely anticipation of a visit, an ad that whispers: "You are just fifteen *gourmet* meals from Europe on the world's fastest ship" is gross and repugnant. Advertisements of travel by plane are worse: "Dinner in New York, indigestion in Paris." Moreover, the photograph has reversed the purpose of travel, which until now had been to encounter the strange and unfamiliar. Descartes, in the early seventeenth century, had observed that traveling was almost like conversing with men of other centuries, a point of view quite unknown before his time. For those who cherish such quaint experience, it is necessary today to go back very many centuries by the art and archaeology route. Professor Boorstin seems unhappy that so many Americans travel so much and are changed by it so little. He feels that the entire travel experience has become "diluted, contrived, prefabricated." He is not concerned to find out why the photograph has done this to us. But in the same way intelligent people in the past always deplored the way in which the book had become a substitute for inquiry, conversation, and reflection, and never troubled to reflect on the nature of the printed book. The book reader has always tended to be passive, because that is the best way to read.

Today, the traveler has become passive. Given travelers checks, a passport, and a toothbrush, the world is your oyster. The macadam road, the railroad, and the steamship have taken the *travail* out of travel. People moved by the silliest whims now clutter the foreign places, because travel differs very little from going to a movie or turning the pages of a magazine. The "Go Now, Pay Later" formula of the travel agencies might as well read: Go now, arrive later," for it could be argued that such people never really leave their beaten paths of impercipience, nor do they ever arrive at any new place. They can have Shanghai or Berlin or Venice in a package tour that they need never open. In 1961, TWA began to provide new movies for its trans-Atlantic flights so that you could visit Portugal, California, or anywhere else, while en-route to Holland, for example. Thus the world itself becomes a sort of museum of objects that have been encountered before in some other medium. It is well known that even museum curators often prefer colored pictures to the originals of various objects in their own cases. In the same way, the tourist who arrives at the Leaning Tower of Pisa, or the Grand Canyon of Arizona, can now merely check his reactions to something with which he has long been familiar, and take his own pictures of the same.

To lament that the packaged tour, like the photograph, cheapens and degrades by making all places easy of access, is to miss most of the game. It is to make value judgments with *fixed* reference to the fragmentary perspective of literary culture. It is the same position that considers a literary landscape as superior to a movie travelogue. For the untrained awareness, all reading and all movies, like all travel are equally banal and unnourishing as experience. Difficulty of access does not confer adequacy of perception, though it may involve an object in an aura of pseudo-values, as with a gem, a movie star, or an old master. This now brings us to the factual core of the "pseudoevent," a label applied to the new media, in general, because of their power to give new patterns to our lives by acceleration of older patterns. It is necessary to reflect that this same insidious power was once felt in the *old* media, including languages. All media exist to invest our lives with artificial perception and arbitrary values.

All meaning alters with acceleration, because all patterns of personal and political interdependence change with any acceleration of information. Some feel keenly that speedup has impoverished the world they knew by changing its forms of human interassociation. There is nothing new or strange in a parochial preference for those pseudo-events that happened to enter into the composition of society just before the electric revolution of this century. The student of media soon comes to expect the new media of any period whatever to be classed as *pseudo* by those who have acquired the patterns of earlier media, whatever they may happen to be. This would seem to be a normal, and even amiable, trait ensuring a maximal degree of social continuity and permanence amidst change and innovation. But all the conservatism in the world does not afford even a token resistance to the ecological sweep of the new electric media. On a moving highway the vehicle that backs up is accelerating in relation to the highway situation. Such would seem to be the ironical status of the cultural reactionary. When the trend is one way his resistance insures a greater speed of change. Control over change would seem to consist in moving not with it but ahead of it. Anticipation gives the power to deflect and control force. Thus we may feel like a man who has been hustled away from his favorite knothole in the ball park by a frantic rout of fans eager to see the arrival of a movie star. We are no sooner in position to look at one kind of event than it is obliterated by another, just as our Western lives seem to native cultures to be one long series of *preparations for living*. But the favorite stance of literary man has long been "to view with alarm" or "to point with pride," while scrupulously ignoring what's going on.

One immense area of photographic influence that affects our lives is the world of packaging and display and, in general, the organization of shops and stores of every kind. The newspaper that could advertise every sort of product on one page quickly gave rise to department stores that provided every kind of product under one roof. Today the decentralizing of such institutions into a multiplicity of small shops in shopping plazas is partly the creation of the car, partly the result of TV. But the photograph still exerts some centralist pressure in the mail-order catalogue. Yet the mail-order houses originally felt not only the centralist forces of railway and postal services, but also, and at the same time, the decentralizing power of the telegraph. The Sears Roebuck enterprise was directly owing to stationmaster use of the telegraph. These men saw that the waste of goods on railway sidings could be ended by the speed of the telegraph to reroute and concentrate.

The complex network of media, other than the photograph that appears in the world of merchandising, is easier to observe in the world of sports. In one instance, the press camera contributed to radical changes in the game of football. A press photo of battered players in a 1905 game between Pennsylvania and Swarthmore came to the attention of President Teddy Roosevelt. He was so angered at the picture of Swarthmore's mangled Bob Maxwell that he issued an immediate ultimatum—that if rough play continued, he would abolish the game by executive edict. The effect was the same as that of the harrowing telegraph reports of Russell from the Crimea, which created the image and role of Florence Nightingale.

No less drastic was the effect of the press photo coverage of the lives of the rich. "Conspicuous consumption" owed less to the phrase of Veblen than to the press photographer, who began to invade the entertainment spots of the very rich.

The sights of men ordering drinks from horseback at the bars of clubs quickly caused a public revulsion that drove the rich into the ways of timid mediocrity and obscurity in America, which they have never abandoned. The photograph made it quite unsafe to come out and play, for it betrayed such blatant dimensions of power as to be self-defeating. On the other hand, the movie phase of photography created a new aristocracy of actors and actresses, who dramatized, on and off the screen, the fantasia of conspicuous consumption that the rich could never achieve. The movie demonstrated the magic power of the photo by providing a consumer package of plutocratic dimension for all the Cinderellas in the world.

The Gutenberg Galaxy provides the necessary background for studying the rapid rise of new visual values after the advent of printing from movable types. "A place for everything and everything in its place" is a feature not only of the compositor's arrangement of his type fonts, but of the entire range of human organization of knowledge and action from the sixteenth century onward. Even the inner life of the feelings and emotions began to be structured and ordered and analyzed according to separate pictorial landscapes, as Christopher Hussey explained in his fascinating study of *The Picturesque*. More than a century of this pictorial analysis of the inner life preceded Talbot's 1839 discovery of photography. Photography, by carrying the pictorial delineation of natural objects much further than paint or language could do, had a *reverse* effect. By conferring a means of self-delineation of objects, of "statement without syntax," photography gave the impetus to a delineation of the inner world. Statement without syntax or verbalization was really statement by gesture, by mime, and by *gestalt*, This new dimension opened for human inspection by poets like Baudelaire and Rimbaud *le paysage intérieur*, or the countries of the mind. Poets and painters invaded this inner landscape world long before Freud and Jung brought their cameras and notebooks to capture states of mind. Perhaps most spectacular of all was Claude Bernard, whose *Introduction to the Study of Experimental Medicine* ushered science into *le milieu intérieur* of the body exactly at the time when the poets did the same for the life of perception and feeling.

It is important to note that this ultimate stage of pictorialization was a reversal of pattern. The world of body and mind observed by Baudelaire and Bernard was not photographical at all, but a nonvisual set of relations such as the physicist, for example, had encountered by means of the new mathematics and statistics. The photograph might be said, also, to have brought to human attention the subvisual world of bacteria that caused Louis Pasteur to be driven from the medical profession by his indignant colleagues. Just as the painter Samuel Morse had unintentionally projected himself into the nonvisual world of the telegraph, so the photograph really transcends the pictorial by capturing the inner gestures and postures of both body and mind, yielding the new worlds of endocrinology and psychopathology.

To understand the medium of the photograph is quite impossible, then, without grasping its relations to other media, both old and new. For media, as extensions of our physical and nervous systems, constitute a world of biochemical interactions that must ever seek new equilibrium as new extensions occur. In America, people can tolerate their images in mirror or photo, but they are made uncomfortable by the recorded sound of their own voices. The photo and visual worlds are secure areas of anesthesia.

21

Press:
Government by News Leak

McLuhan's opening here provides an explicit definition of a phrase so often associated with news stories, human interest: *"a technical term meaning that which happens when multiple book pages or multiple information items are arranged in a mosaic on one sheet." As for the press service itself, he defines it as a contradiction, because it is "an individualistic technology dedicated to shaping and revealing group attitudes." As elsewhere, the link between mankind's first technology, language, and the medium under discussion is emphasized with respect to the inevitable transformations of the former: "Addison and Steele brought written discourse into line with the printed word and away from the variety of pitch and tone of the spoken, and of even the hand-written, word." A century before Addison and Steele, Thomas Nashe studied at Cambridge and travelled abroad, before going to London "to earn a precarious living by his pen"* (Chambers Biographical Dictionary). *The received view of Nashe put him among his journalistic cousins, until McLuhan turned his attention to the full scope of Nashe's work in writing his doctoral dissertation for Cambridge. The germ of McLuhan's media studies is clearly detectable in the dissertation; in the present chapter, the reader finds the sole mention of Nashe in* Understanding Media.

— (Editor)

The headline for an Associated Press release (February 25, 1963) read:

PRESS BLAMED FOR SUCCESS

KENNEDY MANAGES NEWS BOLDLY, CYNICALLY, SUBTLY, KROCK CLAIMS

Arthur Krock is quoted as saying that "the principle onus rests on the printed and electronic process itself." That may seem like another way of saying that "history is to blame." But it is the instant consequences of electrically moved information that makes necessary a deliberate artistic aim in the placing and management of news. In diplomacy the same electric speed causes the decisions to be announced before they are made in order to ascertain the varying responses that might occur when such decisions actually are made. Such procedure, quite inevitable at the electric speed that involves the entire society in the decision-making process, shocks the old press men because it abdicates any definite point of view. As the speed of information increases, the tendency is for politics to move away from representation and delegation of constituents toward immediate involvement of the entire community in the central acts of decision. Slower speeds of information make delegation and representation mandatory. Associated with such delegation are the points of view of the different sectors of public interest that are expected to be put forward for processing and consideration by the rest of the community. When the electric speed is introduced into such a delegated and representational organization, this obsolescent organization can only be made to function by a series of subterfuges and makeshifts. These strike some observers as base betrayals of the original aims and purposes of the established forms.

The massive theme of the press can be managed only by

direct contact with the formal patterns of the medium in question. It is thus necessary to state at once that "human interest" is a technical term meaning that which happens when multiple book pages or multiple information items are arranged in a mosaic on one sheet. The book is a private confessional form that provides a "point of view." The press is a group confessional form that provides communal participation. It can "color" events by using them or by not using them at all. But it is the daily communal exposure of multiple items in juxtaposition that gives the press its complex dimension of human interest.

The book form is not a communal mosaic or corporate image but a private voice. One of the unexpected effects of TV on the press has been a great increase in the popularity of *Time* and *Newsweek*. Quite inexplicably to themselves and without any new effort at subscription, their circulations have more than doubled since TV. These news magazines are preeminently mosaic in form, offering not windows on the world like the old picture magazines, but presenting corporate images of society in action. Whereas the spectator of a picture magazine is passive, the reader of a news magazine becomes much involved in the making of meanings for the corporate image. Thus the TV habit of involvement in mosaic image has greatly strengthened the appeal of these news magazines, but at the same time has diminished the appeal of the older pictorial feature magazines.

Both book and newspaper are confessional in character, creating the effect of *inside story* by their mere form, regardless of content. As the book page yields the inside story of the author's mental adventures, so the press page yields the inside story of the community in action and interaction. It is for this reason that the press seems to be performing its function most when revealing the seamy side. Real news is bad news—bad news *about* somebody, or bad news *for* somebody. In 1962, when Minneapolis had been for months without a newspaper, the chief of police said: "Sure, I miss the news, but so far as my job goes I hope the papers never come back. There is less crime around without a newspaper to pass around the ideas."

Even before the telegraph speedup, the newspaper of the nineteenth century had moved a long way toward a mosaic form. Rotary steam presses came into use decades before electricity, but typesetting by hand remained more satisfactory than any mechanical means until development of linotype about 1890. With linotype, the press could adjust its form more fully to the news-gathering of the telegraph and the newsprinting of the rotary presses. It is typical and significant that the linotype answer to the long-standing slowness of typesetting did not come from those directly engaged with the problem. Fortunes had been vainly spent on typesetting machines before James Clephane, seeking a fast way of writing out and duplicating shorthand notes, found a way to combine the typewriter and the typesetter. It was the *typewriter* that solved the utterly different *typesetting* problem. Today the publishing of book and newspaper both depends on the typewriter.

The speedup of information gathering and publishing naturally created new forms of arranging material for readers. As early as 1830 the French poet Lamartine had said, "The book arrives too late," drawing attention to the fact that the book and the newspaper are quite different forms. Slow down typesetting and news-gathering, and there occurs a change, not only in the physical appearance of the press, but also in the prose style of those writing for it. The first great change in style came early in the eighteenth century, when the famous *Tatler* and *Spectator* of Addison and Steele discovered a new prose technique to match the form of the printed word. It was the technique of equitone It consisted in maintaining a single level of tone and attitude to the reader throughout the entire composition. By this discovery,

Addison and Steele brought written discourse into line with the printed word and away from the variety of pitch and tone of the spoken, and of even the hand-written, word. This way of bringing language into line with print must be clearly understood. The telegraph broke language away again from the printed word, and began to make erratic noises called headlines, journalese, and telegraphese—phenomena that still dismay the literary community with its mannerisms of supercilious equitone that mime typographic uniformity. Headlines produce such effects as

BARBER HONES TONSILS
FOR OLD-TIMER'S EVENT

referring to Sal (the Barber) Maglie, the swarthy curve-ball artist with the old Brooklyn Dodgers, when he was to be guest speaker at a Ball Club dinner. The same community admires the varied tonality and vigor of Aretino, Rabelais, and Nashe, all of whom wrote prose before the print pressure was strong enough to reduce the language gestures to uniform lineality. Talking with an economist who was serving on an unemployment commission, I asked him whether he had considered newspaper reading as a form of paid employment. I was not wrong in supposing that he would be incredulous. Nevertheless, all media that mix ads with other programming are a form of "paid learning." In years to come, when the child will be paid to learn, educators will recognize the sensational press as the forerunner of paid learning. One reason that it was difficult to see this fact earlier is that the processing and moving of information had not been the main business of a mechanical and industrial world. It is, however, easily the dominant business and means of wealth in the electric world. At the end of the mechanical age people still imagined that press and radio and even TV were merely forms of information paid for by the makers and users of "hardware" like cars and soap and gasoline. As automation takes hold, it becomes obvious that *information* is the crucial commodity, and that solid products are merely incidental to information movement. The early stages by which information itself became the basic economic commodity of the electric age were obscured by the ways in which advertising and entertainment put people off the track. Advertisers pay for space and time in paper and magazine, on radio and TV; that is, they buy a piece of the reader, listener, or viewer as definitely as if they hired our homes for a public meeting. They would gladly pay the reader, listener, or viewer directly for his time and attention if they knew how to do so. The only way so far devised is to put on a free show. Movies in America have not developed advertising intervals simply because the movie itself is the greatest of all forms of advertisement for consumer goods.

Those who deplore the frivolity of the press and its natural form of group exposure and communal cleansing simply ignore the nature of the medium and demand that it be a book, as it tends to be in Europe. The book arrived in western Europe long before the newspaper; but Russia and middle Europe developed the book and newspaper almost together, with the result that they have never unscrambled the two forms. Their journalism exudes the private point of view of the literary mandarin. British and American journalism, however, have always tended to exploit the mosaic form of the newspaper format in order to present the discontinuous variety and incongruity of ordinary life. The monotonous demands of the literary community — that the newspaper use its mosaic form to present a fixed point of view on a single plane of perspective — represent a failure to see the form of the press at all. It is as if the public were suddenly to demand that department stores have only one department.

The classified ads (and stock-market quotations) are the bedrock of the press. Should an alternative source of easy access to such diverse daily information be found, the press will fold. Radio and TV can handle the sports, news, comics, and pictures. The editorial, which is the one book-feature of the newspaper, has been ignored for many years, unless put in the form of news or paid advertisement.

If our press is in the main a free entertainment service paid for by advertisers who want to buy readers, the Russian press is *in toto* the basic mode of industrial promotion. If we use news, political and personal, as entertainment to capture ad readers, the Russians use it as a means of promotion for their economy. Their political news has the same aggressive earnestness and posture as the voice of the sponsor in an American ad. A culture that gets the newspaper late (for the same reasons that industrialization is delayed) and one that accepts the press as a form of the book and regards industry as group political action, is not likely to seek entertainment in the news. Even in America, literate people have small skill in understanding the iconographic varieties of the ad world. Ads are ignored or deplored, but seldom studied and enjoyed.

Anybody who could think that the press has the same function in America and Russia, or in France and China, is not really in touch with the medium. Are we to suppose that this kind of media illiteracy is characteristic only of Westerners, and that Russians know how to correct the bias of the medium in order to read it right? Or do people vaguely suppose that the heads of state in the various countries of the world know that the newspaper has totally diverse effects in different cultures? There is no basis for such assumptions. Unawareness of the nature of the press in its subliminal or latent action is as common among politicians as among political scientists. For example, in oral Russia both *Pravda* and *Izvestia* handle domestic news, but the big international themes come to the West over Radio Moscow. In visual America, radio and television handle the domestic stories, and international affairs get their formal treatment in *Time* magazine and *The New York Times*. As a foreign service, the bluntness of Voice of America in no way compares to the sophistication of the BBC and Radio Moscow, but what it lacks in verbal content it makes up in the entertainment value of its American jazz. The implications of this difference of stress are important for an understanding of the kinds of opinions and decisions natural to an oral, as opposed to a visual, culture.

A friend of mine who tried to teach something about the forms of media in secondary school was struck by one unanimous response. The students could not for a moment accept the suggestion that the press or any other public means of communication could be used with base intent. They felt that this would be akin to polluting the air or the water supply, and they didn't feel that their friends and relatives employed in these media would sink to such corruption. Failure in perception occurs precisely in giving attention to the program "content" of our media while ignoring the form, whether it be radio or print or the English language itself. There have been countless Newton Minows (formerly head of the Federal Communications Commission) to talk about the Wasteland of the Media, men who know nothing about the form of any medium whatever. They imagine that a more earnest tone and a more austere theme would pull up the level of the book, the press, the movie, and TV. They are wrong to a farcical degree. They have only to try out their theory for fifty consecutive words in the mass medium of the English language. What would Mr. Minow do, what would any advertiser do, without the well-worn and corny clichés of popular speech? Suppose that we were to try for a few sentences to raise the level of our daily English conversation by a series of sober and serious sentiments? Would this be a way of getting at the problems of improving the medium? If all English were enunciated at a Mandarin level of uniform elegance and sententiousness, would the language and its users be better served? There comes to mind the remark of Artemus Ward that "Shakespeare wrote good plays but he wouldn't have succeeded as the Washington correspondent of a New York daily newspaper. He lacked the reckisit fancy and imagination."

The book-oriented man has the illusion that the press would be better without ads and without the pressure from the advertiser. Reader surveys have astonished even publishers with the revelation that the roving eyes of newspaper readers take equal satisfaction in ads and news copy. During the Second War, the U.S.O. sent special issues of the principal American magazines to the Armed Forces, with the ads omitted. The men insisted on having the ads back again. Naturally. The ads are by far the best part of any magazine or newspaper. More pains and thought, more wit and art go into the making of an ad than into any prose feature of press or magazine. Ads are *news*. What is wrong with them is that they are always *good* news. In order to balance off the effect and to sell good news, it is necessary to have a lot of bad news. Moreover, the newspaper is a hot medium. It has to have bad news for the sake of intensity and reader participation. *Real* news is *bad* news, as already noted, and as any newspaper from the beginning of print can testify. Floods, fires, and other communal disasters by land and sea and air outrank any kind of private horror or villainy as *news*. Ads, in contrast, have to shrill their happy message loud and clear in order to match the penetrating power of bad news.

Commentators on the press and the American Senate have noted that since the Senate began its prying into unsavory subjects it has assumed a role superior to Congress. In fact, the great disadvantage of the Presidency and the Executive arm in relation to public opinion is that it tries to be a source of good news and noble directive. On the other hand, Congressmen and Senators have the free of the seamy side so necessary to the vitality of the press.

Superficially, this may seem cynical, especially to those who imagine that the content of a medium is a matter of policy and personal preference, and for whom all corporate media, not only radio and the press but ordinary popular speech as well, are debased forms of human expression and experience. Here I must repeat that the newspaper, from its beginnings, has tended, not to the book form, but to the mosaic or participational form. With the speedup of print-ing and news-gathering, this mosaic form has become a dominant aspect of human association; for the mosaic form means, not a detached "point of view," but participation in process. For that reason, the press is inseparable from the democratic process, but quite expendable from a literary or book point of view.

Again, the book-oriented man misunderstands the col-lective mosaic form of the press when he complains about its endless reports on the seamy underside of the social garment. Both book and press are, in their very format, dedicated to the job of revealing the inside story, whether it is Montaigne giving to the private reader the delicate contours of his mind, or Hearst and Whitman resonating their barbaric yawps over the roofs of the world. It is the printed form of public address and high intensity with its precise uniformity of repetition that gives to book and press alike the special character of public confessional.

The first items in the press to which all men turn are the ones about which they already know. If we have witnessed some event, whether a ball game or a stock crash or a snow-storm, we turn to the report of that happening, first. Why? The answer is central to any understanding of media. Why does a child like to chatter about the events of its day, how-ever jerkily? Why do we prefer novels and movies about familiar scenes and characters? Because for rational beings to see or re-cognize their experience in a new material form is an unbought grace of life. Experience translated into a new

283

medium literally bestows a delightful playback of earlier awareness. The press repeats the excitement we have in using our wits, and by using our wits we can translate the outer world into the fabric of our own beings. This excitement of translation explains why people quite naturally wish to use their senses all the time. Those external extensions of sense and faculty that we call media we use as constantly as we do our eyes and ears, and from the same motives. On the other hand, the book-oriented man considers this nonstop use of media as debased; it is unfamiliar to him in the book-world.

Up to this point we have discussed the press as a mosaic successor to the book-form. The mosaic is the mode of the corporate or collective image and commands deep participation. This participation is communal rather than private, inclusive rather than exclusive. Further features of its form can best be grasped by a few random views taken from outside the present form of the press. Historically, for example, newspapers had waited for news to come to them. The first American newspaper, issued in Boston by Benjamin Harris on September 25, 1690, announced that it was to be "furnished once a month (or if any Glut of Occurrences happen, oftener)." Nothing could more plainly indicate the idea that news was something outside and beyond the newspaper. Under such rudimentary conditions of awareness, a principal function of the newspaper was to correct rumors and oral reports, as a dictionary might provide "correct" spellings and meanings for words that had long existed without the benefit of dictionaries. Fairly soon the press began to sense that news was not only to be reported but also gathered, and, indeed, to be made. What went *into* the press was news. The rest was not news. "He made the news" is a strangely ambiguous phrase, since to be in the newspaper is both to be news and to make news. Thus "making the news," like "making good," implies a world of actions and fictions alike. But the press is a daily action and fiction or thing made, and it is made out of just about everything in the community. By the mosaic means, it is made into a communal image or cross-section.

When a conventional critic like Daniel Boorstin complains that modern ghost-writing, teletype, and wire services create an insubstantial world of "pseudo-events," he declares, in effect, that he has never examined the nature of any medium prior to those of the electric age. For the pseudo or fictitious character has always permeated the media, not just those of recent origin.

Long before big business and corporations became aware of the image of their operation as a fiction to be carefully tattooed upon the public sensorium, the press had created the image of the community as a series of on-going actions unified by datelines. Apart from the vernacular used, the dateline is the only organizing principle of the newspaper image of the community. Take off the dateline, and one day's paper is the same as the next. Yet to read a week-old newspaper without noticing that it is not today's is a disconcerting experience. As soon as the press recognized that news presentation was not a repetition of occurrences and reports but a direct cause of events, many things began to happen. Advertising and promotion, until then restricted, broke onto the front page, with the aid of Barnum, as sensational stories. Today's press agent regards the newspaper as a ventriloquist does his dummy. He can make it say what he wants. He looks on it as a painter does his palette and tubes of pigment; from the endless resources of available events, an endless variety of managed mosaic effects can be attained. Any private client can be ensconced in a wide range of different patterns and tones of public affairs or human interest and depth items.

If we pay careful attention to the fact that the press is a mosaic, participant kind of organization and a do-it-yourself kind of world, we can see why it is so necessary to democratic government. Throughout his study of the press in *The Fourth Branch of Government*, Douglas Cater is baffled by the fact that amidst the extreme fragmentation of government departments and branches, the press somehow manages to keep them in relation to each other and to the nation. He emphasizes the paradox that the press is dedicated to the process of cleansing by publicity, and yet that, in the electronic world of the seamless web of events, most affairs must be kept secret. Top secrecy is translated into public participation and responsibility by the magic flexibility of the controlled news leak.

It is by this kind of ingenious adaptation from day to day that Western man is beginning to accommodate himself to the electric world of total interdependence. Nowhere is this transforming process of adaptation more visible than in the press. The press, in itself, presents the contradiction of an individualistic technology dedicated to shaping and revealing group attitudes.

It might be well now to observe how the press has been modified by the recent developments of telephone, radio, and TV. We have seen already that the telegraph is the factor that has done most to create the mosaic image of the modern press, with its mass of discontinuous and unconnected features. It is this group-image of the communal life, rather than any editorial outlook or slanting, that constitutes the participant of this medium. To the book-man of detached private culture, this is the scandal of the press: its shameless involvement in the depths of human interest and sentiment. By eliminating time and space in news presentation, the telegraph dimmed the privacy of the book-form, and heightened, instead, the new public image in the press.

The first harrowing experience for the press man visiting Moscow is the absence of telephone books. A further horrifying revelation is the absence of central switchboards in government departments. You know the number, or else. The student of media is happy to read a hundred volumes to discover two facts such as these. They floodlight a vast murky area of the pressworld, and illuminate the role of telephone as seen through another culture. The American newspaperman in large degree assembles his stories and processes his data by telephone because of the speed and immediacy of the oral process. Our popular press is a near approximation to the grapevine. The Russian and European newspaperman is, by comparison, a littérateur. It is a paradoxical situation, but the press in literate America has an intensely oral character, while in oral Russia and Europe the press has a strongly literary character and function.

The English dislike the telephone so much that they substitute numerous mail deliveries for it. The Russians use the telephone for a status symbol, like the alarm clock worn by tribal chiefs as an article of attire in Africa. The mosaic of the press image in Russia is felt as an immediate form of tribal unity and participation. Those features of the press that we find most discordant with austere individual standards of literary culture are just the ones that recommend it to the Communist Party. "A newspaper," Lenin once declared, "is not only a collective propagandist and collective agitator; it is also a collective organizer." Stalin called it "the most powerful weapon of our Party." Khrushchev cites it as "our chief ideological weapon." These men had more an eye to the collective form of the press mosaic, with its magical power to impose its own assumptions, than to the printed word as expressing a private point of view. In oral Russia, fragmentation of government powers is unknown. Not for them our function of the press as unifier of fragmented departments. The Russian monolith has quite different uses for the press mosaic. Russia now needs the press (as we formerly did the book) to translate a tribal and oral community into some degree of visual, uniform culture able to sustain a market system.

In Egypt the press is needed to effect nationalism, that visual kind of unity that springs men out of local and tribal patterns. Paradoxically, radio has come to the fore in Egypt as the rejuvenator of the ancient tribes. The battery radio carried on the camel gives to the Bedouin tribes a power and vitality unknown before, so that to use the word "nationalism" for the fury of oral agitation that the Arabs have felt by radio is to conceal the situation from ourselves. Unity of the Arab-speaking world can only come by the press. Nationalism was unknown to the Western world until the Renaissance, when Gutenberg made it possible to *see* the mother tongue in uniform dress. Radio does nothing for this uniform visual unity so necessary to nationalism. In order to restrict radio-listening to national programs, some Arab governments have passed a law forbidding the use of private headphones, in effect enforcing a tribal collectivism in their radio audiences. Radio restores tribal sensitivity and exclusive involvement in the web of kinship. The press, on the other hand, creates a visual, not-too-involved kind of unity that is hospitable to the inclusion of many tribes, and to diversity of private outlook.

If telegraph shortened the sentence, radio shortened the news story, and TV injected the interrogative mood into journalism. In fact, the press is now not only a telephoto mosaic of the human community hour by hour, but its technology is also a mosaic of all the technologies of the community. Even in its selection of the newsworthy, the press prefers those persons who have already been accorded some notoriety existence in movies, radio, TV, and drama. By this fact, we can test the nature of the press medium, for anybody who appears only in the newspapers is, by that token, an ordinary citizen.

Wallpaper manufacturers have recently begun to issue wallpaper that presents the appearance of a French newspaper. The Eskimo sticks magazine pages on the ceiling of his igloo to deter drip. But even an ordinary newspaper on a kitchen floor will reveal news items that one had missed when the paper was in hand. Yet whether one uses the press for privacy in public conveyances, or for involvement in the communal while enjoying privacy, the mosaic of the press manages to effect a complex many-leveled function of group-awareness and participation such as the book has never been able to perform.

The format of the press—that is, its structural character-istics—were quite naturally taken over by the poets after Baudelaire in order to evoke an inclusive awareness. Our ordinary newspaper page today is not only symbolist and surrealist in an *avant-garde* way, but it was the earlier *inspiration* of symbolism and surrealism in art and poetry, as anybody can discover by reading Flaubert or Rimbaud. Approached as newspaper form, any part of Joyce's *Ulysses* or any poem of T. S. Eliot's before the *Quartets* is more readily enjoyed. Such, however, is the austere continuity of book culture that it scorns to notice these *liaisons dangéreuses* among the media, especially the scandalous affairs of the book-page with elec-tronic creatures from the other side of the linotype.

In view of the inveterate concern of the press with cleans-ing by publicity, it may be well to ask if it does not set up an inevitable clash with the medium of the book. The press as a collective and communal image assumes a natural posture of opposition to all private manipulation. Any mere individual who begins to stir about as if he were a public something-or-other is going to get into the press. Any individual who manipulates the public for his private good may also feel the cleansing power of publicity. The cloak of invisibility, therefore, would seem to fall most naturally on those who own newspapers or who use them extensively for commercial ends. May not this explain the strange obsession of the book-man with the press-lords as essentially corrupt? The merely private and fragmentary point of view assumed by the book reader and writer finds natural grounds for hostility toward the big communal power of the press. As forms, as media, the book and the newspaper would seem to be as incompatible as any two media could be. The owners of media always endeavor to give the public what it wants, because they sense that their power is in the *medium* and not in the *message* or the program.

22

Motorcar:
The Mechanical Bride

Having found a particularly satisfying pun for the subtitle of the following chapter on advertising, McLuhan transplanted here the title of his first book, and first book-length study of ads, The Mechanical Bride, *to evoke America's love affair with the automobile. He referred at the close of the preceding chapter to reactions in a print-dominated culture to "scandalous affairs of the book-page with electronic creatures from the other side of the linotype." Here he shows clearly that the same print culture passively accepts its own affair with the mechanical creature of the assembly-line type. If "Gutenberg made it possible to see the mother tongue in uniform dress," General Motors made it possible to possess a bride without seeing her as a technological extension of self. Narcissus at the drive-in.*

— *(Editor)*

Here is a news item that captures a good deal of the meaning of the automobile in relation to social life:

> I was terrific. There I was in my white Continental, and I was wearing a pure-silk, pure-white, embroidered cowboy shirt, and black gabardine trousers. Beside me in the car was my jet-black Great Dane imported from Europe, named Dana von Krupp. You just can't do any better than that.

Although it may be true to say that an American is a creature of four wheels, and to point out that American youth attributes much more importance to arriving at driver's-license age than at voting age, it is also true that the car has become an article of dress without which we feel uncertain, unclad, and incomplete in the urban compound. Some observers insist that, as a status symbol, the house has, of late, supplanted the car. If so, this shift from the mobile open road to the manicured roots of suburbia may signify a real change in American orientation. There is a growing uneasiness about the degree to which cars have become the real population of our cities, with a resulting loss of human scale, both in power and in distance. The town planners are plotting ways and means to buy back our cities for the pedestrian from the big transportation interests.

Lynn White tells the story of the stirrup and the heavy-armored knight in his *Medieval Technology and Social Change*. So expensive yet so mandatory was the armored rider for shock combat, that the cooperative feudal system came into existence to pay for his equipment. Renaissance gunpowder and ordnance ended the military role of the knight and returned the city to the pedestrian burgess.

If the motorist is technologically and economically far superior to the armored knight, it may be that electric changes in technology are about to dismount him and return us to the pedestrian scale. "Going to work" may be only a transitory phase, like "going shopping." The grocery interests have long foreseen the possibility of shopping by two-way TV, or video-telephone. William M. Freeman, writing for *The New York Times* Service (Tuesday, October 15, 1963), reports that there will certainly be "a decided transition from today's distribution vehicles ... Mrs. Customer will be able to tune in on various stores. Her credit identification will be picked up automatically via television. Items in full and faithful coloring will be viewed. Distance will hold no problem, since by the end of the century the consumer will be able to make direct television connections regardless of how many miles are involved."

What is wrong with all such prophecies is that they assume a stable framework of fact—in this case, the house and the store—which is usually the first to disappear. The changing relation between customer and shopkeeper is as nothing compared to the changing pattern of work itself, in an age of automation. It is true that going-to and coming-from work are almost certain to lose all of their present character. The car as vehicle, in that sense, will go the way of the horse. The horse has lost its role in transportation but has made a strong comeback in entertainment. So with the motorcar. Its future does not belong in the area of transportation. Had the infant automotive industry, in 1910, seen fit to call a conference to consider the future of the horse, the discussion would have been concerned to discover new jobs for the horse and new kinds of training to extend the usefulness of the horse. The complete revolution in transportation and in housing and city arrangement would have been ignored. The turn of our economy to making and servicing motorcars, and the devotion of much leisure time to their use on a vast new highway system, would not even have been thought of. In other words, it is the framework itself that changes with new technology, and not just the picture within the frame. Instead of thinking of doing our shopping by television, we should become aware that TV intercom means the end of shopping itself, and the end of work as we know it at present. The same fallacy besets our thinking about TV and education. We think of TV as an incidental aid, whereas in fact it has already transformed the learning process of the young, quite independently of home and school alike.

In the 1930s, when millions of comic books were inundating the young with gore, nobody seemed to notice that emotionally the violence of millions of cars in our streets was incomparably more hysterical than anything that could ever be printed. All the rhinos and hippos and elephants in the world, if gathered in one city, could not begin to create the menace and explosive intensity of the hourly and daily experience of the internal-combustion engine. Are people really expected to internalize—live with—all this power and explosive violence, without processing and siphoning it off into some form of fantasy for compensation and balance?

In the silent pictures of the 1920s a great many of the sequences involved the motorcar and policemen. Since the film was then accepted as an optical illusion, the cop was the principal reminder of the existence of ground rules in the game of fantasy. As such, he took an endless beating. The motorcars of the 1920s look to our eyes like ingenious contraptions hastily assembled in a tool shop. Their link with the buggy was still strong and clear. Then came the balloon tires, the massive interior, and the bulging fenders. Some people see the big car as a sort of bloated middle age, following the gawky period of the first love-affair between America and the car. But funny as the Viennese analysts have been able to get about the car as sex object, they have at last, in doing so, drawn attention to the fact that, like the bees in the plant world, men have always been the sex organs of the technological world. The car is no more and no less a sex object than the wheel or the hammer. What the motivation researchers have missed entirely is the fact that the American sense of spatial form has changed much since radio, and drastically since TV. It is misleading, though harmless, to try to grasp this change as middle-age reaching out for the sylph Lolita.

Certainly there have been some strenuous slimming programs for the car in recent years. But if one were to ask, "Will the car last?" or "Is the motorcar here to stay?" there would be confusion and doubt at once. Strangely, in so progressive an age, when change has become the only constant in our lives, we never ask, "Is the car here to stay?" The answer, of course, is "No." In the electric age, the wheel itself is obsolescent. At the heart of the car industry there are men who know that the car is passing, as certainly as the cuspidor was doomed when the lady typist arrived on the business scene. What arrangements have they made to ease the automobile industry off the center of the stage? The mere obsolescence of the wheel does not mean its disappearance. It means only that, like penmanship or typography, the wheel will move into a subsidiary role in the culture.

In the middle of the nineteenth century great success was achieved with steam-engined cars on the open road. Only the heavy toll-taxes levied by local road authorities discouraged steam engines on the highways. Pneumatic tires were fitted to a steam car in France in 1887. The American Stanley Steamer began to flourish in 1899. Ford had already built his first car in 1896, and the Ford Motor Company was founded in 1903. It was the electric spark that enabled the gasoline engine to take over from the steam engine. The crossing of electricity, the biological form, with the mechanical form was never to release a greater force.

It is TV that has dealt the heavy blow to the American car. The car and the assembly line had become the ultimate expression of Gutenberg technology; that is, of uniform and repeatable processes applied to all aspects of work and living. TV brought a questioning of all mechanical assumptions about uniformity and standardization, as of all consumer values. TV brought also obsession with depth study and analysis. Motivation research, offering to hook the ad and the *id*, became immediately acceptable to the frantic executive world that felt the same way about the new American tastes as Al Capp did about his 50,000,000 audience when TV struck. Something had happened. America was not the same.

For forty years the car had been the great leveler of physical space and of social distance as well. The talk about the American car as a status symbol has always overlooked the basic fact that it is the *power* of the motorcar that levels all social differences, and makes the pedestrian a second-class citizen. Many people have observed how the real integrator or leveler of white and Negro in the South was the private car and the truck, not the expression of moral points of view. The simple and obvious fact about the car is that, more than any horse, it is an extension of man that turns the rider into a superman. It is a hot, explosive medium of social communication. And TV, by cooling off the American public tastes and creating new needs for unique wrap-around space, which the European car promptly provided, practically unhorsed the American auto-cavalier. The small European cars reduce him to near-pedestrian status once more. Some people manage to drive them on the sidewalk.

The car did its social leveling by horsepower alone. In turn, the car created highways and resorts that were not only very much alike in all parts of the land, but equally available to all. Since TV, there is naturally frequent complaint about this uniformity of vehicle and vacation scene. As John Keats put it in his attack on the car and the industry in *The Insolent Chariots*, where one automobile can go, all other automobiles do go, and wherever the automobile goes, the automobile version of civilization surely follows. Now this is a TV-oriented sentiment that is not only anti-car and anti-standardization, but anti-Gutenberg, and therefore anti-American as well.

Of course, I know that John Keats doesn't *mean* this. He had never thought about media or the way in which Gutenberg created Henry Ford and the assembly line and standardized culture. All he knew was that it was popular to decry the uniform, the standardized, and the hot forms of communication, in general. For that reason, Vance Packard could make hay with *The Hidden Persuaders*. He hooted at the old salesmen and the hot media, just as *MAD* does. Before TV, such gestures would have been meaningless. It wouldn't have paid off. Now, it pays to laugh at the mechanical and the merely standardized. John Keats could question the central glory of classless American society by saying, "If you've seen one part of America, you've seen it all," and that the car gave the American the opportunity, not to travel and experience adventure, but "to make himself more and more common." Since TV, it has become popular to regard the more and more uniform and repeatable products of industry with the same contempt that a Brahmin like Henry James might have felt for a chamber-pot dynasty in 1890. It is true that automation is about to produce the unique and custom-built at assembly-line speed and cheapness. Automation can manage the bespoke car or coat with less fuss than we ever produced the standardized ones. But the unique product cannot circulate in our market or distribution setups. As a result, we are moving into a most revolutionary period in marketing, as in everything else.

When Europeans used to visit America before the Second War they would say, "But you have communism here!" What they meant was that we not only *had* standardized goods, but *everybody* had them. Our millionaires not only ate corn-flakes and hot dogs, but really thought of themselves as middle-class people. What else? How could a millionaire be anything but "middle-class" in America unless he had the creative imagination of an artist to make a unique life for himself? Is it strange that Europeans should associate uniformity of environment and commodities with communism? And that Lloyd Warner and his associates, in their studies of American cities, should speak of the American class system in terms of income? The highest income cannot liberate a North American from his "middle-class" life. The lowest income gives everybody a considerable piece of the same middle-class existence. That is, we really have homogenized our schools and factories and cities and entertainment to a great extent, just because we are literate and do accept the logic of uniformity and homogeneity that is inherent in Gutenberg technology. This logic, which had never been accepted in Europe until very recently, has suddenly been questioned in America, since the tactile mesh of the TV mosaic has begun to permeate the American sensorium. When a popular writer can, with confidence, decry the use of the car for travel as making the driver "more and more common," the fabric of American life has been questioned.

Only a few years back Cadillac announced its "El Dorado Brougham" as having anti-dive control, outriggers, pillarless styling, projectile-shaped gull-wing bumpers, outboard exhaust ports, and various other exotic features borrowed from the non-motorcar world. We were invited to associate it with Hawaiian surf riders, with gulls soaring like sixteen-inch shells, and with the boudoir of Madame de Pompadour. Could *MAD* magazine do any better? In the TV age, any of these tales from the Vienna woods, dreamed up by motivational researchers, could be relied upon to be an ideal comic script for *MAD*. The script was always there, in fact, but not till TV was the audience conditioned to enjoy it.

To mistake the car for a status symbol, just because it is asked to be taken as anything but a car, is to mistake the whole meaning of this very late product of the mechanical age that is now yielding its form to electric technology. The car is a superb piece of uniform, standardized mechanism that is of a piece with the Gutenberg technology and literacy which created the first classless society in the world. The car gave to the democratic cavalier his horse and armor and haughty insolence in one package, transmogrifying the knight into a misguided missile. In fact, the American car did not level downward, but upward, toward the aristocratic idea. Enormous increase and distribution of power had also been the equalizing force of literacy and various other forms of mechanization. The willingness to accept the car as a status symbol, restricting its more expansive form to the use of higher executives, is not a mark of the car and mechanical age, but of the electric forces that are now ending this mechanical age of uniformity and standardization, and recreating the norms of status and role.

When the motorcar was new, it exercised the typical mechanical pressure of explosion and separation of functions. It broke up family life, or so it seemed, in the 1920s. It separated work and domicile, as never before. It exploded each city into a dozen suburbs, and then extended many of the forms of urban life along the highways until the open road seemed to become non-stop cities. It created the asphalt jungles, and caused 40,000 square miles of green and pleasant land to be cemented over. With the arrival of plane travel, the motorcar and truck teamed up together to wreck the railways. Today small children plead for a train ride as if it were a stagecoach or horse and cutter: "Before they're *gone*, Daddy."

The motorcar ended the countryside and substituted a new landscape in which the car was a sort of steeplechaser. At the same time, the motor destroyed the city as a casual environment in which families could be reared. Streets, and even sidewalks, became too intense a scene for the casual interplay of growing up. As the city filled with mobile strangers, even next door neighbors became strangers. This is the story of the motorcar, and it has not much longer to run. The tide of taste and tolerance has turned, since TV, to make the hot-car medium increasingly tiresome. Witness the portent of the crosswalk, where the small child has power to stop a cement truck. The same change has rendered the big city unbearable to many who would no more have felt that way ten years ago than they could have enjoyed reading *MAD*.

The continuing power of the car medium to transform the patterns of settlement appears fully in the way in which the new urban kitchen has taken on the same central and multiple social character as the old farm kitchen. The farm kitchen had been the key point of entry to the farmhouse, and had become the social center, as well. The new suburban home again makes the kitchen the center and, ideally, is localized for access to and from the car. The car has become the carapace, the protective and aggressive shell, of urban and suburban man. Even before the Volkswagen, observers above street level have often noticed the near-resemblance of cars to shiny-backed insects. In the age of the tactile-oriented skin-diver, this hard shiny carapace is one of the blackest marks against the motorcar. It is for motorized man that the shopping plazas have emerged. They are strange islands that make the pedestrian feel friendless and disembodied. The car bugs him.

The car, in a word, has quite refashioned all of the spaces that unite and separate men, and it will continue to do so for a decade more, by which time the electronic successors to the car will be manifest.

23

Ads:
Keeping Upset with
the Joneses

This chapter is the central panel of the triptych dedicated to the study of advertising in McLuhan's writings, beginning with The Mechanical Bride *and concluding with* Culture Is Our Business. *It is rich in observations on the interplay of media that advertising has deliberately exploited to enhance its power and even richer in examples of such interplay proving to be inevitable and uncontrollable, because of the inherent power of any new medium. If McLuhan might seem to be expressing admiration for advertising moguls in saying that "no group of sociologists can approximate the ad teams in the gathering and processing of exploitable social data," it is important to look elsewhere here and discover his references to "the exploitation of the unconscious," "Madison Avenue frog-men-of-the-mind" and the ominous prospect of "programmed harmony." McLuhan practices his own media-centered approach to sociology, observing that "it was easy for the retribalized Nazi to feel superior to the American consumer." As for the overriding project of understanding media and bringing them into the "orderly service" referred to in the author's introduction, the reader is invited to approach advertising in the same spirit of play in which it is created, even more so today than when McLuhan wrote that "any ad put in a new setting is funny."*

— *(Editor)*

The continuous pressure is to create ads more and more in the image of audience motives and desires. The product matters less as the audience participation increases. An extreme example is the corset series that protests that "it is not the corset that you feel." The need is to make the ad include the audience experience. The product and the public response become a single complex pattern. The art of advertising has wondrously come to fulfill the early definition of anthropology as "the science of man embracing woman." The steady trend in advertising is to manifest the product as an integral part of large social purposes and processes. With very large budgets the commercial artists have tended to develop the ad into an icon, and icons are not specialist fragments or aspects but unified and compressed images of complex kind. They focus a large region of experience in tiny compass. The trend in ads, then, is away from the consumer picture of product to the producer image of process. The corporate image of process includes the consumer in the producer role as well.

This powerful new trend in ads toward the iconic image has greatly weakened the position of the magazine industry in general and the picture magazines in particular. Magazine features have long employed the pictorial treatment of themes and news. Side by side with these magazine features that present shots and fragmentary points of view, there are the new massive iconic ads with their compressed images that include producer and consumer, seller and society in a single image. The ads make the features seem pale, weak, and anemic. The features belong to the old pictorial world that preceded TV mosaic imagery.

It is the powerful mosaic and iconic thrust in our experience since TV that explains the paradox of the upsurge of *Time* and *Newsweek* and similar magazines. These magazines present the news in a compressed mosaic form that is a real parallel to the ad world. Mosaic news is neither narrative, nor point of view, nor explanation, nor comment. It is a corporate image in depth of the community in action and invites maximal participation in the social process.

Ads seem to work on the very advanced principle that a small pellet or pattern in a noisy, redundant barrage of repetition will gradually assert itself. Ads push the principle of noise all the way to the plateau of persuasion. They are quite in accord with the procedures of brain-washing. This depth principle of onslaught on the unconscious may be the reason why.

Many people have expressed uneasiness about the advertising enterprise in our time. To put the matter abruptly, the advertising industry is a crude attempt to extend the principles of automation to every aspect of society. Ideally, advertising aims at the goal of a programmed harmony among all human impulses and aspirations and endeavors. Using handicraft methods, it stretches out toward the ultimate electronic goal of a collective consciousness. When all production and all consumption are brought into a pre-established harmony with all desire and all effort, then advertising will have liquidated itself by its own success.

Since the advent of TV, the exploitation of the unconscious by the advertiser has hit a snag. TV experience favors much more consciousness concerning the unconscious than do the hard-sell forms of presentation in the press, the magazine, movie, or radio. The sensory tolerance of the audience has changed, and so have the methods of appeal by the advertisers. In the new cool TV world, the old hot world of hard-selling, earnest-talking salesmen has all the antique charm of the songs and togs of the 1920s. Mort Sahl and Shelley Berman are merely following, not setting, a trend in spoofing the ad world. They discovered that they have only to reel off an ad or news item to have the audience in fits. Will Rogers discovered years ago that any newspaper read aloud

from a theater stage is hilarious. The same is true today of ads.
Any ad put into a new setting is funny. This is a way of saying
that any ad consciously attended to is comical. Ads are not
meant for conscious consumption. They are intended as subliminal pills for the subconscious in order to exercise an hyp-
notic spell, especially on sociologists. That is one of the most
edifying aspects of the huge educational enterprise that we
call advertising, whose twelve-billion-dollar annual budget
approximates the national school budget. Any expensive ad
represents the toil, attention, testing, wit, art, and skill of many
people. Far more thought and care go into the composition of
any prominent ad in a newspaper or magazine than go into
the writing of their features and editorials. Any expensive ad
is as carefully built on the tested foundations of public stereo-
types or "sets" of established attitudes, as any skyscraper is
built on bedrock. Since highly skilled and perceptive teams
of talent cooperate in the making of an ad for any established
line of goods whatever, it is obvious that any acceptable ad is
a vigorous dramatization of communal experience. No group
of sociologists can approximate the ad teams in the gathering
and processing of exploitable social data. The ad teams have
billions to spend annually on research and testing of reactions,
and their products are magnificent accumulations of material
about the shared experience and feelings of the entire community. Of course, if ads were to depart from the center of this
shared experience, they would collapse at once, by losing all
hold on our feelings.

It is true, of course, that ads use the most basic and tested
human experience of a community in grotesque ways. They
are as incongruous, if looked at consciously, as the playing of
"Silver Threads among the Gold" as music for a strip-tease
act. But ads are carefully designed by the Madison Avenue
frog-men-of-the-mind for semiconscious exposure. Their
mere existence is a testimony, as well as a contribution, to
the somnambulistic state of a tired metropolis.

307

After the Second War, an ad-conscious American army officer in Italy noted with misgiving that Italians could tell you the names of cabinet ministers, but not the names of commodities preferred by Italian celebrities. Furthermore, he said, the wall space of Italian cities was given over to political, rather than commercial, slogans. He predicted that there was small hope that Italians would ever achieve any sort of domestic prosperity or calm until they began to worry about the rival claims of cornflakes and cigarettes, rather than the capacities of public men. In fact, he went so far as to say that democratic freedom very largely consists in ignoring politics and worrying, instead, about the threat of scaly scalp, hairy legs, sluggish bowels, saggy breasts, receding gums, excess weight, and tired blood.

The army officer was probably right. Any community that wants to expedite and maximize the exchange of goods and services has simply got to homogenize its social life. The decision to homogenize comes easily to the highly literate population of the English-speaking world. Yet it is hard for oral cultures to agree on this program of homogenization, for they are only too prone to translate the message of radio into tribal politics, rather than into a new means of pushing Cadillacs. This is one reason that it was easy for the retribalized Nazi to feel superior to the American consumer. The tribal man can spot the gaps in the literate mentality very easily. On the other hand, it is the special illusion of literate societies that they are highly aware and individualistic. Centuries of typographic conditioning in patterns of lineal uniformity and fragmented repeatability have, in the electric age, been given increasing critical attention by the artistic world. The lineal process has been pushed out of industry, not only in management and production, but in entertainment, as well. It is the new mosaic form of the TV image that has replaced the Gutenberg structural assumptions. Reviewers of William Burroughs' *The Naked Lunch* have alluded to the prominent use of the "mosaic" term and method in his novel. The TV image renders the world of standard brands and consumer goods merely amusing. Basically, the reason is that the mosaic mesh of the TV image compels so much active participation on the part of the viewer that he develops a nostalgia for pre-consumer ways and days. Lewis Mumford gets serious attention when he praises the cohesive form of medieval towns as relevant to our time and needs.

Advertising got into high gear only at the end of the last century, with the invention of photoengraving. Ads and pictures then became interchangeable and have continued so. More important, pictures made possible great increases in newspaper and magazine circulation that also increased the quantity and profitability of ads. Today it is inconceivable that any publication, daily or periodical, could hold more than a few thousand readers without pictures. For both the pictorial ad or the picture story provide large quantities of instant information and instant humans, such as are necessary for keeping abreast in our kind of culture. Would it not seem natural and necessary that the young be provided with at least as much training of perception in this graphic and photographic world as they get in the typographic? In fact, they need more training in graphics, because the art of casting and arranging actors in ads is both complex and forcefully insidious.

Some writers have argued that the Graphic Revolution has shifted our culture away from private ideals to corporate images. That is really to say that the photo and TV seduce us from the *literate* and private "point of view" to the complex and inclusive world of the group icon. That is certainly what advertising does. Instead of presenting a private argument or vista, it offers a way of life that is for everybody or nobody. It offers this prospect with arguments that concern only irrelevant and trivial matters. For example, a lush car ad features a baby's rattle on the rich rug of the back floor and says that it has removed unwanted car rattles as easily as the user could remove the baby's rattle. This kind of copy has really nothing to do with rattles. The copy is merely a punning gag to distract the critical faculties while the image of the car goes to work on the hypnotized viewer. Those who have spent their lives protesting about "false and misleading ad copy" are godsends to advertisers, as teetotalers are to brewers, and moral censors are to books and films. The protesters are the best acclaimers and accelerators. Since the advent of pictures, the job of the ad copy is as incidental and latent as the "meaning" of a poem is to a poem, or the words of a song are to a song. Highly literate people cannot cope with the nonverbal art of the pictorial, so they dance impatiently up and down to express a pointless disapproval that renders them futile and gives new power and authority to the ads. The unconscious depth-messages of ads are never attacked by the literate, because of their incapacity to notice or discuss nonverbal forms of arrangement and meaning. They have not the art to argue with pictures. When early in TV broadcasting hidden ads were tried out, the literate were in a great panic until they were dropped. The fact that typography is itself mainly subliminal in effect and that pictures are, as well, is a secret that is safe from the book-oriented community.

When the movies came, the entire pattern of American life went on the screen as a nonstop ad. Whatever any actor or actress wore or used or ate was such an ad as had never been dreamed of. The American bathroom, kitchen, and car, like everything else, got the *Arabian Nights* treatment. The result was that all ads in magazines and the press had to look like scenes from a movie. They still do. But the focus has had to become softer since TV.

With radio, ads openly went over to the incantation of the singing commercial. Noise and nausea as a technique of achieving unforgetability became universal. Ad and image making became, and have remained, the one really dynamic and growing part of the economy. Both movie and radio are hot media, whose arrival pepped up everybody to a great degree, giving us the Roaring Twenties. The effect was to provide a massive platform and a mandate for sales promotion as a way of life that ended only with *The Death of A Salesman* and the advent of TV. These two events did not coincide by accident. TV introduced that "experience in depth" and the "do-it-yourself" pattern of living that has shattered the image of the individualist hard-sell salesman and the docile consumer, just as it has blurred the formerly clear figures of the movie stars. This is not to suggest that Arthur Miller was trying to explain TV to America on the eve of its arrival, though he could as appropriately have titled his play "The Birth of the PR Man." Those who saw Harold Lloyd's *World of Comedy* film will remember their surprise at how much of the 1920s they had forgotten. Also, they were surprised to find evidence of how naïve and simple the Twenties really were. That age of the vamps, the sheiks, and the cavemen was a raucous nursery compared to our world, in which children read *MAD* magazine for chuckles. It was a world still innocently engaged in expanding and exploding, in separating and teasing and tearing. Today, with TV, we are experiencing the opposite process of integrating and interrelating that is anything but innocent. The simple faith of the salesman in the irresistibility of his line (both talk and goods) now yields to the complex togetherness of the corporate posture, the process and the organization.

Ads have proved to be a self-liquidating form of community entertainment. They came along just after the Victorian gospel of work, and they promised a Beulah land of perfectibility, where it would be possible to "iron shirts without hating your husband." And now they are deserting the individual consumer-product in favor of the all inclusive and never-ending process that is the Image of any great corporate enterprise. The Container Corporation of America does not feature paper bags and paper cups in its ads, but the container *function*, by means of great art. The historians and archeologists will one day discover that the ads of our times are the richest and most faithful daily reflections that any society ever made of its entire range of activities. The Egyptian hieroglyph lags far behind in this respect. With TV, the smarter advertisers have made free with fur and fuzz, and blur and buzz. They have, in a word, taken a skin-dive. For that is what the TV viewer is. He is a skin-diver, and he no longer likes garish daylight on hard, shiny surfaces, though he must continue to put up with a noisy radio sound track that is painful.

24

Games:
The Extensions of Man

This chapter is no less integrated with McLuhan's framework for media studies than any other. Games, in all their diversity, are grist to McLuhan's mill, and he makes a point of noting that "we are looking at their role as media of communication in society as a whole." Like every other medium, the game functions as a translator or transformer of human experience. McLuhan finds earlier attempts at analysis deficient: "The form of any game is of first importance. Game theory, like information theory, has ignored this aspect of game and information movement."

— *(Editor)*

Alcohol and gambling have very different meanings in different cultures. In our intensely individualist and fragmented Western world, "booze" is a social bond and a means of festive involvement. By contrast, in closely knit tribal society, "booze" is destructive of all social pattern and is even used as a means to mystical experience.

In tribal societies, gambling, on the other hand, is a welcome avenue of entrepreneurial effort and individual initiative. Carried into an individualist society, the same gambling games and sweepstakes seem to threaten the whole social order. Gambling pushes individual initiative to the point of mocking the individualist social structure. The tribal virtue is the capitalist vice.

When the boys came home from the mud and blood baths of the Western Front in 1918 and 1919, they encountered the Volstead Prohibition Act. It was the social and political recognition that the war had fraternalized and tribalized us to the point where alcohol was a threat to an individualist society. When we too are prepared to legalize gambling, we shall, like the English, announce to the world the end of individualist society and the trek back to tribal ways.

We think of humor as a mark of sanity for a good reason: in fun and play we recover the integral person, who in the workaday world or in professional life can use only a small sector of his being. Philip Deane, in *Captive in Korea*, tells a story about games in the midst of successive brainwashings that is to the point.

> There came a time when I had to stop reading those books, to stop practising Russian because with the study of language the absurd and constant assertion began to leave its mark, began to find an echo, and I felt my thinking processes getting tangled, my critical faculties getting blunted … then they made a

mistake. They gave us Robert Louis Stevenson's
Treasure Island in English ... I could read Marx again,
and question myself honestly without fear. Robert
Louis Stevenson made us lighthearted, so we started
dancing lessons.

Games are popular art, collective, social reactions to the
main drive or action of any culture. Games, like institutions,
are extensions of social man and of the body politic, as technologies are extensions of the animal organism. Both games
and technologies are counter-irritants or ways of adjusting to
the stress of the specialized actions that occur in any social
group. As extensions of the popular response to the workaday
stress, games become faithful models of a culture. They incorporate both the action and the reaction of whole populations
in a single dynamic image.

A Reuters dispatch for December 13, 1962, reported from
Tokyo:

BUSINESS IS A BATTLEFIELD

Latest fashion among Japanese businessmen is the
study of classical military strategy and tactics in order
to apply them to business operations ... It has been
reported that one of the largest advertising companies in Japan has even made these books compulsory
reading for all its employees.

Long centuries of tight tribal organization now stand the
Japanese in very good stead in the trade and commerce of
the electric age. A few decades ago they underwent enough
literacy and industrial fragmentation to release aggressive
individual energies. The close teamwork and tribal loyalty
now demanded by electrical intercom again puts the Japanese
in positive relation to their ancient traditions. Our own tribal
ways are much too remote to be of any social avail. We have
begun retribalizing with the same painful groping with which
a preliterate society begins to read and write, and to organize
its life visually in three-dimensional space.

The search for Michael Rockefeller brought the life of a New Guinea tribe into prominent attention in *Life* a year ago. The editors explained the war games of these people:

> The traditional enemies of the Willigiman-Walla-lua are the Wittaia, a people exactly like themselves in language, dress and custom ... Every week or two the Willigiman-Wallalua and their enemies arrange a formal battle at one of the traditional fighting grounds. In comparison with the catastrophic conflicts of "civilized" nations, these frays seem more like a dangerous field sport than true war. Each battle lasts but a single day, always stops before nightfall (because of the danger of ghosts) or if it begins to rain (no one wants to get his hair or ornaments wet). The men are very accurate with their weapons — they have all played war games since they were small boys — but they are equally adept at dodging, and hence are rarely hit by anything.
>
> The truly lethal part of this primitive warfare is not the formal battle but the sneak raid or stealthy ambush in which not only men but women and children are mercilessly slaughtered....
>
> This perpetual bloodshed is carried on for none of the usual reasons for waging war. No territory is won or lost; no goods or prisoners are seized ... They fight because they enthusiastically enjoy it, because it is to them a vital function of the complete man, and because they feel they must satisfy the ghosts of slain companions.

These people, in short, detect in these games a kind of model of the universe, in whose deadly gavotte they can participate through the ritual of war games.

Games are dramatic models of our psychological lives providing release of particular tensions. They are collective and popular art forms with strict conventions. Ancient and non-literate societies naturally regarded games as live

317

dramatic models of the universe or of the outer cosmic drama. The Olympic games were direct enactments of the *agon*, or struggle of the Sun god. The runners moved around a track adorned with the zodiacal signs in imitation of the daily circuit of the sun chariot. With games and plays that were dramatic enactments of a cosmic struggle, the spectator role was plainly religious. The participation in these rituals kept the cosmos on the right track, as well as providing a booster shot for the tribe. The tribe or the city was a dim replica of that cosmos, as much as were the games, the dances, and the icons. How art became a sort of civilized substitute for magical games and rituals is the story of the detribalization which came with literacy. Art, like games, became a mimetic echo of, and relief from, the old magic of total involvement. As the audience for the magic games and plays became more individualistic, the role of art and ritual shifted from the cosmic to the humanly psychological, as in Greek drama. Even the ritual became more verbal and less mimetic or dancelike. Finally, the verbal narrative from Homer and Ovid became a romantic literary substitute for the corporate liturgy and group participation. Much of the scholarly effort of the past century in many fields has been devoted to a minute reconstruction of the conditions of primitive art and ritual, for it has been felt that this course offers the key to understanding the mind of primitive man. The key to this understanding, however, is also available in our new electric technology that is so swiftly and profoundly re-creating the conditions and attitudes of primitive tribal man in ourselves.

The wide appeal of the games of recent times — the popular sports of baseball and football and ice hockey — seen as outer models of inner psychological life, become understandable. As models, they are collective rather than private dramatizations of inner life. Like our vernacular tongues, all games are media of interpersonal communication, and they could have neither existence nor meaning except as extensions of our immediate inner lives. If we take a tennis racket in hand, or thirteen playing cards, we consent to being a part of a dynamic mechanism in an artificially contrived situation. Is this not the reason we enjoy those games most that mimic other situations in our work and social lives? Do not our favorite games provide a release from the monopolistic tyranny of the social machine? In a word, does not Aristotle's idea of drama as a mimetic reenactment and relief from our besetting pressures apply perfectly to all kinds of games and dance and fun? For fun or games to be welcome, they must convey an echo of workaday life. On the other hand, a man or society without games is one sunk in the zombie trance of the automaton. Art and games enable us to stand aside from the material pressures of routine and convention, observing and questioning. Games as popular art forms offer to all an immediate means of participation in the full life of a society, such as no single role or job can offer to any man. Hence the contradiction in "professional" sport. When the games door opening into the free life leads into a merely specialist job, everybody senses an incongruity.

The games of a people reveal a great deal about them. Games are a sort of artificial paradise like Disneyland, or some Utopian vision by which we interpret and complete the meaning of our daily lives. In games we devise means of non-specialized participation in the larger drama of our time. But for civilized man the idea of participation is strictly limited. Not for him the depth participation that erases the boundaries of individual awareness as in the Indian cult of *darshan*, the mystic experience of the physical presence of vast numbers of people.

A game is a machine that can get into action only if the players consent to become puppets for a time. For individualist Western man, much of his "adjustment" to society has the character of a personal surrender to the collective demands. Our games help both to teach us this kind of adjustment and also to provide a release from it. The uncertainty of the outcomes of our contests makes a rational excuse for the mechanical rigor of the rules and procedures of the game.

When the social rules change suddenly, then previously accepted social manners and rituals may suddenly assume the stark outlines and the arbitrary patterns of a game. The *Gamesmanship* of Stephen Potter speaks of a social revolution in England. The English are moving toward social equality and the intense personal competition that goes with equality. The older rituals of long-accepted class behavior now begin to appear comic and irrational, gimmicks in a game. Dale Carnegie's *How to Win Friends and Influence People* first appeared as a solemn manual of social wisdom, but it seemed quite ludicrous to sophisticates. What Carnegie offered as serious discoveries already seemed like a naïve mechanical ritual to those beginning to move in a milieu of Freudian awareness charged with the psychopathology of everyday life. Already the Freudian patterns of perception have become an outworn code that begins to provide the cathartic amusement of a game, rather than a guide to living.

The social practices of one generation tend to get codified into the "game" of the next. Finally, the game is passed on as a joke, like a skeleton stripped of its flesh. This is especially true of periods of suddenly altered attitudes, resulting from some radically new technology. It is the inclusive mesh of the TV image, in particular, that spells for a while, at least, the doom of baseball. For baseball is a game of one-thing-at-a-time, fixed positions and visibly delegated specialist jobs such as belonged to the now passing mechanical age, with its fragmented tasks and its staff and line in management organization. TV, as the very image of the new corporate and participant ways of electric living, fosters habits of unified awareness and social interdependence that alienate us from the peculiar style of baseball, with its specialist and positional stress. When cultures change, so do games. Baseball, that had become the elegant abstract image of an industrial society living by split-second timing, has in the new TV decade lost its psychic and social relevance for our new way of life. The ball game has been dislodged from the social center and been conveyed to the periphery of American life.

In contrast, American football is nonpositional, and any or all of the players can switch to any role during play. It is, therefore, a game that at the present is supplanting baseball in general acceptance. It agrees very well with the new needs of decentralized team play in the electric age. Offhand, it might be supposed that the tight tribal unity of football would make it a game that the Russians would cultivate. Their devotion to ice hockey and soccer, two very individualist forms of game, would seem little suited to the psychic needs of a collectivist society. But Russia is still in the main an oral, tribal world that is undergoing detribalization and just now discovering individualism as a novelty. Soccer and ice hockey have for them, therefore, an exotic and Utopian quality of promise that they do not convey to the West. This is the quality that we tend to call "snob value," and *we* might derive some similar "value" from owning race horses, polo ponies, or twelve-meter yachts.

Games, therefore, can provide many varieties of satisfaction. Here we are looking at their role as media of communication in society as a whole. Thus, poker is a game that has often been cited as the expression of all the complex attitudes and unspoken values of a competitive society. It calls for shrewdness, aggression, trickery, and unflattering appraisals of character. It is said women cannot play poker well because it stimulates their curiosity, and curiosity is fatal in poker. Poker is intensely individualist, allowing no place for kindness or consideration, but only for the greatest good of the greatest number—the number one. It is in this perspective that it is easy to see why war has been called the sport of kings. For kingdoms are to monarchs what patrimonies and private income are to the private citizen. Kings can play poker with kingdoms, as the generals of their armies do with the troops. They can bluff and deceive the opponent about their resources and their intentions. What disqualifies war from being a true game is probably what also disqualifies the stock market and business — the rules are not fully known nor accepted by all the players. Furthermore, the audience is too fully participant in war and business, just as in a native society there is no true art because *everybody* is engaged in making art. Art and games need rules, conventions, and spectators. They must stand forth from the over-all situation as models of it in order for the quality of play to persist. For "play," whether in life or in a wheel, implies *interplay*. There must be give and take, or dialogue, as between two or more persons and groups. This quality can, however, be diminished or lost in any kind of situation. Great teams often play practice games without any audience at all. This is not sport in our sense, because much of the quality of interplay, the very medium of interplay, as it were, is the feeling of the audience. Rocket Richard, the Canadian hockey player, used to comment on the poor acoustics of some arenas. He felt that the puck off his stick rode on the roar of the crowd. Sport, as a popular art form, is not just self-expression but is deeply and necessarily a means of interplay within an entire culture.

Art is not just play but an extension of human awareness in contrived and conventional patterns. Sport as popular art is a deep reaction to the typical action of the society. But high art, on the other hand, is not a reaction but a profound reappraisal of a complex cultural state. Jean Genet's *The Balcony* appeals to some people as a shatteringly logical appraisal of mankind's madness in its orgy of self-destruction. Genet offers a brothel enveloped by the holocaust of war and revolution as an inclusive image of human life. It would be easy to argue that Genet is hysterical, and that football offers a more serious criticism of life than he does. Seen as live models of complex social situations, games may lack moral earnestness, it has to be admitted. Perhaps there is, just for this reason, a desperate need for games in a highly specialized industrial culture, since they are the only form of art accessible to many minds. Real interplay is reduced to nothing in a specialist world of delegated tasks and fragmented jobs. Some backward or tribal societies suddenly translated into industrial and specialist forms of mechanization cannot easily devise the antidote of sports and games to create countervailing force. They bog down into grim earnest. Men without art, and men without the popular arts of games, tend toward automatism.

A comment on the different kinds of games played in the British Parliament and the French Chamber of Deputies will rally the political experience of many readers. The British had the luck to get the two-team pattern into the House benches, whereas the French, trying for centralism by seating the deputies in a semicircle facing the chair, got instead a multiplicity of teams playing a great variety of games. By trying for unity, the French got anarchy. The British, by setting up diversity, achieved, if anything, too much unity. The British representative, by playing his "side," is not tempted into private mental effort, nor does he have to follow the debates until the ball is passed to him. As one critic said, if the benches did not face each other the British could not tell truth from falsehood, nor wisdom from folly, unless they listened to it *all*. And since most of the debate must be nonsense, it would be stupid to listen to all.

The form of any game is of first importance. Game theory, like information theory, has ignored this aspect of game and information movement. Both theories have dealt with the information content of systems, and have observed the "noise" and "deception" factors that divert data. This is like approaching a painting or a musical composition from the point of view of its content. In other words, it is guaranteed to miss the central structural core of the experience. For as it is the *pattern* of a game that gives it relevance to our inner lives, and not who is playing nor the outcome of the game, so it is with information movement. The selection of our human senses employed makes all the difference, say, between photo and telegraph. In the arts the particular mix of our senses in the medium employed is all-important. The ostensible program content is a lulling distraction needed to enable the structural form to get through the barriers of conscious attention.

Any game, like any medium of information, is an extension of the individual or the group. Its effect on the group or individual is a reconfiguring of the parts of the group or individual that are *not* so extended. A work of art has no existence or function apart from its *effects* on human observers. And art, like games or popular arts, and like media of communication, has the power to impose its own assumptions by setting the human community into new relationships and postures.

Art, like games, is a translator of experience. What we have already felt or seen in one situation we are suddenly given in a new kind of material. Games, likewise, shift familiar experience into new forms, giving the bleak and the blear side of things sudden luminosity. The telephone companies make tapes of the blither of boors, who inundate defenseless telephone operators with various kinds of revolting expressions. When played back this becomes salutary fun and play, and helps the operators to maintain equilibrium.

The world of science has become quite self-conscious about the play element in its endless experiments with models of situations otherwise unobservable. Management training centers have long used games as a means of developing new business perception. John Kenneth Galbraith argues that business must now study art, for the artist makes models of problems and situations that have not yet emerged in the larger matrix of society, giving the artistically perceptive businessman a decade of leeway in his planning.

In the electric age, the closing of the gaps between art and business, or between campus and community, are part of the overall implosion that closes the ranks of specialists at all levels. Flaubert, the French novelist of the nineteenth century, felt that the Franco-Prussian War could have been avoided if people had heeded his *Sentimental Education*. A similar feeling has since come to be widely held by artists. They know that they are engaged in making live models of situations that have not yet matured in the society at large. In their artistic play, they discovered what is actually happening, and thus they appear to be "ahead of their time." Non-artists always look at the present through the spectacles of the preceding age. General staffs are always magnificently prepared to fight the previous war.

Games, then, are contrived and controlled situations, extensions of group awareness that permit a respite from customary patterns. They are a kind of talking to itself on the part of society as a whole. And talking to oneself is a recognized form of play that is indispensable to any growth of self-confidence. The British and Americans have enjoyed during recent times an enormous self-confidence born of the playful spirit of fun and games. When they sense the absence of this spirit in their rivals, it causes embarrassment. To take mere worldly things in dead earnest betokens a defect of awareness that is pitiable. From the first days of Christianity there grew a habit, in some quarters, of spiritual clowning, of "playing the fool in Christ," as St. Paul put it. Paul also associated this sense of spiritual confidence and Christianity play with the games and sports of his time. Play goes with an awareness of huge disproportion between the ostensible situation and the real stakes. A similar sense hovers over the game situation, as such. Since the game, like any art form, is a mere tangible model of another situation that is less accessible, there is always a tingling sense of oddity and fun in play or games that renders the very earnest and very serious person or society laughable. When the Victorian Englishman began to lean toward the pole of seriousness, Oscar Wilde and Bernard Shaw and G. K. Chesterton moved in swiftly as countervailing force. Scholars have often pointed out that Plato conceived of play dedicated to the Deity as the loftiest reach of man's religious impulse.

Bergson's famous treatise on laughter sets forth the idea of mechanism taking over life-values as the key to the ludicrous. To see a man slip on a banana skin is to see a rationally structured system suddenly translated into a whirling machine. Since industrialism had created a similar situation in the society of his time, Bergson's idea was readily accepted. He seems not to have noticed that he had mechanically turned up a mechanical metaphor in a mechanical age in order to explain the very unmechanical thing, laughter, or "the mind sneezing," as Wyndham Lewis described it.

The game spirit suffered a defeat a few years ago over the rigged TV quiz shows. For one thing, the big prize seemed to make fun of money. Money as store of power and skill, and expediter of exchange, still has for many people the ability to induce a trance of great earnestness. Movies, in a sense, are also rigged shows. Any play or poem or novel is, also, rigged *to produce an effect*. So was the TV quiz show. But with the TV effect there is deep audience *participation*. Movie and drama do not permit as much participation as that afforded by the mosaic mesh of the TV image. So great was the audience participation in the quiz shows that the directors of the show were prosecuted as con men. Moreover press and radio ad interests, bitter about the success of the new TV medium, were delighted to lacerate the flesh of their rivals. Of course, the riggers had been blithely unaware of the nature of their medium, and had given it the movie treatment of intense realism, instead of the softer mythic focus proper to TV. Charles Van Doren merely got clobbered as an innocent bystander, and the whole investigation elicited no insight into the nature or effects of the TV medium. Regrettably, it simply provided a field day for the earnest moralizers. A moral point of view too often serves as a substitute for understanding in technological matters.

That games are extensions, not of our private but of our social selves, and that they are media of communication, should now be plain. If, finally, we ask, "Are games mass media?" the answer has to be "Yes." Games are situations contrived to permit simultaneous participation of many people in some significant pattern of their own corporate lives.

25

Telegraph:
The Social Hormone

With this chapter, McLuhan resumes chronicling the succession of new media that transformed mankind's first technology for getting a new grasp on the environment — language. It contains a passage missed by all those critics who declared that McLuhan was "anti-book" (see Critical Reception of Understanding Media*): "We are in great danger of wiping out our entire investment in the preelectric technology of the literate and mechanical kind by means of an indiscriminate use of electrical energy." The conversion of mechanical processes under electricity, discussed throughout this book under the name of* automation, *here receives an alternate term,* cybernation, *along with a new dimension to its meaning: "a way of thinking, as much as a way of doing." Moving into the final chapters, McLuhan reminds the reader of the consequences of failing to understand technological change, its character, its effects — we may all become "media victims."*

— (Editor)

The wireless telegraph was given spectacular publicity in 1910 when it led to the arrest at sea of Dr. Hawley H. Crippen, a U.S. physician who had been practicing in London, murdered his wife, buried her in the cellar of their home, and fled the country with his secretary aboard the liner *Montrose*. The secretary was dressed as a boy, and the pair traveled as Mr. Robinson and son. Captain George Kendall of the *Montrose* became suspicious of the Robinsons, having read in the English papers about the Crippen case.

The *Montrose* was one of the few ships then equipped with Marconi's wireless. Binding his wireless operator to secrecy, Captain Kendall sent a message to Scotland Yard, and the Yard sent Inspector Dews on a faster liner to race the *Montrose* across the Atlantic. Inspector Dews, dressed as a pilot, boarded the *Montrose* before it reached port, and arrested Crippen. Eighteen months after Crippen's arrest, an act was passed in the British Parliament making it compulsory for all passenger ships to carry wireless.

The Crippen case illustrates what happens to the best-laid plans of mice and men in any organization when the instant speed of information movement begins. There is a collapse of delegated authority and a dissolution of the pyramid and management structures made familiar in the organization chart. The separation of functions, and the division of stages, spaces, and tasks are characteristic of literate and visual society and of the Western world. These divisions tend to dissolve through the action of the instant and organic interrelations of electricity.

Former German Armaments minister Albert Speer, in a speech at the Nuremberg trials, made some bitter remarks about the effects of electric media on German life: "The telephone, the teleprinter and the wireless made it possible for orders from the highest levels to be given direct to the lowest

331

levels, where, on account of the absolute authority behind them, they were carried out uncritically...."

The tendency of electric media is to create a kind of organic interdependence among all the institutions of society, emphasizing de Chardin's view that the discovery of electromagnetism is to be regarded as "a prodigious biological event." If political and commercial institutions take on a biological character by means of electric communications, it is also common now for biologists like Hans Selye to think of the physical organism as a communication network: "Hormone is a specific chemical messenger-substance, made by an endocrine gland and secreted into the blood, to regulate and co-ordinate the functions of distant organs."

This peculiarity about the electric form, that it ends the mechanical age of individual steps and specialist functions, has a direct explanation. Whereas all previous technology (save speech, itself) had, in effect, extended some part of our bodies, electricity may be said to have outered the central nervous system itself, including the brain. Our central nervous system is a unified field quite without segments. As J. Z. Young writes in *Doubt and Certainty in Science: A Biologist's Reflections on the Brain* (Galaxy, Oxford University Press, New York, 1960):

> It may be that a great part of the secret of the brain's power is the enormous opportunity provided for interaction between the effects of stimulating each part of the receiving fields. It is this provision of interacting-places or mixing-places that allows us to react to the world *as a whole* to a much greater degree than most other animals can do.

Failure to understand the organic character of electric technology is evident in our continuing concern with the dangers of mechanizing the world. Rather, we are in great danger of wiping out our entire investment in the preelectric technology of the literate and mechanical kind by means of an indiscriminate use of electrical energy. What makes a mechanism is the separation and extension of separate parts of our body as hand, arm, feet, in pen, hammer, wheel. And the mechanization of a task is done by segmentation of each part of an action in a series of uniform, repeatable, and movable parts. The exact opposite characterizes cybernation (or automation), which has been described as a way of thinking, as much as a way of doing. Instead of being concerned with separate machines, cybernation looks at the production problem as an integrated system of information handling.

It is this same provision of interacting places in the electric media that now compels us to react to the world as a whole. Above all, however, it is the speed of electric involvement that creates the integral whole of both private and public awareness. We live today in the Age of Information and of Communication because electric media instantly and constantly create a total field of interacting events in which all men participate. Now, the world of public interaction has the same inclusive scope of integral interplay that has hitherto characterized only our private nervous systems. That is because electricity is organic in character and confirms the organic social bond by its technological use in telegraph and telephone, radio, and other forms. The simultaneity of electric communication, also characteristic of our nervous system, makes each of us present and accessible to every other person in the world. To a large degree our co-presence everywhere at once in the electric age is a fact of passive, rather than active, experience. Actively, we are more likely to have this awareness when reading the newspaper or watching a TV show.

One way to grasp the change from the mechanical to the electric age is by noticing the difference between the layout of a literary and a telegraph press, say between the London *Times* and the *Daily Express*, or between *The New York Times*

and the New York *Daily News*. It is the difference between
columns representing points of view, and a mosaic of unrelated scraps in a field unified by a dateline. Whatever else
there is, there can be no point of view in a mosaic of simultaneous items. The world of impressionism, associated with
painting in the late nineteenth century, found its more extreme
form in the *pointillisme* of Seurat and the refractions of light
in the world of Monet and Renoir. The stipple of points of
Seurat is close to the present technique of sending pictures
by telegraph, and close to the form of the TV image or mosaic
made by the scanning finger. All of these anticipate later electric forms because, like the digital computer with its multiple
yes-no dots and dashes, they caress the contours of every kind
of being by the multiple touches of these points. Electricity
offers a means of getting in touch with every facet of being
at once, like the brain itself. Electricity is only incidentally
visual and auditory; it is primarily tactile.

As the age of electricity began to establish itself in the
later nineteenth century, the entire world of the arts began
to reach again for the iconic qualities of touch and sense
interplay (synesthesia, as it was called) in poetry, as in painting. The German sculptor Adolf von Hildebrand inspired
Berenson's remark that "the painter can accomplish his task
only by giving tactile values to retinal impressions." Such a
program involves the endowing of each plastic form with a
kind of nervous system of its own.

The electric form of pervasive impression is profoundly
tactile and organic, endowing each object with a kind of
unified sensibility, as the cave painting had done. The unconscious task of the painter in the new electric age was to raise
this fact to the level of conscious awareness. From this time
on, the mere specialist in any field was doomed to the sterility
and inanity that echoed an archaic form of the departing
mechanical age. Contemporary awareness had to become
integral and inclusive again, after centuries of dissociated

sensibilities. The Bauhaus school became one of the great centers of effort tending toward an inclusive human awareness; but the same task was accepted by a race of giants that sprang up in music and poetry, architecture, and painting. They gave the arts of this century an ascendancy over those of other ages comparable to that which we have long recognized as true of modern science.

During its early growth, the telegraph was subordinate to railway and newspaper, those immediate extensions of industrial production and marketing. In fact, once the railways began to stretch across the continent, they relied very much on the telegraph for their coordination, so that the image of the stationmaster and the telegraph operator were easily superimposed in the American mind.

It was in 1844 that Samuel Morse opened a telegraph line from Washington to Baltimore with $30,000 obtained from Congress. Private enterprise, as usual, waited for bureaucracy to clarify the image and goals of the new operation. Once it proved profitable, the fury of private promotion and initiative became impressive, leading to some savage episodes. No new technology, not even the railroad, manifested a more rapid growth than the telegraph. By 1858 the first cable had been laid across the Atlantic, and by 1861 telegraph wires had reached across America. That each new method of transporting commodity or information should have to come into existence in a bitter competitive battle against previously existing devices is not surprising. Each innovation is not only commercially disrupting, but socially and psychologically corroding, as well.

It is instructive to follow the embryonic stages of any new growth, for during this period of development it is much misunderstood, whether it be printing or the motorcar or TV. Just because people are at first oblivious of its nature, the new form deals some revealing blows to the zombie-eyed spectators. The original telegraph line between Baltimore and

Washington promoted chess games between experts in the two cities. Other lines were used for lotteries and play in general, just as early radio existed in isolation from any commercial commitments and was, in fact, fostered by the amateur "hams" for years before it was seized by big interests.

A few months ago John Crosby wrote to the *New York Herald Tribune* from Paris in a way that well illustrates why the "content" obsession of the man of print culture makes it difficult for him to notice any facts about the *form* of a new medium:

> Telstar, as you know, is that complicated ball that whirls through space, transmitting television broadcasts, telephone messages, and everything except common sense. When it was first cast aloft, trumpets sounded. Continents would share each other's intellectual pleasures. Americans would enjoy Brigitte Bardot. Europeans would partake of the heady intellectual stimulation of "Ben Casey" ... The fundamental flaw in this communications miracle is the same one that has bugged every communications miracle since they started carving hieroglyphics on stone tablets. What do you say on it? Telstar went into operation in August when almost nothing of importance was happening anywhere in Europe. All the networks were ordered to say something, any-thing, on this miracle instrument. "It was a new toy and they just had to use it," the men here say. CBS combed Europe for hot news and came up with a sausage-eating contest, which was duly sent back via the miracle ball, although that particular news event could have gone by camelback without losing any of its essence.

Any innovation threatens the equilibrium of existing organization. In big industry new ideas are invited to rear their heads so that they can be clobbered at once. The idea department of a big firm is a sort of lab for isolating dangerous viruses. When one is found, it is assigned to a group for neutralizing and immunizing treatment. It is comical, therefore, when anybody applies to a big corporation with a new idea that would result in a great "increase of production and sales." Such an increase would be a disaster for the existing management. They would have to make way for new management. Therefore, no new idea ever starts from within a big operation. It must assail the organization from outside, through some small but competing organization. In the same way, the outering or extension of our bodies and senses in a "new invention" compels the whole of our bodies and senses to shift into new positions in order to maintain equilibrium. A new "closure" is effected in all our organs and senses, both private and public, by any new invention. Sight and sound assume new postures, as do all the other faculties. With the telegraph, the entire method, both of gathering and of presenting news, was revolutionized. Naturally, the effects on language and on literary style and subject matter were spectacular.

In the same year, 1844, then, that men were playing chess and lotteries on the first American telegraph, Søren Kierkegaard published *The Concept of Dread*. The Age of Anxiety had begun. For with the telegraph, man had initiated that outering or extension of his central nervous system that is now approaching an extension of consciousness with satellite broadcasting. To put one's nerves outside, and one's physical organs inside the nervous system, or the brain, is to initiate a situation—if not a concept—of dread.

Having glanced at the major trauma of the telegraph on conscious life, noting that it ushers in the Age of Anxiety and of Pervasive Dread, we can turn to some specific instances of this uneasiness and growing jitters. Whenever any new medium or human extension occurs, it creates a new myth for itself, usually associated with a major figure: Aretino, the

Scourge of Princes and the Puppet of Printing; Napoleon and the trauma of industrial change; Chaplin, the public conscience of the movie; Hitler, the tribal totem of radio; and Florence Nightingale, the first singer of human woe by telegraph wire.

Florence Nightingale (1820–1910), wealthy and refined member of the powerful new English group engendered by industrial power, began to pick up human-distress signals as a young lady. They were quite undecipherable at first. They upset her entire way of life, and couldn't be adjusted to her image of parents or friends or suitors. It was sheer genius that enabled her to translate the new diffused anxiety and dread of life into the idea of deep human involvement and hospital reform. She began to think, as well as to live, her time, and she discovered the new formula for the electronic age: Medicare. Care of the body became balm for the nerves in the age that had extended its nervous system outside itself for the first time in human history.

To put the Florence Nightingale story in new media terms is quite simple. She arrived on a distant scene where controls from the London center were of the common pre-electric hierarchical pattern. Minute division and delegation of functions and separation of powers, normal in military and industrial organization then and long afterward, created an imbecile system of waste and inefficiency which for the first time got reported daily by telegraph. The legacy of literacy and visual fragmentation came home to roost every day on the telegraph wire:

> In England fury succeeded fury. A great storm of rage, humiliation, and despair had been gathering through the terrible winter of 1854–55. For the first time in history, through reading the dispatches of Russell, the public had realized "with what majesty the British soldier fights." And these heroes were dead. The men who had stormed the heights of Alma, charged with the Light Brigade at Balaclava ... had

perished of hunger and neglect. Even horses which had taken part in the Charge of the Light Brigade had starved to death. (*Lonely Crusader*, Cecil Woodham-Smith, McGraw-Hill)

The horrors that William Howard Russell relayed by wire to *The Times* were normal in British military life. He was the first war correspondent, because the telegraph gave that immediate and inclusive dimension of "human interest" to news that does not belong to a "point of view." It is merely a comment on our absentmindedness and general indifference that after more than a century of telegraph news reporting, nobody has seen that "human interest" is the electronic or depth dimension of immediate involvement in news. With telegraph, there ended that separation of interests and division of faculties that are certainly not without their magnificent monuments of toil and ingenuity. But with telegraph, came the integral insistence and wholeness of Dickens, and of Florence Nightingale, and of Harriet Beecher Stowe. The electric gives powerful voices to the weak and suffering, and sweeps aside the bureaucratic specialisms and job descriptions of the mind tied to a manual of instructions. The "human interest" dimension is simply that of immediacy of participation in the experience of others that occurs with instant information. People become instant, too, in their response of pity or of fury when they must share the common extension of the central nervous system with the whole of mankind. Under these conditions, "conspicuous waste" or "conspicuous consumption" become lost causes, and even the hardiest of the rich dwindle into modest ways of timid service to mankind.

At this point, some may still inquire why the telegraph should create "human interest," and why the earlier press did not. The section on The Press may help these readers. But there may also be a lurking obstacle to perception. The instant all-at-onceness and total involvement of the telegraphic form still repels some literary sophisticates. For them, visual continuity and fixed "point-of-view" render the immediate participation of the instant media as distasteful and unwelcome as popular sports. These people are as much media victims, unwittingly mutilated by their studies and toil, as children in a Victorian blacking factory. For many people, then, who have had their sensibilities irremediably skewed and locked into the fixed postures of mechanical writing and printing, the iconic forms of the electric age are as opaque, or even as invisible, as hormones to the unaided eye. It is the artist's job to try to dislocate older media into postures that permit attention to the new. To this end, the artist must ever play and experiment with new means of arranging experience, even though the majority of his audience may prefer to remain fixed in their old perceptual attitudes. The most that can be done by the prose commentator is to capture the media in as many characteristic and revealing postures as he can manage to discover. Let us examine a series of these postures of the telegraph, as this new medium encounters other media like the book and the newspaper.

By 1848 the telegraph, then only four years old, compelled several major American newspapers to form a collective organization for newsgathering. This effort became the basis of the Associated Press, which, in turn, sold news service to subscribers. In one sense, the real meaning of this form of the electric, instant coverage was concealed by the mechanical overlay of the visual and industrial patterns of print and printing. The specifically electric effect may seem to appear in this instance as a centralizing and compressional force. By many analysts, the electric revolution has been regarded as a continuation of the process of the mechanization of mankind. Closer inspection reveals quite a different character. For example, the regional press, that had had to rely on postal service and political control through the post office, quickly escaped from this center-margin type of monopoly by means of the new telegraph services. Even in England, where short distances and concentrated population made the railway a powerful agent of centralism, the monopoly of London was dissolved by the invention of the telegraph, which now encouraged provincial competition. The telegraph freed the marginal provincial press from dependence on the big metropolitan press. In the whole field of the electric revolution, this pattern of decentralization appears in multiple guises. It is Sir Lewis Namier's view that telephone and airplane are the biggest single cause of trouble in the world today. Professional diplomats with delegated powers have been supplanted by prime ministers, presidents, and foreign secretaries, who think they could conduct all important negotiations personally. This is also the problem encountered in big business, where it has been found impossible to exercise delegated authority when using the telephone. The very nature of the telephone, as all electric media, is to compress and unify that which had previously been divided and specialized. Only the "authority of knowledge" works by telephone because of the speed that creates a total and inclusive field of relations. Speed requires that the decisions made be inclusive, not fragmentary or partial, so that literate people typically resist the telephone. But radio and TV, we shall see, have the same power of imposing an inclusive order, as of an oral organization. Quite in contrast is the center-margin form of visual and written structures of authority.

Many analysts have been misled by electric media because of the seeming ability of these media to extend man's spatial powers of organization. Electric media, however, abolish the spatial dimension, rather than enlarge it. By electricity, we everywhere resume person-to-person relations as if on the smallest village scale. It is a relation in depth, and without delegation of functions or powers. The organic everywhere supplants the mechanical. Dialogue supersedes the lecture. The greatest dignitaries hobnob with youth. When a group of Oxford undergraduates heard that Rudyard Kipling received ten shillings for every word he wrote, they sent him ten shillings by telegram during their meeting: "Please send us one of your very best words." Back came the word a few minutes later: "Thanks."

The hybrids of electricity and the older mechanics have been numerous. Some of them, such as the phonograph and the movie, are discussed elsewhere in this volume. Today the wedding of mechanical and electric technology draws to a close, with TV replacing the movie and Telstar threatening the wheel. A century ago the effect of the telegraph was to send the presses racing faster, just as the application of the electric spark was to make possible the internal-combustion engine with its instant precision. Pushed further, however, the electric principle everywhere dissolves the mechanical technique of visual separation and analysis of functions. Electronic tapes with exactly synchronized information replace the old lineal sequence of the assembly line.

Acceleration is a formula for dissolution and breakdown in any organization. Since the entire mechanical technology of the Western world has been wedded to electricity, it has pushed toward higher speeds. All the mechanical aspects of our world seem to probe toward self-liquidation. The United States had built up a large degree of central political controls through the interplay of the railway, the post office, and the newspaper. In 1848 the Postmaster General wrote, in his report, that newspapers have "always been esteemed of so much importance to the public, as the best means of disseminating intelligence among the people, that the lowest rate has always been afforded for the purpose of encouraging their circulation." The telegraph quickly weakened this center-margin pattern and, more important, by intensifying the volume of news, it greatly weakened the role of editorial opinions. News has steadily overtaken views as a shaper of public attitudes, though few examples of this change are quite so striking as the sudden growth of the Florence Nightingale image in the British world. And yet nothing has been more misunderstood than the power of the telegraph in this matter. Perhaps the most decisive feature is this. The natural dynamic of the book and, also, newspaper is to create a unified national outlook on a centralized pattern. All literate people, therefore, experience a desire for an extension of the most enlightened opinions in a uniform horizontal and homogeneous pattern to the "most backward areas," and to the least literate minds. The telegraph ended that hope. It decentralized the newspaper world so thoroughly that uniform national views were quite impossible, even before the Civil War. Perhaps, an even more important consequence of the telegraph was that in America the literary talent was drawn into journalism, rather than into the book medium. Poe and Twain and Hemingway are examples of writers who could find neither training nor outlet save through the newspaper. In Europe, on the other hand, the numerous small national groups presented a discontinuous mosaic that the telegraph merely intensified. The result was that the telegraph in Europe strengthened the position of the book, and forced even the press to assume a literary character.

Not least of the developments since the telegraph has been the weather forecast, perhaps the most popularly participative of all the human interest items in the daily press. In the early days of telegraph, rain created problems in the grounding of wires. These problems drew attention to weather dynamics. One report in Canada in 1883 stated: "It was early discovered that when the wind at Montreal was from the East or North-East, rain storms traveled from the West, and the stronger the land current, the faster came the rain from the opposite direction." It is clear that telegraph, by providing a wide sweep of instant information, could reveal meteorological patterns of force quite beyond observation by pre-electric man.

26

The Typewriter:
Into the Age of the
Iron Whim

The chapter is rich in delineating the explicit effects of the invention of the typewriter for poets, to whose verse it gave a technological support comparable to the musician's bars and staves; women, to whom it gave careers as typists; the work place, to which it gave typists. The typewriter was not exempt from the reversal principle that manifests itself in all media effects, witness the oral quality it conferred on the poetry of Eliot and Pound, with its characteristic "colloquial freedom of the world of jazz and ragtime." And the transformations wrought by the typewriter provide yet another example of how the medium is the message: "In any given structure, the rate of staff accumulation is not related to the work done but to the intercommunication among the staff, itself."

— *(Editor)*

The comments of Robert Lincoln O'Brien, writing in the *Atlantic Monthly* in 1904, indicate a rich field of social material that still remains unexplored. For example:

> The invention of the typewriter has given a tremendous impetus to the dictating habit ... This means not only greater diffuseness ... but it also brings forward the point of view of the one who speaks. There is the disposition on the part of the talker to explain, as if watching the facial expression of his hearers to see how far they are following. This attitude is not lost when his audience is following. It is no uncommon thing in the typewriting booths at the Capitol in Washington to see Congressmen in dictating letters use the most vigorous gestures as if the oratorical methods of persuasion could be transmitted to the printed page.

In 1882, ads proclaimed that the typewriter could be used as an aid in learning to read, write, spell, and punctuate. Now, eighty years later, the typewriter is used only in experimental classrooms. The ordinary classroom still holds the typewriter at bay as a merely attractive and distractive toy. But poets like Charles Olson are eloquent in proclaiming the power of the typewriter to help the poet to indicate exactly the breath, the pauses, the suspension, even, of syllables, the juxtaposition, even, of parts of phrases which he intends, observing that, for the first time, the poet has the stave and the bar that the musician has had.

The same kind of autonomy and independence which Charles Olson claims that the typewriter confers on the voice of the poet was claimed for the typewriter by the career woman of fifty years ago. British women were reputed to have developed a "twelve-pound look" when typewriters became available for sixty dollars or so. This look was in some way related to the Viking gesture of Ibsen's Nora Helmer, who slammed the door of her doll's house and set off on a quest of vocation and soul-testing. The age of the iron whim had begun.

The reader will recall earlier mention that when the first wave of female typists hit the business office in the 1890s, the cuspidor manufacturers read the sign of doom. They were right. More important, the uniform ranks of fashionable lady typists made possible a revolution in the garment industry. What she wore, every farmer's daughter wanted to wear, for the typist was a popular figure of enterprise and skill. She was a style-maker who was also eager to follow styles. As much as the typewriter, the typist brought into business a new dimension of the uniform, the homogeneous, and the continuous that has made the typewriter indispensable to every aspect of mechanical industry. A modern battleship needs dozens of typewriters for ordinary operations. An army needs more typewriters than medium and light artillery pieces, even in the field, suggesting that the typewriter now fuses the functions of the pen and sword.

But the effect of the typewriter is not all of this kind. If the typewriter has contributed greatly to the familiar forms of the homogenized specialism and fragmentation that is print culture, it has also caused an integration of functions and the creation of much private independence. G. K. Chesterton demurred about this new independence as a delusion, remarking that "women refused to be dictated to and went out and became stenographers." The poet or novelist now composes on the typewriter. The typewriter fuses composition and publication, causing an entirely new attitude to the written and printed word. Composing on the typewriter has altered the forms of the language and of literature in ways best seen in the later novels of Henry James that were dictated to Miss Theodora Bosanquet, who took them down, not by shorthand, but on a typewriter. Her memoir, *Henry James at Work*, should have been followed by other studies of how the typewriter has altered English verse and prose, and, indeed, the very mental habits, themselves, of writers.

With Henry James, the typewriter had become a confirmed habit by 1907, and his new style developed a sort of free, incantatory quality. His secretary tells of how he found dictating not only easier but more inspiring than composing by hand: "It all seems to be so much more effectively and unceasingly *pulled* out of me in speech than in writing," he told her. Indeed, he became so attached to the sound of his typewriter that, on his deathbed, Henry James called for his Remington to be worked near his bedside.

Just how much the typewriter has contributed by its unjustified right-hand margin to the development of *vers libre* would be hard to discover, but free verse was really a recovery of spoken, dramatic stress in poetry, and the typewriter encouraged exactly this quality. Seated at the typewriter, the poet, much in the manner of the jazz musician, has the experience of performance as composition. In the nonliterate world, this had been the situation of the bard or minstrel. He had themes, but no text. At the typewriter, the poet commands the resources of the printing press. The machine is like a public-address system immediately at hand. He can shout or whisper or whistle, and make funny typographic faces at the audience, as does e. e. cummings in this sort of verse:

> In Just-
> spring   when the world is mud-
> luscious the little
> lame baloonman
> whistles   far   and wee
>
> and eddieandbill come
> running from marbles and

piracies and it's
spring
when the world is puddle wonderful

the queer
old baloonman whistles
far   and   wee
and bettyandisbel come dancing
from hop-scotch and jump-rope and

it's spring
and
            the
                        goat footed
baloonman whistles
far
and
wee

e. e. cummings is here using the typewriter to provide a poem with a musical score for choral speech. The older poet, separated from the print form by various technical stages, could enjoy none of the freedom of oral stress provided by the typewriter. The poet at the typewriter can do Njinsky leaps or Chaplin-like shuffles and wiggles. Because he is an audience for his own mechanical audacities, he never ceases to react to his own performance. Composing on the typewriter is like flying a kite.

The e. e. cummings poem, when read aloud with widely varying stresses and paces, will duplicate the perceptual process of its typewriting creator. How Gerard Manley Hopkins would have loved to have had a typewriter to compose on! People who feel that poetry is for the eye and is to be read silently can scarcely get anywhere with Hopkins or cummings. Read aloud, such poetry becomes quite natural. Putting first names in lower case, as "eddieandbill," bothered the literate people of forty years ago. It was supposed to.

Eliot and Pound used the typewriter for a great variety of central effects in their poems. And with them, too, the typewriter was an oral and mimetic instrument that gave them the colloquial freedom of the world of jazz and ragtime. Most colloquial and jazzy of all Eliot's poems, *Sweeney Agonistes*, in its first appearance in print, carried the note: "From Wanna Go Home Baby?"

That the typewriter, which carried the Gutenberg technology into every nook and cranny of our culture and economy, should, also, have given out with these opposite oral effects is a characteristic reversal. Such a reversal of form happens in all extremes of advanced technology, as with the wheel today.

As expediter, the typewriter brought writing and speech and publication into close association. Although a merely mechanical form, it acted in some respects as an implosion, rather than an explosion.

In its explosive character, confirming the existing procedures of movable types, the typewriter had an immediate effect in regulating spelling and grammar. The pressure of Gutenberg technology toward "correct" or uniform spelling and grammar was felt at once. Typewriters caused an enormous expansion in the sale of dictionaries. They also created the innumerable overstuffed files that led to the rise of the file-cleaning companies in our time. At first, however, the typewriter was not seen as indispensable to business. The personal touch of the hand-penned letter was considered so important that the typewriter was ruled out of commercial use by the pundits. They thought, however, that it might be of use to authors, clergymen, and telegraph operators. Even newspapers were lukewarm about this machine for some time.

Once any part of the economy feels a step-up in pace, the rest of the economy has to follow suit. Soon, no business

351

could be indifferent to the greatly increased pace set by the typewriter. It was the telephone, paradoxically, that sped the commercial adoption of the typewriter. The phrase "Send me a memo on that," repeated into millions of phones daily, helped to create the huge expansion of the typist function. Northcote Parkinson's law that "work expands so as to fill the time available for its completion" is precisely the zany dynamic provided by the telephone. In no time at all, the telephone expanded the work to be done on the typewriter to huge dimensions. Pyramids of paperwork rise on the basis of a small telephone network inside a single business. Like the typewriter, the telephone fuses functions, enabling the call-girl, for example, to be her own procurer and madam.

Northcote Parkinson had discovered that any business or bureaucratic structure functions by itself, independently of "the work to be done." The number of personnel and "the quality of the work are not related to each other at all." In any given structure, the rate of staff accumulation is not related to the work done but to the intercommunication among the staff, itself. (In other words, the medium is the message.) Mathematically stated, Parkinson's Law says that the rate of accumulation of office staff per annum will be between 5.17 per cent and 6.56 per cent, "irrespective of any variation in the amount of work (if any) to be done."

"Work to be done," of course, means the transformation of one kind of material energy into some new form, as trees into lumber or paper, or clay into bricks or plates, or metal into pipe. In terms of this kind of work, the accumulation of office personnel in a navy, for example, goes up as the number of ships goes down. What Parkinson carefully hides from himself and his readers is simply the fact that in the area of information movement, the main "work to be done" is actually the movement of information. The mere interrelating of people by selected information is now the principal source of wealth in the electric age. In the preceding mechanical age, work had not been like that at all. Work had meant the processing of various materials by assembly-line fragmentation of operations and hierarchically delegated authority. Electric power circuits, in relation to the same processing, eliminate both the assembly line and the delegated authority. Especially with the computer, the work effort is applied at the "programming" level, and such effort is one of information and knowledge. In the decision-making and "make happen" aspect of the work operation, the telephone and other such speedups of information have ended the divisions of delegated authority in favor of the "authority of knowledge." It is as if a symphony composer, instead of sending his manuscript to the printer and thence to the conductor and to the individual members of the orchestra, were to compose directly on an electronic instrument that would render each note or theme as if on the appropriate instrument. This would end at once all the delegation and specialism of the symphony orchestra that makes it such a natural model of the mechanical and industrial age. The typewriter, with regard to the poet or novelist, comes very close to the promise of electronic music, insofar as it compresses or unifies the various jobs of poetic composition and publication.

The historian Daniel Boorstin was scandalized by the fact that celebrity in our information age was not due to a person's having done anything but simply to his being known for being well known. Professor Parkinson is scandalized that the structure of human work now seems to be quite independent of any job to be done. As an economist, he reveals the same incongruity and comedy, as between the old and the new, that Stephen Potter does in his *Gamesmanship*. Both have revealed the hollow mockery of "getting ahead in the world," in its old sense. Neither honest toil nor clever ploy will serve to advance the eager executive. The reason is simple. Positional warfare is finished, both in private and corporate action. In business, as in society, "getting on" may mean getting out. There is no "ahead" in a world that is an echo chamber of instantaneous celebrity.

The typewriter, with its promise of careers for the Nora Helmers of the West, has really turned out to be an elusive pumpkin coach, after all.

27

The Telephone: Sounding Brass or Tinkling Symbol?

In a pattern to which the reader will be accustomed by now, McLuhan pulls media together without confining himself to the principal subject at hand. Here, an illuminating passage discusses how the movie rivals the book. A pattern of prose style too, if already evident much earlier, shows itself in high definition: it is the 1-2-3-4 punch of McLuhan's sentences: "The reversal of that one-way movement outward from center-to-margin is as clearly owing to electricity as the great Western explosion had, in the first place, been due to electricity."

— *(Editor)*

The readers of the New York *Evening Telegram* were told in 1904: "Phony implies that a thing so qualified has no more substance than a telephone talk with a supposititious friend." The folklore of the telephone in song and story has been augmented in the memoirs of Jack Paar, who writes that his resentment toward the telephone began with the singing telegram. He tells how he got a call from a woman who said she was so lonesome she had been taking a bath three times a day in hopes that the phone would ring.

James Joyce in *Finnegans Wake* headlined TELEVISION KILLS TELEPHONY IN BROTHERS BROIL, introducing a major theme in the battle of the technologically extended senses that has, indeed, been raging through our culture for more than a decade. With the telephone, there occurs the extension of ear and voice that is a kind of extrasensory perception. With television came the extension of the sense of touch or of sense interplay that even more intimately involves the entire sensorium.

The child and the teenager understand the telephone, embracing the cord and the ear-mike as if they were beloved pets. What we call "the French phone," the union of mouth-piece and earphone in a single instrument, is a significant indication of the French liaison of the senses that English-speaking people keep firmly separate. French is "the language of love" just because it unites voice and ear in an especially close way, as does the telephone. So it is quite natural to kiss via phone, but not easy to visualize while phoning.

No more unexpected social result of the telephone has been observed than its elimination of the red-light district and its creation of the call-girl. To the blind, all things are unexpected. The form and character of the telephone, as of all electric technology, appear fully in this spectacular development. The prostitute was a specialist, and the call-girl is not. A "house" was not a home; but the call-girl not only lives at home, she may be a matron. The power of the telephone to decentralize every operation and to end positional warfare, as well as localized prostitution, has been felt but not understood by every business in the land.

The telephone, in the case of the call-girl, is like the typewriter that fuses the functions of composition and publication. The call-girl dispenses with the procurer and the madam. She has to be an articulate person of varied conversation and social accomplishments since she is expected to be able to join any company on a basis of social equality. If the typewriter has splintered woman from the home and turned her into a specialist in the office, the telephone gave her back to the executive world as a general means of harmony, an invitation to happiness, and a sort of combined confessional-and-wailing wall for the immature American executive.

The typewriter and the telephone are most unidentical twins that have taken over the revamping of the American girl with technological ruthlessness and thoroughness.

Since all media are fragments of ourselves extended into the public domain, the action upon us of any one medium tends to bring the other senses into play in a new relation. As we read, we provide a sound track for the printed word; as we listen to the radio, we provide a visual accompaniment. Why can we not visualize while telephoning? At once the reader will protest, "But I do visualize on the telephone!" When he has a chance to try the experiment deliberately, he will find that he simply can't visualize while phoning, though all literate people try to do so and, therefore, believe they are succeeding. But that is not what most irritates the literate and visualizing Westerner about the telephone. Some people can scarcely talk to their best friends on the phone without becoming angry. The telephone demands complete participation, unlike the written and printed page. Any literate man resents such a heavy demand for his total attention, because he has long been accustomed to fragmentary attention. Similarly, literate man can learn to speak other languages only with great difficulty, for learning a language calls for participation of *all* the senses at once. On the other hand, our habit of visualizing renders the literate Westerner helpless in the nonvisual world of advanced physics. Only the visceral and audile-tactile Teuton and Slav have the needed immunity to visualization for work in the non-Euclidean math and quantum physics. Were we to teach our math and physics by telephone, even a highly literate and abstract Westerner could eventually compete with the European physicists. This fact does not interest the Bell Telephone research department, for like any other book-oriented group they are oblivious to the telephone as a *form*, and study only the content aspect of wire service. As already mentioned, the Shannon and Weaver [1] hypothesis about Information Theory, like the Morgenstern Game Theory, tends to ignore the function of the form as form. Thus both Information Theory and Game Theory have bogged down into sterile banalities, but the psychic and social changes resulting from these forms have altered the whole of our lives.

Many people feel a strong urge to "doodle" while telephoning. This fact is very much related to the characteristic of this medium, namely that it demands participation of our senses and faculties. Unlike radio, it cannot be used as background. Since the telephone offers a very poor auditory image, we strengthen and complete it by the use of all the other senses. When the auditory image is of high definition, as with radio, we visualize the experience or complete it with

1 Information theory and game theory are mentioned briefly in Chapter 24. The specific works referred to here are Claude E. Shannon and Warren Weaver, *The Mathematical Theory of Communication*, first published in 1949, and John von Neumann and Oskar Morgenstern, *Theory of Games and Economic Behavior*, first published in 1953.

the sense of sight. When the visual image is of high definition or intensity, we complete it by providing sound. That is why, when the movies added sound track, there was such deep artistic upset. In fact the disturbance was almost equal to that caused by the movie itself. For the movie is a rival of the book, tending to provide a visual track of narrative description and statement that is much fuller than the written word.

In the 1920s a popular song was "All Alone by the Telephone, All Alone Feeling Blue." Why should the phone create an intense feeling of loneliness? Why should we feel compelled to answer a ringing public phone when we know the call cannot concern us? Why does a phone ringing on the stage create instant tension? Why is that tension so very much less for an unanswered phone in a movie scene? The answer to all of these questions is simply that the phone is a participant form that demands a partner, with all the intensity of electric polarity. It simply will not act as a background instrument like radio.

A standard practical joke of the small town in the early days of the telephone draws attention to the phone as a form of communal participation. No back fence could begin to rival the degree of heated participation made possible by the partyline. The joke in question took the form of calling several people, and, in an assumed voice, saying that the engineering department was going to clean out the telephone lines: "We recommend that you cover your telephone with a sheet or pillow case to prevent your room from being filled with dirt and grease." The jokester would then make the rounds of his friends in question to enjoy their preparations and their momentary expectation of a hiss and roar that was sure to come when the lines were blown out. The joke now serves to recall that not long ago the phone was a new contraption, used more for entertainment than for business.

The invention of the telephone was an incident in the

larger effort of the past century to render speech visible. Melville Bell, the father of Alexander Graham Bell, spent his life devising a universal alphabet that he published in 1867 under the title *Visible Speech*. Besides the aim to make all the languages of the world immediately present to each other in a simple visual form, the Bells, father and son, were much concerned to improve the state of the deaf. Visible speech seemed to promise immediate means of release for the deaf from their prison. Their struggle to perfect visible speech for the deaf led the Bells to a study of the new electrical devices that yielded the telephone. In much the same way, the Braille system of dots-for-letters had begun as a means of reading military messages in darkness, then was transferred to music, and finally to reading for the blind. Letters had been codified as dots for the fingers long before the Morse Code was developed for telegraph use. And it is relevant to note how electric technology, in like manner, had converged on the world of speech and language, from the beginning of electricity. That which had been the first great extension of our central nervous system — the mass media of the spoken word — was soon wedded to the second great extension of the central nervous system — electric technology.

The New York *Daily Graphic* for March 15, 1877, portrayed on its front page "The Terrors of the Telephone — The Orator of the Future." A disheveled Svengali stands before a microphone haranguing in a studio. The same mike is shown in London, San Francisco, on the Prairies, and in Dublin. Curiously, the newspaper of that time saw the telephone as a rival to the press as P.A. system, such as radio was in fact to be fifty years later. But the telephone, intimate and personal, is the most removed of any medium from the P.A. form. Thus wire-tapping seems even more odious than the reading of other people's letters.

The word "telephone" came into existence in 1840, before Alexander Graham Bell was born. It was used to describe a device made to convey musical notes through wooden rods. By the 1870s, inventors in many places were trying to achieve the electrical transmission of speech, and the American Patent Office received Elisha Gray's design for a telephone on the same day as Bell's, but an hour or two later. The legal profession benefited enormously from this coincidence. But Bell got the fame, and his rivals became footnotes. The telephone presumed to offer service to the public in 1877, paralleling wire telegraphy. The new telephone group was puny beside the vast telegraph interests, and Western Union moved at once to establish control over the telephone service.

It is one of the ironies of Western man that he has never felt any concern about invention as a threat to his way of life. The fact is that, from the alphabet to the motorcar, Western man has been steadily refashioned in a slow technological explosion that has extended over 2,500 years. From the time of the telegraph onward, however, Western man began to live an implosion. He began suddenly with Nietzschean insouciance to play the movie of his 2,500-year explosion backward. But he still enjoys the results of the extreme fragmentation of the original components of his tribal life. It is this fragmentation that enables him to ignore cause-and-effect in all interplay of technology and culture. It is quite different in Big Business. There, tribal man is on the alert for stray seeds of change. That was why William H. Whyte could write *The Organization Man* as a horror story. Eating people is wrong. Even grafting people into the ulcer of a big corporation seems wrong to anybody brought up in literate visual fragmented freedom. "I call them up at night when their guard's down," said one senior executive.

In the 1920s, the telephone spawned a good deal of dialogue humor that sold as gramophone records. But radio and the talking pictures were not kind to the monologue, even when it was made by W. C. Fields or Will Rogers. These hot media pushed aside the cooler forms that TV has now brought back on a large scale. The new race of night-club entertainers (Newhart, Nichols and May) have a curious early-telephone flavor that is very welcome, indeed. We can thank TV, with its call for such high participation, that mime and dialogue are back. Our Mort Sahls and Shelley Bermans and Jack Paars are almost a variety of "living newspaper," such as was provided for the Chinese revolutionary masses by dramatic teams in the 1930s and 1940s. Brecht's plays have the same participational quality of the world of the comic strip and the newspaper mosaic that TV has made acceptable, as pop art.

The mouthpiece of the telephone was a direct outgrowth of a prolonged attempt beginning in the seventeenth century to mimic human physiology by mechanical means. It is very much in the nature of the electric telephone, therefore, that it has such natural congruity with the organic. On the advice of a Boston surgeon, Dr. C. J. Blake, the receiver of the phone was directly modeled on the bone and diaphragm structure of the human ear. Bell paid much attention to the work of the great Helmholtz, whose work covered many fields. Indeed, it was because of his conviction that Helmholtz had sent vowels by telegraph that Bell was encouraged to persevere in his efforts. It turned out that it was his inadequate German that had fostered this optimistic impression. Helmholtz had failed to achieve any speech effects by wire. But Bell argued, if vowels could be sent, why not consonants? "I thought that Helmholtz, himself, had done it, and that my failure was due only to my ignorance of electricity. It was a very valuable blunder. It gave me confidence. If I had been able to read German in those days I might never have commenced my experiments!"

One of the most startling consequences of the telephone was its introduction of a "seamless web" of interlaced patterns in management and decision-making. It is not feasible to exercise delegated authority by telephone. The pyramidal structure of job-division and description and delegated powers cannot withstand the speed of the phone to by-pass all hierarchical arrangements, and to involve people in depth. In the same way, mobile panzer divisions equipped with radio telephones upset the traditional army structure. And we have seen how the news reporter linking the printed page to the telephone and the telegraph created a unified corporate image out of the fragmented government departments.

Today the junior executive can get on a first-name basis with seniors in different parts of the country. "You just start telephoning. Anybody can walk into any manager's office by telephone. By ten o'clock of the day I hit the New York office I was calling everybody by their first names."

The telephone is an irresistible intruder in time or place, so that high executives attain immunity to its call only when dining at head tables. In its nature the telephone is an intensely personal form that ignores all the claims of visual privacy prized by literate man. One firm of stockbrokers recently abolished all private offices for its executives, and settled them around a kind of seminar table. It was felt that the instant decisions that had to be made based on the continuous flow of teletype and other electric media could only receive group approval fast enough if private space were abolished. When on the alert, even the grounded crews of military aircraft cannot be out of sight of one another at any time. This is merely a time factor. More relevant is the need for total involvement in *role* that goes with this instant structure. The two pilots of one Canadian jet fighter are matched with all the care used in a marriage bureau. After many tests and long experience together they are officially *married* by their commanding officer "till death do you part." There is no tongue-in-cheek about this. It is this same kind of total integration into a role that raises the hackles of any literate man faced by the implosive demands of the seamless web of electric decision-making. Freedom in the Western world has always taken the form of the explosive, the divisive, advancing the separation of the individual from the state. The reversal of that one-way movement outward from center-to-margin is as clearly owing to electricity as the great Western explosion had, in the first place, been due to phonetic literacy.

If delegated chain-of-command authority won't work by telephone but only by written instruction, what sort of authority does come into play? The answer is simple, but not easy to convey. On the telephone only the authority of knowledge will work. Delegated authority is lineal, visual, hierarchical. The authority of knowledge is nonlineal, nonvisual, and inclusive. To act, the delegated person must always get clearance from the chain-of-command. The electric situation eliminates such patterns; such "checks and balances" are alien to the inclusive authority of knowledge. Consequently, restraints on electric absolutist power can be achieved, not by the separation of powers, but by a pluralism of centers. This problem has arisen apropos of the direct private line from the Kremlin to the White House. President Kennedy stated his preference for teletype over telephone, with a natural Western bias.

The separation of powers had been a technique for restraining action in a centralist structure radiating out to remote margins. In an electric structure there are, so far as the time and space of this planet are concerned, no margins. There can, therefore, be dialogue only among centers and among equals. The chain-of-command pyramids cannot obtain support from electric technology. But in place of delegated power, there tends to appear again with electric media, the *role*. A person can now be reinvested with all kinds of nonvisual character. King and emperor were legally endowed to act as the collective ego of all the private egos of their subjects. So far, Western man has encountered the restoration of the role only tentatively. He still manages to keep individuals in delegated *jobs*. In the cult of the movie star, we have allowed ourselves somnambulistically to abandon our Western traditions, conferring on these jobless images a mystic role. They are collective embodiments of the multitudinous private lives of their subjects.

An extraordinary instance of the power of the telephone to involve the whole person is recorded by psychiatrists, who report that neurotic children lose all neurotic symptoms when telephoning. *The New York Times* of September 7, 1949, printed an item that provides bizarre testimony to the cooling participational character of the telephone:

> On September 6, 1949, a psychotic veteran, Howard B. Unruh, in a mad rampage on the streets of Camden, New Jersey, killed thirteen people, and then returned home. Emergency crews, bringing up machine guns, shotguns, and tear gas bombs, opened fire. At this point an editor on the *Camden Evening Courier* looked up Unruh's name in the telephone directory and called him. Unruh stopped firing and answered, "Hello."
> "This Howard?"
> "Yes...."
> "Why are you killing people?"
> "I don't know. I can't answer that yet. I'll have to talk to you later. I'm too busy now."

Art Seidenbaum, in a recent article in the *Los Angeles Times*, "Dialectics of Unlisted Telephone Numbers," said:

> Celebrities have been hiding for a long time. Paradoxically, as their names and images are bloated on ever widening screens, they take increasing pains to be unapproachable in the flesh or phone ... Many big names never answer up to their numbers; a service takes every call, and only upon request, delivers

the accumulated messages ... "Don't call us" could become the real area code for Southern California.

"All Alone by the Telephone" has come full circle. It will soon be the telephone that is "all alone, and feeling blue."

28

The Phonograph:
The Toy that Shrank the
National Chest

McLuhan's frequent disquisitions on automation in the electric age are nowhere as extensive as here. He begins by stressing the crucial role in the eventual automation of human song and dance played by mechanical technologies such as film and the phonograph, noting that that development was paralleled by automation in the work place. The subsequent commentary ranges over the assembly-line, the disappearance of the human hand, fragmentation under mechanization, and computer programming. Culture and technology, typically dissociated in commentaries on the former, are explicitly linked in the preceding chapter and come under McLuhan's attention again: "The Romantics knew as little about real savages as they did about assembly lines." There is also a definition of consciousness, tailored to McLuhan's imperative for media study: "an inclusive process, not at all dependent on content."

— (Editor)

The phonograph, which owes its origin to the electrical telegraph and the telephone, had not manifested its basically
electric form and function until the tape recorder released it
from its mechanical trappings. That the world of sound is essentially a unified field of instant relationships lends it a near
resemblance to the world of electromagnetic waves. This fact
brought the phonograph and radio into early association.

Just how obliquely the phonograph was at first received
is indicated in the observation of John Philip Sousa, the
brassband director and composer. He commented: "With
the phonograph vocal exercises will be out of vogue! Then
what of the national throat? Will it not weaken? What of the
national chest? Will it not shrink?"

One fact Sousa had grasped: The phonograph is an
extension and amplification of the voice that may well have
diminished individual vocal activity, much as the car had
reduced pedestrian activity.

Like the radio that it still provides with program content,
the phonograph is a hot medium. Without it, the twentieth
century as the era of tango, ragtime, and jazz would have
had a different rhythm. But the phonograph was involved in
many misconceptions, as one of its early names—gramophone
—implies. It was conceived as a form of auditory writing
(*gramma*-letters). It was also called "graphophone," with the
needle in the role of pen. The idea of it as a "talking machine"
was especially popular. Edison was delayed in his approach
to the solution of its problems by considering it at first as a
"telephone repeater"; that is, a storehouse of data from the
telephone, enabling the telephone to "provide invaluable
records, instead of being the recipient of momentary and
fleeting communication." These words of Edison, published
in the *North American Review* of June, 1878, illustrate how the
then recent telephone invention already had the power to color

thinking in other fields. So, the record player had to be seen as a kind of phonetic record of telephone conversation. Hence, the names "phonograph" and "gramophone."

Behind the immediate popularity of the phonograph was the entire electric implosion that gave such new stress and importance to actual speech rhythms in music, poetry, and dance alike. Yet the phonograph was a machine merely. It did not at first use an electric motor or circuit. But in providing a mechanical extension of the human voice and the new ragtime melodies, the phonograph was propelled into a central place by some of the major currents of the age. The fact of acceptance of a new phrase, or a speech form, or a dance rhythm is already direct evidence of some actual development to which it is significantly related. Take, for example, the shift of English into an interrogative mood, since the arrival of "How about that?" Nothing could induce people to begin suddenly to use such a phrase over and over, unless there was some new stress, rhythm, or nuance in interpersonal relations that gave it relevance.

It was while handling paper tape, impressed by Morse Code dots and dashes, that Edison noticed the sound given off when the tape moved at high speed resembled "human talk heard indistinctly." It then occurred to him that indented tape could record a telephone message. Edison became aware of the limits of lineality and the sterility of specialism as soon as he entered the electric field. "Look," he said, "it's like this. I start here with the intention of reaching here in an experiment, say, to increase the speed of the Atlantic cable; but when I've arrived part way in my straight line, I meet with a phenomenon, and it leads me off in another direction and develops into a phonograph." Nothing could more dramatically express the turning point from mechanical explosion to electrical implosion. Edison's own career embodied that very change in our world, and he himself was often caught in the confusion between the two forms of procedure.

It was just at the end of the nineteenth century that the psychologist Lipps revealed by a kind of electric audiograph that the single clang of a bell was an intensive manifold containing all possible symphonies. It was somewhat on the same lines that Edison approached his problems. Practical experience had taught him that embryonically all problems contained all answers when one could discover a means of rendering them explicit. In his own case, his determination to give the phonograph, like the telephone, a direct practical use in business procedures led to his neglect of the instrument as a means of entertainment. Failure to foresee the phonograph as a means of entertainment was really a failure to grasp the meaning of the electric revolution in general. In our time we are reconciled to the phonograph as a toy and a solace; but press, radio, and TV have also acquired the same dimension of entertainment. Meantime, entertainment pushed to an extreme becomes the main form of business and politics. Electric media, because of their total "field" character, tend to eliminate the fragmented specialties of form and function that we have long accepted as the heritage of alphabet, printing, and mechanization. The brief and compressed history of the phonograph includes all phases of the written, the printed, and the mechanized word. It was the advent of the electric tape recorder that only a few years ago released the phonograph from its temporary involvement in mechanical culture. Tape and the LP record suddenly made the phonograph a means of access to all the music and speech of the world.

Before turning to the LP and tape recording revolution, we should note that the earlier period of mechanical recording and sound reproduction had one large factor in common with the silent picture. The early phonograph produced a brisk and raucous experience not unlike that of a Mack Sennett movie. But the undercurrent of mechanical music is strangely sad. It was the genius of Charles Chaplin to have captured for film this sagging quality of a deep blues, and to have overlaid it with jaunty jive and bounce. The poets and painters and musicians of the later nineteenth century all insist on a sort of metaphysical melancholy as latent in the great industrial world of the metropolis. The Pierrot figure is as crucial in the poetry of Laforgue as it is in the art of Picasso or the music of Satie. Is not the mechanical at its best a remarkable approximation to the organic? And is not a great industrial civilization able to produce anything in abundance for everybody? The answer is "Yes." But Chaplin and the Pierrot poets and painters and musicians pushed this logic all the way to reach the image of Cyrano de Bergerac, who was the greatest lover of all, but who was never permitted the return of his love. This weird image of Cyrano, the unloved and unlovable lover, was caught up in the phonograph cult of the blues. Perhaps it is misleading to try to derive the origin of the blues from Negro folk music; however, Constant Lambert, English conductor-composer, in his *Music Ho!*, provides an account of the blues that preceded the jazz of the post-World War I. He concludes that the great flowering of jazz in the twenties was a popular response to the highbrow richness and orchestral subtlety of the Debussy-Delius period. Jazz would seem to be an effective bridge between highbrow and lowbrow music, much as Chaplin made a similar bridge for pictorial art. Literary people eagerly accepted these bridges, and Joyce got Chaplin into *Ulysses* as Bloom, just as Eliot got jazz into the rhythms of his early poems.

Chaplin's clown-Cyrano is as much a part of a deep melancholy as Laforgue's or Satie's Pierrot art. Is it not inherent in the very triumph of the mechanical and its omission of the human? Could the mechanical reach a higher level than the talking machine with its mime of voice and dance? Do not T. S. Eliot's famous lines about the typist of the jazz age capture the entire pathos of the age of Chaplin and the ragtime blues?

> When lovely woman stoops to folly and
> Paces about her room again, alone,
> She smoothes her hair with automatic hand,
> And puts a record on the gramophone.

Read as a Chaplin-like comedy, Eliot's Prufrock makes ready sense. Prufrock is the complete Pierrot, the little puppet of the mechanical civilization that was about to do a flip into its electric phase.

It would be difficult to exaggerate the importance of complex mechanical forms such as film and phonograph as the prelude to the automation of human song and dance. As this automation of human voice and gesture had approached perfection, so the human work force approached automation. Now in the electric age the assembly line with its human hands disappears, and electric automation brings about a withdrawal of the work force from industry. Instead of being automated themselves—fragmented in task and function— as had been the tendency under mechanization, men in the electric age move increasingly to involvement in diverse jobs simultaneously, and to the work of learning, and to the programming of computers.

This revolutionary logic inherent in the electric age was made fairly clear in the early electric forms of telegraph and telephone that inspired the "talking machine." These new forms that did so much to recover the vocal, auditory, and mimetic world that had been repressed by the printed word, also inspired the strange new rhythms of "the jazz age," the various forms of syncopation and symbolist discontinuity that, like relativity and quantum physics, heralded the end of the Gutenberg era with its smooth, uniform lines of type and organization.

The word "jazz" comes from the French *jaser*, to chatter. Jazz is, indeed, a form of dialogue among instrumentalists and dancers alike. Thus it seemed to make an abrupt break with the homogeneous and repetitive rhythms of the smooth waltz. In the age of Napoleon and Lord Byron, when the waltz was a new form, it was greeted as a barbaric fulfillment of the Rousseauistic dream of the noble savage. Grotesque as this idea now appears, it is really a most valuable clue to the dawning mechanical age. The impersonal choral-dancing of the older, courtly pattern was abandoned when the waltzers held each other in a personal embrace. The waltz is precise, mechanical, and military, as its history manifests. For a waltz to yield its full meaning, there must be military dress. "There was a sound of revelry by night" was how Lord Byron referred to the waltzing before Waterloo. To the eighteenth century and to the age of Napoleon, the citizen armies seemed to be an individualistic release from the feudal framework of courtly hierarchies. Hence the association of waltz with noble savage, meaning no more than freedom from status and hierarchic deference. The waltzers were all uniform and equal, having free movement in any part of the hall. That this was the Romantic idea of the life of the noble savage now seems odd, but the Romantics knew as little about real savages as they did about assembly lines.

In our own century the arrival of jazz and ragtime was also heralded as the invasion of the bottom-wagging native. The indignant tended to appeal from jazz to the beauty of the mechanical and repetitive waltz that had once been greeted as pure native dancing. If jazz is considered as a break with mechanism in the direction of the discontinuous, the partic-ipant, the spontaneous and improvisational, it can also be seen as a return to a sort of oral poetry in which performance is both creation and composition. It is a truism among jazz performers that recorded jazz is "as stale as yesterday's newspaper." Jazz is alive, like conversation; and like conver-sation it depends upon a repertory of available themes. But performance is composition. Such performance insures maxi-mal participation among players and dancers alike. Put in this way, it becomes obvious at once that jazz belongs in that family of mosaic structures that reappeared in the Western world with the wire services. It belongs with symbolism in poetry, and with the many allied forms in painting and in music.

The bond between the phonograph and song and dance is no less deep than its earlier relation to telegraph and telephone. With the first printing of musical scores in the sixteenth century, words and music drifted apart. The separate virtuosity of voice and instruments became the basis of the great musical developments of the eighteenth and nineteenth centuries. The same kind of fragmentation and specialism in the arts and sciences made possible mammoth results in industry and in military enterprise, and in massive cooperative enterprises such as the newspaper and symphony orchestra.

Certainly the phonograph as a product of industrial, assembly-line organization and distribution showed little of the electric qualities that had inspired its growth in the mind of Edison. There were prophets who could foresee the great day when the phonograph would aid medicine by providing a medical means of discrimination between "the sob of hysteria and the sigh of melancholia ... the ring of whooping cough and the hack of the consumptive. It will be an expert in insanity, distinguishing between the laugh of the maniac and the drivel of the idiot ... It will accomplish this feat in the anteroom, while the physician is busying himself with his last patient." In practice, however, the phonograph stayed with the voices of the Signor Foghornis, the basso-tenores, robusto-profundos.

Recording facilities did not presume to touch anything so subtle as an orchestra until after the First War. Long before this, one enthusiast looked to the record to rival the photograph album and to hasten the happy day when "future generations will be able to condense within the space of twenty minutes a tone-picture of a single lifetime: five minutes of a child's prattle, five of the boy's exultations, five of the man's reflections, and five from the feeble utterances of the death-bed." James Joyce, somewhat later, did better. He made *Finnegans Wake* a tone poem that condensed in a single sentence all the prattlings, exultations, observations, and remorse of the entire human race. He could not have conceived this work in any other age than the one that produced the phonograph and the radio.

It was radio that finally injected a full electric charge into the world of the phonograph. The radio receiver of 1924 was already superior in sound quality, and soon began to depress the phonograph and record business. Eventually, radio restored the record business by extending popular taste in the direction of the classics.

The real break came after the Second War with the availability of the tape recorder. This meant the end of the incision recording and its attendant surface noise. In 1949 the era of electric hi-fi was another rescuer of the phonograph business. The hi-fi quest for "realistic sound" soon merged with the TV image as part of the recovery of tactile experience. For the sensation of having the performing instruments "right in the room with you" is a striving toward the union of the audile and tactile in a finesse of fiddles that is in large degree the sculptural experience. To be in the presence of performing musicians is to experience their touch and handling of instruments as tactile and kinetic, not just as resonant. So it can be said that hi-fi is not any quest for abstract effects of sound in separation from the other senses. With hi-fi, the phonograph meets the TV tactile challenge.

Stereo sound, a further development, is "all-around" or "wrap-around" sound. Previously sound had emanated from a single point in accordance with the bias of visual culture with its fixed point of view. The hi-fi changeover was really for music what cubism had been for painting, and what symbolism had been for literature; namely, the acceptance of multiple facets and planes in a single experience. Another way to put it is to say that stereo is sound in depth, as TV is the visual in depth.

Perhaps it is not very contradictory that when a medium becomes a means of depth experience the old categories of "classical" and "popular" or of "highbrow" and "lowbrow" no longer obtain. Watching a blue-baby heart operation on TV is an experience that will fit none of the categories. When LP and hi-fi and stereo arrived, a depth approach to musical experience also came in. Everybody lost his inhibitions about "highbrow," and the serious people lost their qualms about popular music and culture. Anything that is approached in depth acquires as much interest as the greatest matters. Because "depth" means "in interrelation," not in isolation. Depth means insight, not point of view; and insight is a kind of mental involvement in process that makes the content of the item seem quite secondary. Consciousness itself is an inclusive process not at all dependent on content. Consciousness does not postulate consciousness of anything in particular.

With regard to jazz, LP brought many changes, such as the cult of "real cool drool," because the greatly increased length of a single side of a disk meant that the jazz band could really have a long and casual chat among its instruments. The repertory of the 1920s was revived and given new depth and complexity by this new means. But the tape recorder in combination with LP revolutionized the repertory of classical music. Just as tape meant the new study of spoken rather than written languages, so it brought in the entire musical culture of many centuries and countries. Where before there had been a narrow selection from periods and composers, the tape recorder, combined with LP, gave a full musical spectrum that made the sixteenth century as available as the nineteenth, and Chinese folk song as accessible as the Hungarian.

A brief summary of technological events relating to the phonograph might go this way:

The telegraph translated writing into sound, a fact directly related to the origin of both the telephone and phonograph. With the telegraph, the only walls left are the vernacular walls that the photograph and movie and wirephoto overleap so easily. The electrification of writing was almost as big a step into the nonvisual and auditory space as the later steps soon taken by telephone, radio, and TV.

The telephone: speech without walls.
The phonograph: music hall without walls.
The photograph: museum without walls.
The electric light: space without walls.
The movie, radio, and TV: classroom without walls.

Man the food-gatherer reappears incongruously as information-gatherer. In this role, electronic man is no less a nomad than his paleolithic ancestors.

29

Movies:
The Reel World

In his comments on closure *(see also Glossary) in the early pages here, McLuhan is referring to closure of sensory* input, *whereas the closure frequently referred to elsewhere, in connection with media-induced somnambulism or the Narcissus-trance, is closure of one of the physical senses* itself. *Structure is revealed here as sculptural in quality, and manuscript culture as oral in quality. As for film itself, it proves not to reduce to a single medium, but it is the "final fulfillment of the great potential of typographic fragmentation."*

— *(Editor)*

In England the movie theater was originally called "The Bioscope," because of its visual presentation of the actual movements of the forms of life (from Greek *bios*, way of life). The movie, by which we roll up the real world on a spool in order to unroll it as a magic carpet of fantasy, is a spectacular wedding of the old mechanical technology and the new electric world. In the chapter on The Wheel, the story was told of how the movie had a kind of symbolic origin in an attempt to photograph the flying hooves of galloping horses, for to set a series of cameras to study animal movement is to merge the mechanical and the organic in a special way. In the medieval world, curiously, the idea of change in organic beings was that of the substitution of one static form for another, in sequence. They imagined the life of a flower as a kind of cinematic strip of phases or essences. The movie is the total realization of the medieval idea of change, in the form of an entertaining illusion. Physiologists had very much to do with the development of film, as they did with the telephone. On film the mechanical appears as organic, and the growth of a flower can be portrayed as easily and as freely as the movement of a horse.

If the movie merges the mechanical and organic in a world of undulating forms, it also links with the technology of print. The reader in projecting words, as it were, has to follow the black and white sequences of stills that is typography, providing his own sound track. He tries to follow the contours of the author's mind, at varying speeds and with various illusions of understanding. It would be difficult to exaggerate the bond between print and movie in terms of their power to generate fantasy in the viewer or reader. Cervantes devoted his *Don Quixote* entirely to this aspect of the printed word and its power to create what James Joyce throughout *Finnegans Wake* designates as "the ABCED-minded," which can be taken as "ab-said" or "ab-sent," or just alphabetically controlled.

The business of the writer or the film-maker is to transfer the reader or viewer from one world, his *own*, to another, the world created by typography and film. That is so obvious, and happens so completely, that those undergoing the experience accept it subliminally and without critical awareness. Cervantes lived in a world in which print was as new as movies are in the West, and it seemed obvious to him that print, like the images now on the screen, had usurped the real world. The reader or spectator had become a dreamer under their spell, as René Clair said of film in 1926.

Movies as a nonverbal form of experience are like photography, a form of statement without syntax. In fact, however, like the print and the photo, movies assume a high level of literacy in their users and prove baffling to the nonliterate. Our literate acceptance of the mere movement of the camera eye as it follows or drops a figure from view is not acceptable to an African film audience. If somebody disappears off the side of the film, the African wants to know what happened to him. A literate audience, however, accustomed to following printed imagery line by line without questioning the logic of lineality, will accept film sequence without protest.

It was René Clair who pointed out that if two or three people were together on a stage, the dramatist must ceaselessly motivate or explain their being there at all. But the film audience, like the book reader, accepts mere sequence as rational. Whatever the camera turns to, the audience accepts. We are transported to another world. As René Clair observed, the screen opens its white door into a harem of beautiful visions and adolescent dreams, compared to which the loveliest real body seems defective. Yeats saw the movie as a world of Platonic ideals with the film projector playing "a spume upon a ghostly paradigm of things." This was the world that haunted Don Quixote, who found it through the folio door of the newly printed romances.

The close relation, then, between the reel world of film and the private fantasy experience of the printed word is indispensable to our Western acceptance of the film form. Even the film industry regards all of its greatest achievements as derived from novels, nor is this unreasonable. Film, both in its reel form and in its scenario or script form, is completely involved with book culture. All one need do is to imagine for a moment a film based on newspaper form in order to see how close film is to book. Theoretically, there is no reason why the camera should not be used to photograph complex groups of items and events in dateline configurations, just as they are presented on the page of a newspaper. Actually, poetry tends to do this configuring or "bunching" more than prose. Symbolist poetry has much in common with the mosaic of the newspaper page, yet very few people can detach themselves from uniform and connected space sufficiently to grasp symbolist poems. Natives, on the other hand, who have very little contact with phonetic literacy and lineal print, have to learn to "see" photographs or film just as much as we have to learn our letters. In fact, after having tried for years to teach Africans their letters by film, John Wilson of London University's African Institute found it easier to teach them their letters as a means to film literacy. For even when natives have learned to "see" pictures, they cannot accept our ideas of time and space "illusions." On seeing Charlie Chaplin's *The Tramp*, the African audience concluded that Europeans were magicians who could restore life. They saw a character who survived a mighty blow on the head without any indication of being hurt. When the camera shifts, they think they see trees moving, and buildings growing or shrinking, because they cannot make the literate assumption that space is continuous and uniform. Nonliterate people simply don't get perspective or distancing effects of light and shade that we assume are innate human equipment. Literate people think of cause and effect as sequential, as if one thing pushed another along by physical force. Nonliterate people register very little interest in this kind of "efficient" cause and effect, but are fascinated by hidden forms that produce magical results. Inner, rather than outer, causes interest the nonliterate and nonvisual cultures. And that is why the literate West sees the rest of the world as caught in the seamless web of superstition.

Like the oral Russian, the African will not accept sight and sound together. The talkies were the doom of Russian film-making because, like any backward or oral culture, Russians have an irresistible need for participation that is defeated by the addition of sound to the visual image. Both Pudovkin and Eisenstein denounced the sound film but considered that if sound were used symbolically and contrapuntally, rather than realistically, there would result less harm to the visual image. The African insistence on group participation and on chanting and shouting during films is wholly frustrated by sound track. Our own talkies were a further completion of the visual package as a mere consumer commodity. For with silent film we automatically provide sound for ourselves by way of "closure" or completion. And when it is filled in for us there is very much less participation in the work of the image.

Again, it has been found that nonliterates do not know how to fix their eyes, as Westerners do, a few feet in front of the movie screen, or some distance in front of a photo. The result is that they move their eyes over photo or screen as they might their hands. It is this same habit of using the eyes as hands that makes European men so "sexy" to American women. Only an extremely literate and abstract society learns to fix the eyes, as we must learn to do in reading the printed page. For those who thus fix their eyes, perspective results. There is great subtlety and synesthesia in native art, but no perspective. The old belief that everybody really saw in perspective, but only that Renaissance painters had learned how to paint it, is erroneous. Our own first TV generation is rapidly losing this habit of visual perspective as a sensory modality, and along with this change comes an interest in words, not as visually uniform and continuous, but as unique worlds in depth. Hence the craze for puns and word-play, even in sedate ads.

In terms of other media such as the printed page, film has the power to store and to convey a great deal of information. In an instant it presents a scene of landscape with figures that would require several pages of prose to describe. In the next instant it repeats, and can go on repeating, this detailed information. The writer, on the other hand, has no means of holding a mass of detail before his reader in a large bloc or *gestalt*. As the photograph urged the painter in the direction of abstract, sculptural art, so the film has confirmed the writer in verbal economy and depth symbolism where the film cannot rival him.

Another facet of the sheer quantity of data possible in a movie shot is exemplified in historical films like *Henry V* or *Richard III*. Here extensive research went into the making of the sets and costumes that any six-year-old can now enjoy as readily as any adult. T. S. Eliot reported how, in the making of the film of his *Murder in the Cathedral*, it was not only necessary to have costumes of the period, but — so great is the precision and tyranny of the camera eye — these costumes had to be woven by the same techniques as those used in the twelfth century. Hollywood, amidst much illusion, had also to provide authentic scholarly replicas of many past scenes. The stage and TV can make do with very rough approximations, because they offer an image of low definition that evades detailed scrutiny.

At first, however, it was the detailed realism of writers like Dickens that inspired movie pioneers like D. W. Griffith, who carried a copy of a Dickens novel on location. The realistic novel, that arose with the newspaper form of communal cross-section and human-interest coverage in the eighteenth century, was a complete anticipation of film form. Even the poets took up the same panoramic style, with human interest vignettes and close-ups as variant. Gray's *Elegy*, Burns' *The Cotter's Saturday Night*, Wordsworth's *Michael*, and Byron's *Childe Harold* are all like shooting scripts for some contemporary documentary film.

"The kettle began it ..." Such is the opening of Dickens' *The Cricket on the Hearth*. If the modern novel came out of Gogol's *The Overcoat*, the modern movie, says Eisenstein, boiled up out of that kettle. It should be plain that the American and even British approach to film is much lacking in that free interplay among the senses and the media that seems so natural to Eisenstein or René Clair. For the Russian, especially, it is easy to approach any situation structurally, which is to say, sculpturally. To Eisenstein, the overwhelming fact of film was that it is an "act of juxtaposition." But to a culture in an extreme reach of typographic conditioning, the juxtaposition must be one of uniform and connected characters and qualities. There must be no leaps from the unique space of the tea kettle to the unique space of the kitten or the boot. If such objects appear, they must be leveled off by some continuous narrative, or be "contained" in some uniform pictorial space. All that Salvador Dali had to do to create a furor was to allow the chest of drawers or the grand piano to exist in its own space against some Sahara or Alpine backdrop. Merely by releasing objects from the uniform continuous space of typography we got modern art and poetry. We can measure the psychic pressure of typography by the uproar generated by that release. For most people, their own ego image seems to have been typographically conditioned, so that the electric age with its return to inclusive experience threatens their idea of self. These are the fragmented ones, for whom specialist toil renders the mere prospect of leisure or jobless security a nightmare. Electric simultaneity ends specialist learning and activity, and demands interrelation in depth, even of the personality.

The case of Charlie Chaplin films helps to illumine this problem. His *Modern Times* was taken to be a satire on the fragmented character of modern tasks. As clown, Chaplin presents the acrobatic feat in a mime of elaborate incompetence, for any specialist task leaves out most of our faculties. The clown reminds us of our fragmented state by tackling acrobatic or special jobs in the spirit of the whole or integral man. This is the formula for helpless incompetence. On the street, in social situations, on the assembly line, the worker continues his compulsive twitchings with an imagery wrench. But the mime of this Chaplin film and others is precisely that of the robot, the mechanical doll whose deep pathos it is to approximate so closely to the condition of human life. Chaplin, in all his work, did a puppetlike ballet of the Cyrano de Bergerac kind. In order to capture this puppetlike pathos, Chaplin (a devotee of ballet and a personal friend of Pavlova), adopted from the first the foot postures of classical ballet. Thus he could have the aura of *Spectre de la Rose* shimmering around his clown getup. From the British music hall, his first training ground, with a sure touch of genius he took images like that of Mr. Charles Pooter, the haunting figure of a nobody. This shoddy-genteel image he invested with an envelope of fairy romance by means of adherence to the classic ballet postures. The new film form was perfectly adapted to this composite image, since film is itself a jerky mechanical ballet of flicks that yields a sheer dream world of romantic illusions. But the film form is not just a puppetlike dance of arrested still shots, for it manages to approximate and even to surpass real life by means of illusion. That is why Chaplin, in his silent pictures at least, was never tempted to abandon the Cyrano role of the puppet who could never really be a lover. In this stereotype Chaplin discovered the heart of the film illusion, and he manipulated that heart with easy mastery, as the key to the pathos of a mechanized civilization. A mechanized world is always in the process of getting ready to live, and to this end it brings to bear the most appalling pomp of skill and method and resourcefulness.

The film pushed this mechanism to the utmost mechanical verge and beyond, into a surrealism of dreams that money can buy. Nothing is more congenial to the film form than this pathos of superabundance and power that is the dower of a puppet for whom they can never be real. This is the key to *The Great Gatsby* that reaches its moment of truth when Daisy breaks down in contemplating Gatsby's superb collection of shirts. Daisy and Gatsby live in a tinsel world that is both corrupted by power, yet innocently pastoral in its dreaming.

The movie is not only a supreme expression of mechanism, but paradoxically it offers as product the most magical of consumer commodities, namely dreams. It is, therefore, not accidental that the movie has excelled as a medium that offers poor people roles of riches and power beyond the dreams of avarice. In the chapter on The Photograph, it was pointed out how the press photo in particular had discouraged the really rich from the paths of conspicuous consumption. The life of display that the photo had taken from the rich, the movie gave to the poor with lavish hand:

> Oh lucky, lucky me,
> I shall live in luxury,
> For I've got a pocketful of dreams.

The Hollywood tycoons were not wrong in acting on the assumption that movies gave the American immigrant a means of self-fulfillment without any delay. This strategy, however deplorable in the light of the "absolute ideal good," was perfectly in accord with film form. It meant that in the 1920s the American way of life was exported to the entire world in cans. The world eagerly lined up to buy canned dreams. The film not only accompanied the first great consumer age, but was also incentive, advertisement and, in itself, a major commodity. Now in terms of media study it is clear that the power of film to store information in accessible form is unrivaled. Audio tape and video tape were to excel film eventually as information storehouses. But film remains a major information resource, a rival of the book whose technology it did so much to continue and also to surpass. At the present time, film is still in its manuscript phase, as it were; shortly it will, under TV pressure, go into its portable, accessible, printed-book phase. Soon everyone will be able to have a small, inexpensive film projector that plays an 8-mm sound cartridge as if on a TV screen. This type of development is part of our present technological implosion. The present dissociation of projector and screen is a vestige of our older mechanical world of explosion and separation of functions that is now ending with the electrical implosion.

Typographic man took readily to film just because, like books, it offers an inward world of fantasy and dreams. The film viewer sits in psychological solitude like the silent book reader. This was not the case with the manuscript reader, nor is it true of the watcher of television. It is not pleasant to turn on TV just for oneself in a hotel room, nor even at home. The TV mosaic image demands social completion and dialogue. So with the manuscript before typography, since manuscript culture is oral and demands dialogue and debate, as the entire culture of the ancient and medieval worlds demonstrates. One of the major pressures of TV has been to encourage the "teaching machine." In fact, these devices are adaptations of the book in the direction of dialogue. These teaching machines are really private tutors, and their being misnamed on the principle that produced the names "wireless" and "horseless carriage" is another instance in that long list that illustrates how every innovation must pass through a primary phase in which the new effect is secured by the old method, amplified or modified by some new feature.

Film is not really a single medium like song or the written word, but a collective art form with different individuals directing color, lighting, sound, acting, speaking. The press, radio and TV, and the comics are also art forms dependent upon entire teams and hierarchies of skill in corporate action. Prior to the movies, the most obvious example of such corporate artistic action had occurred early in the industrialized world, with the large new symphony orchestras of the nineteenth century. Paradoxically, as industry went its ever more specialized fragmented course, it demanded more and more teamwork in sales and supplies. The symphony orchestra became a major expression of the ensuing power of such coordinated effort, though for the players themselves this effect was lost, both in the symphony and in industry.

When the magazine editors recently introduced film scenario procedures to the constructing of idea articles, the idea article supplanted the short story. The film is the rival of the book in that sense. (TV in turn is the rival of the magazine because of its mosaic power.) Ideas presented as a sequence of shots or pictorialized situations, almost in the manner of a teaching machine, actually drove the short story out of the magazine field.

Hollywood has fought TV mainly by becoming a subsidiary of TV. Most of the film industry is now engaged in supplying TV programs. But one new strategy has been tried, namely the big-budget picture. The fact is that Technicolor is the closest the movie can get to the effect of the TV image. Technicolor greatly lowers photographic intensity and creates, in part, the visual conditions for participant viewing. Had Hollywood understood the reasons for *Marty*'s success, TV might have given us a revolution in film. *Marty* was a TV show that got onto the screen in the form of low definition or low-intensity visual realism. It was not a success story, and it had no stars, because the low-intensity TV image is quite incompatible with the high-intensity star image. *Marty*, which in fact looked like an early silent movie or an old Russian picture, offered the film industry all the clues it needed for meeting the TV challenge.

This kind of casual, cool realism has given the new British films easy ascendancy. *Room at the Top* features the new cool realism. Not only is it not a success story, it is as much an announcement of the end of the Cinderella package as Marilyn Monroe was the end of the star system. *Room at the Top* is the story of how the higher a monkey climbs, the more you see of his backside. The moral is that success is not only wicked but also the formula for misery. It is very hard for a hot medium like film to accept the cool message of TV. But the Peter Sellers movies *I'm All Right, Jack* and *Only Two Can Play* are perfectly in tune with the new temper created by the cool TV image. Such is also the meaning of the ambiguous success of *Lolita*. As a novel, its acceptance announced the antiheroic approach to romance. The film industry had long beaten out a royal road to romance in keeping with the crescendo of the success story. *Lolita* announced that the royal road was only a cowtrack, after all, and as for success, it shouldn't happen to a dog.

In the ancient world and in medieval times, the most popular of all stories were those dealing with *The Falls of Princes*. With the coming of the very hot print medium, the preference changed to a rising rhythm and to tales of success and sudden elevation in the world. It seemed possible to achieve anything by the new typographic method of minute, uniform segmentation of problems. It was by this method, eventually, that film was made. Film was, as a form, the final fulfillment of the great potential of typographic fragmentation. But the electric implosion has now reversed the entire process of expansion by fragmentation. Electricity has brought back the cool, mosaic world of implosion, equilibrium, and stasis. In our electric age, the one-way expansion of the berserk individual on his way to the top now appears as a gruesome image of trampled lives and disrupted harmonies. Such is the subliminal message of the TV mosaic with its total field of simultaneous impulses. Film strip and sequence cannot but bow to this superior power. Our own youngsters have taken the TV message to heart in their beatnik rejection of consumer mores and of the private success story.

Since the best way to get to the core of a form is to study its effect in some unfamiliar setting, let us note what President Sukarno of Indonesia announced in 1956 to a large group of Hollywood executives. He said that he regarded them as political radicals and revolutionaries who had greatly hastened political change in the East. What the Orient saw in a Hollywood movie was a world in which all the *ordinary people* had cars and electric stoves and refrigerators. So the Oriental now regards himself as an ordinary person who has been deprived of the ordinary man's birthright.

That is another way of getting a view of the film medium as monster ad for consumer goods. In America this major aspect of film is merely subliminal. Far from regarding our pictures as incentives to mayhem and revolution, we take them as solace and compensation, or as a form of deferred payment by daydreaming. But the Oriental is right, and we are wrong about this. In fact, the movie is a mighty limb of the industrial giant. That it is being amputated by the TV image reflects a still greater revolution going on at the center of American life. It is natural that the ancient East should feel the political pull and industrial challenge of our movie industry. The movie, as much as the alphabet and the printed word, is an aggressive and imperial form that explodes outward into other cultures. Its explosive force was significantly greater in silent pictures than in talkies, for the electromagnetic sound track already forecast the substitution of electric implosion for mechanical explosion. The silent pictures were immediately acceptable across language barriers as the talkies were not. Radio teamed up with film to give us the talkie and to carry us further on our present reverse course of implosion or re-integration after the mechanical age of explosion and expansion. The extreme form of this implosion or contraction is the image of the astronaut locked into his wee bit of wrap-around space. Far from enlarging our world, he is announcing its contraction to village size. The rocket and the space capsule are ending the rule of the wheel and the machine, as much as did the wire services, radio, and TV.

We may now consider a further instance of the film's influence in a most conclusive aspect. In modern literature there is probably no more celebrated technique than that of the stream of consciousness or interior monologue. Whether in Proust, Joyce, or Eliot, this form of sequence permits the reader an extraordinary identification with personalities of the utmost range and diversity. The stream of consciousness is really managed by the transfer of film technique to the printed page, where, in a deep sense it really originated; for as we have seen, the Gutenberg technology of movable types is quite indispensable to any industrial or film process. As much as the infinitesimal calculus that *pretends* to deal with motion and change by minute fragmentation, the film *does* so by making motion and change into a series of static shots. Print does likewise while pretending to deal with the whole mind in action. Yet film and the stream of consciousness alike seemed to provide a deeply desired release from the mechanical world of increasing standardization and uniformity. Nobody ever felt oppressed by the monotony or uniformity of the Chaplin ballet or by the monotonous, uniform musings of his literary twin, Leopold Bloom.

In 1911 Henri Bergson in *Creative Evolution* created a sensation by associating the thought process with the form of the movie. Just at the extreme point of mechanization represented by the factory, the film, and the press, men seemed by the stream of consciousness, or interior film to obtain release into a world of spontaneity, of dreams, and of unique personal experience. Dickens perhaps began it all with his Mr. Jingle in *Pickwick Papers*. Certainly in *David Copperfield* he made a great technical discovery, since for the first time the world unfolds realistically through the use of the eyes of a growing child as camera. Here was the stream of consciousness, perhaps, in its original form before it was adopted by Proust and Joyce and Eliot. It indicates how the enrichment of human experience can occur unexpectedly with the crossing and interplay of the life of media forms.

The film imports of all nations, especially those from the United States, are very popular in Thailand, thanks in part to a deft Thai technique for getting round the foreign-language obstacle. In Bangkok, in place of subtitles, they use what is called "Adam-and-Eving." This takes the form of live Thai dialogue read through a loudspeaker by Thai actors concealed from the audience. Split-second timing and great endurance enable these actors to demand more than the best-paid movie stars of Thailand.

Everyone has at some time wished he were equipped with his own sound system during a movie performance, in order to make appropriate comments. In Thailand, one might achieve great heights of interpretive interpolation during the inane exchanges of great stars.

30

Radio:
The Tribal Drum

*When McLuhan says of radio in Nazi Germany that it worked
"to create depth involvement for everybody," it would seem to be
a contradiction of the definitions of hot vs. cool media (Chapter
2). But he stresses, in what follows, that radio produces different effects in different cultures. Its classification as a hot or
cool medium depends entirely on the literacy or lack of it in
the culture to which it is introduced. Literacy itself McLuhan
characterizes as a typographic technology, observing that "learning
to read and write is a minor facet of literacy in the uniform,
continuous environments of the English-speaking world." He
ascribes attitudes in young people to media effects: "So much
do-it-yourself, or completion and 'closure' of action, develops
a kind of independent isolation in the young that makes them
remote and inaccessible." Recalling the place that language
occupies for McLuhan as a primary technology, his further
comments on technological effects as parallel to numbing under
stress and shock offer an intriguing possibility for explaining
the language crisis evoked by writers, beginning in the years
immediately after World War I.*

— (Editor)

England and America had had their "shots" against radio in the form of long exposure to literacy and industrialism. These forms involve an intense visual organization of experience. The more earthy and less visual European cultures were not immune to radio. Its tribal magic was not lost on them, and the old web of kinship began to resonate once more with the note of fascism. The inability of literate people to grasp the language and message of the media as such is involuntarily conveyed by the comments of sociologist Paul Lazarsfeld in discussing the effects of radio:

> The last group of effects may be called the monopolistic effects of radio. Such have attracted most public attention because of their importance in the totalitarian countries. If a government monopolizes the radio, then by mere repetition and by exclusion of conflicting points of view it can determine the opinions of the population. We do not know much about how this monopolistic effect really works, but it is important to note its singularity. No inference should be drawn regarding the effects of radio as such. It is often forgotten that Hitler did not achieve control through radio but almost despite it, because at the time of his rise to power radio was controlled by his enemies. The monopolistic effects have probably less social importance than is generally assumed.

Professor Lazarsfeld's helpless unawareness of the nature and effects of radio is not a personal defect, but a universally shared ineptitude.

In a radio speech in Munich, March 14, 1936, Hitler said, "I go my way with the assurance of a somnambulist." His victims and his critics have been equally somnambulistic. They danced entranced to the tribal drum of radio that extended their central nervous system to create depth involvement for everybody. "I live right inside radio when I listen. I more easily lose myself in radio than in a book," said a voice from a radio poll. The power of radio to involve people in depth is manifested in its use during homework by youngsters and by many other people who carry transistor sets in order to provide a private world for themselves amidst crowds. There is a little poem by the German dramatist Berthold Brecht:

> You little box, held to me when escaping
> So that your valves should not break,
> Carried from house to ship from ship to train,
> So that my enemies might go on talking to me
> Near my bed, to my pain
> The last thing at night, the first thing in the morning,
> Of their victories and of my cares,
> Promise me not to go silent all of a sudden.

One of the many effects of television on radio has been to shift radio from an entertainment medium into a kind of nervous information system. News bulletins, time signals, traffic data, and, above all, weather reports now serve to enhance the native power of radio to involve people in one another. Weather is that medium that involves all people equally. It is the top item on radio, showering us with fountains of auditory space or *lebensraum*.

It was no accident that Senator McCarthy lasted such a very short time when he switched to TV. Soon the press decided, "He isn't news anymore." Neither McCarthy nor the press ever knew what had happened. TV is a cool medium. It rejects hot figures and hot issues and people from the hot press media. Fred Allen was a casualty of TV. Was Marilyn Monroe? Had TV occurred on a large scale during Hitler's reign he would have vanished quickly. Had TV come first there would have been no Hitler at all. When Khrushchev appeared on American TV he was more acceptable than Nixon, as a clown and a lovable sort of old boy. His appearance is rendered by TV as a comic cartoon. Radio, however, is a hot medium and takes cartoon characters seriously. Mr. K. on radio would be a different proposition.

In the Kennedy-Nixon debates, those who heard them on radio received an overwhelming idea of Nixon's superiority. It was Nixon's fate to provide a sharp, high-definition image and action for the cool TV medium that translated that sharp image into the impression of a phony. I suppose "phony" is something that resonates wrong, that doesn't *ring* true. It might well be that F.D.R. would not have done well on TV. He had learned, at least, how to use the hot radio medium for his very cool job of fireside chatting. He first, however, had had to hot up the press media against himself in order to create the right atmosphere for his radio chats. He learned how to use the press in close relation to radio. TV would have presented him with an entirely different political and social mix of components and problems. He would possibly have enjoyed solving them, for he had the kind of playful approach necessary for tackling new and obscure relationships.

Radio affects most people intimately, person-to-person, offering a world of unspoken communication between writer-speaker and listener. That is the immediate aspect of radio. A private experience. The subliminal depths of radio are charged with the resonating echoes of tribal horns and antique drums. This is inherent in the very nature of this medium, with its power to turn the psyche and society into a single echo chamber. The resonating dimension of radio is unheeded by the script writers, with few exceptions. The famous Orson Welles broadcast about the invasion from Mars was a simple demonstration of the all-inclusive, completely involving scope of the auditory image of radio. It was Hitler who gave radio the Orson Welles treatment for *real*.

That Hitler came into political existence at all is directly

401

owing to radio and public-address systems. This is not to say that these media relayed his thoughts effectively to the German people. His thoughts were of very little consequence. Radio provided the first massive experience of electronic implosion, that reversal of the entire direction and meaning of literate Western civilization. For tribal peoples, for those whose entire social existence is an extension of family life, radio will continue to be a violent experience. Highly literate societies, that have long subordinated family life to individualist stress in business and politics, have managed to absorb and to neutralize the radio implosion without revolution. Not so, those communities that have had only brief or superficial experience of literacy. For them, radio is utterly explosive.

To understand such effects, it is necessary to see literacy as typographic technology, applied not only to the rationalizing of the entire procedures of production and marketing, but to law and education and city planning, as well. The principles of continuity, uniformity, and repeatability derived from print technology have, in England and America, long permeated every phase of communal life. In those areas a child learns literacy from traffic and street, from every car and toy and garment. Learning to read and write is a minor facet of literacy in the uniform, continuous environments of the English-speaking world. Stress on literacy is a distinguishing mark of areas that are striving to initiate that process of standardization that leads to the visual organization of work and space. Without psychic transformation of the inner life into segmented visual terms by literacy, there cannot be the economic "take-off" that insures a continual movement of augmented production and perpetually accelerated change-and-exchange of goods and services.

Just prior to 1914, the Germans had become obsessed with the menace of "encirclement." Their neighbors had all developed elaborate railway systems that facilitated mobilization of manpower resources. Encirclement is a highly

visual image that had great novelty for this newly industrialized nation. In the 1930s, by contrast, the German obsession
was with *lebensraum*. This is not a visual concern, at all. It is
a claustrophobia, engendered by the radio implosion and
compression of space. The German defeat had thrust them
back from visual obsession into brooding upon the resonating
Africa within. The tribal past has never ceased to be a reality
for the German psyche.

It was the ready access of the German and middle-Euro-
pean world to the rich nonvisual resources of auditory and
tactile form that enabled them to enrich the world of music
and dance sculpture. Above all their tribal mode gave them
easy access to the new nonvisual world of subatomic physics,
in which long-literate and long-industrialized societies are
decidedly handicapped. The rich area of preliterate vitality
felt the hot impact of radio. The message of radio is one of
violent, unified, implosion and resonance. For Africa, India,
China, and even Russia, radio is a profound archaic force,
a time bond with the most ancient past and long-forgotten
experience.

Tradition, in a word, is the sense of the total past as
now. Its awakening is a natural result of radio impact and
of electric information, in general. For the intensely literate
population, however, radio engendered a profound unlocal-
izable sense of guilt that sometimes expressed itself in the
fellow-traveler attitude. A newly found human involvement
bred anxiety and insecurity and unpredictability. Since literacy had fostered an extreme of individualism, and radio had
done just the opposite in reviving the ancient experience of
kinship webs of deep tribal involvement, the literate West
tried to find some sort of compromise in a larger sense of
collective responsibility. The sudden impulse to this end was
just as subliminal and obscure as the earlier literary pressure
toward individual isolation and irresponsibility; therefore,
nobody was happy about any of the positions arrived at. The

Gutenberg technology had produced a new kind of visual, national entity in the sixteenth century that was gradually meshed with industrial production and expansion. Telegraph and radio neutralized nationalism but evoked archaic tribal ghosts of the most vigorous brand. This is exactly the meeting of eye and ear, of explosion and implosion, or as Joyce puts it in the *Wake*, "In that earopean end meets Ind." The opening of the European ear brought to an end the open society and reintroduced the Indic world of tribal man to West End woman. Joyce puts these matters not so much in cryptic, as in dramatic and mimetic, form. The reader has only to take any of his phrases such as this one, and mime it until it yields the intelligible. Not a long or tedious process, if approached in the spirit of artistic playfulness that guarantees "lots of fun at Finnegan's wake."

Radio is provided with its cloak of invisibility, like any other medium. It comes to us ostensibly with person-to-person directness that is private and intimate, while in more urgent fact, it is really a subliminal echo chamber of magical power to touch remote and forgotten chords. All technological extensions of ourselves must be numb and subliminal, else we could not endure the leverage exerted upon us by such extension. Even more than telephone or telegraph, radio is that extension of the central nervous system that is matched only by human speech itself. Is it not worthy of our meditation that radio should be specially attuned to that primitive extension of our central nervous system, that aboriginal mass medium, the vernacular tongue? The crossing of these two most intimate and potent of human technologies could not possibly have failed to provide some extraordinary new shapes for human experience. So it proved with Hitler, the somnambulist. But does the detribalized and literate West imagine that it has earned immunity to the tribal magic of radio as a permanent possession? Our teenagers in the 1950s began to manifest many of the tribal stigmata. The adolescent, as opposed to the teenager, can now be classified as a phenomenon of literacy. Is it not significant that the adolescent was indigenous only to those areas of England and America where literacy had invested even food with abstract visual values? Europe never had adolescents. It had chaperones. Now, to the teenager, radio gives privacy, and at the same time it provides the tight tribal bond of the world of the common market, of song, and of resonance. The ear is hyperesthetic compared to the neutral eye. The ear is intolerant, closed, and exclusive, whereas the eye is open, neutral, and associative. Ideas of tolerance came to the West only after two or three centuries of literacy and visual Gutenberg culture. No such saturation with visual values had occurred in Germany by 1930. Russia is still far from any such involvement with visual order and values.

If we sit and talk in a dark room, words suddenly acquire new meanings and different textures. They become richer, even, than architecture, which Le Corbusier rightly says can best be felt at night. All those gestural qualities that the printed page strips from language come back in the dark, and on the radio. Given only the *sound* of a play, we have to fill in *all* of the senses, not just the sight of the action. So much do-it-yourself, or completion and "closure" of action, develops a kind of independent isolation in the young that makes them remote and inaccessible. The mystic screen of sound with which they are invested by their radios provides the privacy for their homework, and immunity from parental behest.

With radio came great changes to the press, to advertising, to drama, and to poetry. Radio offered new scope to practical jokers like Morton Downey at CBS. A sportscaster had just begun his fifteen-minute reading from a script when he was joined by Mr. Downey, who proceeded to remove his shoes and socks. Next followed coat and trousers and then underwear, while the sportscaster helplessly continued his broadcast, testifying to the compelling power of the mike to command loyalty over modesty and the self-protective impulse.

Radio created the disk jockey, and elevated the gag writer into a major national role. Since the advent of radio, the gag has supplanted the joke, not because of gag writers, but because radio is a fast hot medium that has also rationed the reporter's space for stories.

Jean Shepherd of WOR in New York regards radio as a new medium for a new kind of novel that he writes nightly. The mike is his pen and paper. His audience and their knowledge of the daily events of the world provide his characters, his scenes, and moods. It is his idea that, just as Montaigne was the first to use the page to record his reactions to the new world of printed books, he is the first to use radio as an essay and novel form for recording our common awareness of a totally new world of universal human participation in all human events, private or collective.

To the student of media, it is difficult to explain the human indifference to social effects of these radical forces. The phonetic alphabet and the printed word that exploded the closed tribal world into the open society of fragmented functions and specialist knowledge and action have never been studied in their roles as a magical transformer. The antithetic electric power of instant information that reverses social explosion into implosion, private enterprise into organization man, and expanding empires into common markets, has obtained as little recognition as the written word. The power of radio to retribalize mankind, its almost instant reversal of individualism into collectivism, Fascist or Marxist, has gone unnoticed. So extraordinary is this unawareness that *it* is what needs to be explained. The transforming power of media is easy to explain, but the ignoring of this power is not at all easy to explain. It goes without saying that the universal ignoring of the psychic action of technology bespeaks some inherent function, some essential numbing of consciousness such as occurs under stress and shock conditions.

The history of radio is instructive as an indicator of the bias and blindness induced in any society by its pre-existent technology. The word "wireless," still used for radio in Britain, manifests the negative "horseless-carriage" attitude toward a new form. Early wireless was regarded as a form of telegraph, and was not seen even in relation to the telephone. David Sarnoff in 1916 sent a memo to the Director of the American Marconi Company that employed him, advocating the idea of a music box in the home. It was ignored. That was the year of the Irish Easter rebellion and of the first radio *broadcast*. Wireless had already been used on ships as ship-to-shore "telegraph." The Irish rebels used a ship's wireless to make, not a point-to-point message, but a diffused broadcast in the hope of getting word to some ship that would relay their story to the American press. And so it proved. Even after broadcasting had been in existence for some years, there was no commercial interest in it. It was the amateur operators or hams and their fans, whose petitions finally got some action in favor of the setting up of facilities. There was reluctance and opposition from the world of the press, which, in England, led to the formation of the BBC and the firm shackling of radio by newspaper and advertising interests. This is an obvious rivalry that has not been openly discussed. The restrictive pressure by the press on radio and TV is still a hot issue in Britain and in Canada. But, typically, misunderstanding of the nature of the medium rendered the restraining policies quite futile. Such has always been the case, most notoriously in government censorship of the press and of the movies. Although the medium is the *message*, the controls go beyond programming. The restraints are always directed to the "content," which is always another medium. The content of the press is literary statement, as the content of the book is speech, and the content of the movie is the novel. So the

407

effects of radio are quite independent of its programming. To those who have never studied media, this fact is quite as baffling as literacy is to natives, who say, "Why do you write? Can't you remember?"

Thus, the commercial interests who think to render media universally acceptable, invariably settle for "entertainment" as a strategy of neutrality. A more spectacular mode of the ostrich-head-in-sand could not be devised, for it ensures maximal pervasiveness for any medium whatever. The literate community will always argue for a controversial or point-of-view use of press, radio, and movie that would in effect diminish the operation, not only of press, radio and movie, but of the book as well. The commercial entertainment strategy automatically ensures maximum speed and force of impact for any medium, on psychic and social life equally. It thus becomes a comic strategy of unwitting self-liquidation, conducted by those who are dedicated to permanence, rather than to change. In the future, the only effective media controls must take the thermostatic form of quantitative rationing. Just as we now try to control atom-bomb fallout, so we will one day try to control media fallout. Education will become recognized as civil defense against media fallout. The only medium for which our education now offers some civil defense is the print medium. The educational establishment, founded on print, does not yet admit any other responsibilities.

Radio provides a speedup of information that also causes acceleration in other media. It certainly contracts the world to village size, and creates insatiable village tastes for gossip, rumor, and personal malice. But while radio contracts the world to village dimensions, it hasn't the effect of homogenizing the village quarters. Quite the contrary. In India, where radio is the supreme form of communication, there are more than a dozen official languages and the same number of official radio networks. The effect of radio as a reviver of archaism and ancient memories is not limited to Hitler's Germany. Ireland, Scotland, and Wales have undergone resurgence of their ancient tongues since the coming of radio, and the Israeli present an even more extreme instance of linguistic revival. They now speak a language which has been dead in books for centuries. Radio is not only a mighty awakener of archaic memories, forces, and animosities, but a decentralizing, pluralistic force, as is really the case with all electric power and media.

Centralism of organization is based on the continuous, visual, lineal structuring that arises from phonetic literacy. At first, therefore, electric media merely followed the established patterns of literate structures. Radio was released from these centralist network pressures by TV. TV then took up the burden of centralism, from which it may be released by Telstar. With TV accepting the central network burden derived from our centralized industrial organization, radio was free to diversify, and to begin a regional and local community service that it had not known, even in the earliest days of the radio "hams." Since TV, radio has turned to the individual needs of people at different times of the day, a fact that goes with the multiplicity of receiving sets in bedrooms, bathrooms, kitchens, cars, and now in pockets. Different programs are provided for those engaged in diverse activities. Radio, once a form of group listening that emptied churches, has reverted to private and individual uses since TV. The teenager withdraws from the TV group to his private radio.

This natural bias of radio to a close tie-in with diversified community groups is best manifested in the disk-jockey cults, and in radio's use of the telephone in a glorified form of the old trunkline wire-tapping. Plato, who had old-fashioned tribal ideas of political structure, said that the proper size of a city was indicated by the number of people who could hear the voice of a public speaker. Even the printed book, let alone radio, renders the political assumptions of

409

Plato quite irrelevant for practical purposes. Yet radio, because of its ease of decentralized intimate relation with both private and small communities, could easily implement the Platonic political dream on a world scale.

The uniting of radio with phonograph that constitutes the average radio program yields a very special pattern quite superior in power to the combination of radio and telegraph press that yields our news and weather programs. It is curious how much more arresting are the weather reports than the news, on both radio and TV. Is not this because "weather" is now entirely an electronic form of information, whereas news retains much of the pattern of the printed word? It is probably the print and book bias of the BBC and the CBC that renders them so awkward and inhibited in radio and TV presentation. Commercial urgency, rather than artistic insight, fostered by contrast a hectic vivacity in the corresponding American operation.

31

Television:
The Timid Giant

It is intriguing to speculate as to whether McLuhan might have thought the epithet of his subtitle appropriate forty years later. The chapter is absorbing for its anecdotes, prescient for its comments on HDTV, valuable for its paraphrases ("the medium is the message or the basic source of effects"), definitions ("the seeking of multi-uses for rooms and things and objects, in a single word—the iconic"; what is deep *in* depth structure *is set in opposition to* fragmentary*) and explanations ("The mosaic can be* seen *as dancing can, but it is not* structured *visually; nor is is it an extension of the visual power. For the mosaic is not uniform, continuous, or repetitive."). McLuhan reveals why he suspends value judgment in studying media ("their effects are not capable of being isolated") and why he is an optimist ("uniqueness and diversity ... can be fostered under electric conditions as never before"). His signature juxtapositions continue (TV, Bauhaus, and Montessori), as does his compressed history of the media ("This mosaic TV image had already been adumbrated in the popular press that grew up with the telegraph"). The "group of simulcasts of several media done in Toronto" is a reference to the Ryerson experiment, published here as an appendix. Named in the same appendix is political scientist and economist Harold Innis, who here merely waits in the wings: "Political scientists have been quite unaware of the effects of media ..." There is a rich echo of the tradition of training of perception that McLuhan absorbed during his Cambridge years ("Everybody experiences far more than he understands. Yet it is experience, rather than understanding, that influences behavior, especially in collective matters of media and technology, where the individual is almost inevitably unaware of their effects upon him.") McLuhan expanded and translated that tradition of training into the program that marks his entire career and all his writings, summarized by him here in reference only to* Understanding Media: *"It is the theme of this book that not even the most lucid understanding of the peculiar force of a medium can head off the ordinary 'closure' of the senses that causes us to conform to the pattern of experience presented."*

— *(Editor)*

Perhaps the most familiar and pathetic effect of the TV image is the posture of children in the early grades. Since TV, children—regardless of eye condition—average about six and a half inches from the printed page. Our children are striving to carry over to the printed page the all-involving sensory mandate of the TV image. With perfect psycho-mimetic skill, they carry out the commands of the TV image. They pore, they probe, they slow down and involve themselves in depth. This is what they had learned to do in the cool iconography of the comic-book medium. TV carried the process much further. Suddenly they are transferred to the hot print medium with its uniform patterns and fast lineal movement. Pointlessly they strive to read print in depth. They bring to print all their senses, and print rejects them. Print asks for the isolated and stripped-down visual faculty, not for the unified sensorium.

The Mackworth head-camera, when worn by children watching TV, has revealed that their eyes follow, not the actions, but the reactions. The eyes scarcely deviate from the faces of the actors, even during scenes of violence. This head-camera shows by projection both the scene and the eye movement simultaneously. Such extraordinary behavior is another indication of the very cool and involving character of this medium.

On the Jack Paar show for March 8, 1963, Richard Nixon was Paared down and remade into a suitable TV image. It turns out that Mr. Nixon is both a pianist and a composer. With sure tact for the character of the TV medium, Jack Paar brought out this *pianoforte* side of Mr. Nixon, with excellent effect. Instead of the slick, glib, legal Nixon, we saw the doggedly creative and modest performer. A few timely touches like this would have quite altered the result of the Kennedy-Nixon campaign. TV is a medium that rejects the sharp personality and favors the presentation of processes rather than of products.

The adaptation of TV to processes, rather than to the neatly packaged products, explains the frustration many people experience with this medium in its political uses. An article by Edith Efron in *TV Guide* (May 18–24, 1963) labeled TV "The Timid Giant," because it is unsuited to hot issues and sharply defined controversial topics: "Despite official freedom from censorship, a self-imposed silence renders network documentaries almost mute on many great issues of the day." As a cool medium TV has, some feel, introduced a kind of *rigor mortis* into the body politic. It is the extraordinary degree of audience participation in the TV medium that explains its failure to tackle hot issues. Howard K. Smith observed: "The networks are delighted if you go into a controversy in a country 14,000 miles away. They don't want real controversy, real dissent, at home." For people conditioned to the hot newspaper medium, which is concerned with the clash of *views*, rather than involvement in *depth* in a situation, the TV behavior is inexplicable.

Such a hot news item that concerns TV directly was headlined "It finally happened—a British film with English subtitles to explain the dialects." The film in question is the British comedy "Sparrows Don't Sing." A glossary of Yorkshire, Cockney, and other slang phrases has been printed for the customers so that they can figure out just what the subtitles mean. Sub subtitles are as handy an indicator of the depth effects of TV as the new "rugged" styles in feminine attire. One of the most extraordinary developments since TV in England has been the upsurge of regional dialects. A regional brogue or "burr" is the vocal equivalent of gaiter stockings. Such brogues undergo continual erosion from literacy. Their sudden prominence in England in areas in which previously one had heard only standard English is one of the most significant cultural events of our time. Even in the classrooms of

Oxford and Cambridge, the local dialects are heard again. The undergraduates of those universities no longer strive to achieve a uniform speech. Dialectal speech since TV has been found to provide a social bond in depth, not possible with the artificial "standard English" that began only a century ago.

An article on Perry Como bills him as "Low-pressure king of high-pressure realm." The success of any TV performer depends on his achieving a low-pressure style of presentation although getting his act on the air may require much high-pressure organization. Castro may be a case in point. According to Tad Szulc's story on "Cuban Television's One-man Show" (*The Eighth Art*), "in his seemingly improvised 'as-I-go-along' style he can evolve politics and govern his country — right on camera." Now, Tad Szulc is under the illusion that TV is a hot medium, and suggests that in the Congo "television might have helped Lumumba to incite the masses to even greater turmoil and bloodshed." But he is quite wrong. Radio is the medium for frenzy, and it has been the major means of hotting up the tribal blood of Africa, India, and China, alike. TV has cooled Cuba down, as it is cooling down America. What the Cubans are getting by TV is the experience of being directly engaged in the making of political decisions. Castro presents himself as a teacher, and as Szulc says, "manages to blend political guidance and education with propaganda so skillfully that it is often difficult to tell where one begins and the other ends." Exactly the same mix is used in entertainment in Europe and America alike. Seen outside the United States, any American movie looks like subtle political propaganda. Acceptable entertainment has to flatter and exploit the cultural and political assumptions of the land of its origin. These unspoken presuppositions also serve to blind people to the most obvious facts about a new medium like TV.

In a group of simulcasts of several media done in Toronto

a few years back, TV did a strange flip. Four randomized groups of university students were given the same information at the same time about the structure of preliterate languages. One group received it via radio, one from TV, one by lecture, and one read it. For all but the reader group, the information was passed along in straight verbal flow by the same speaker without discussion or questions or use of blackboard. Each group had half an hour of exposure to the material. Each was asked to fill in the same quiz afterward. It was quite a surprise to the experimenters when the students performed better with TV-channeled information and with radio than they did with lecture and print—and the TV group stood *well* above the radio group. Since nothing had been done to give special stress to any of these four media, the experiment was repeated with other randomized groups. This time each medium was allowed full opportunity to do its stuff. For radio and TV, the material was dramatized with many auditory and visual features. The lecturer took full advantage of the blackboard and class discussion. The printed form was embellished with an imaginative use of typography and page layout to stress each point in the lecture. All of these media had been stepped up to high intensity for this repeat of the original performance. Television and radio once again showed results high above lecture and print. Unexpectedly to the testers, however, radio now stood significantly above television. It was a long time before the obvious reason declared itself, namely that TV is a cool, participant medium. When hotted up by dramatization and stingers, it performs less well because there is less opportunity for participation. Radio is a hot medium. When given additional intensity, it performs better. It doesn't invite the same degree of participation in its users. Radio will serve as background-sound or as noise-level control, as when the ingenious teenager employs it as a means of privacy. TV will not work as background. It engages you. You have to be *with* it. (The phrase has gained acceptance since TV.)

A great many things will not work since the arrival of TV. Not only the movies, but the national magazines as well, have been hit very hard by this new medium. Even the comic books have declined greatly. Before TV, there had been much concern about why Johnny couldn't read. Since TV, Johnny has acquired an entirely new set of perceptions. He is not at all the same. Otto Preminger, director of *Anatomy of a Murder* and other hits, dates a great change in movie making and viewing from the very first year of general TV programming. "In 1951, " he wrote, "I started a fight to get the release in motion-picture theaters of *The Moon Is Blue* after the production code approval was refused. It was a small fight and I won it." (*Toronto Daily Star*, October 19, 1963)

He went on to say, "The very fact that it was the word 'virgin' that was objected to in *The Moon Is Blue* is today laughable, almost incredible." Otto Preminger considers that American movies have advanced toward maturity owing to the influence of TV. The cool TV medium promotes depth structures in art and entertainment alike, and creates audience involvement in depth as well. Since nearly all our technologies and entertainment since Gutenberg have been not cool, but hot; and not deep, but fragmentary; not producer-oriented, but consumer-oriented, there is scarcely a single area of established relationships, from home and church to school and market, that has not been profoundly disturbed in its pattern and texture.

The psychic and social disturbance created by the TV image, and not the TV programming, occasions daily comment in the press. Raymond Burr, who plays Perry Mason, spoke to the National Association of Municipal Judges, reminding them that, "Without our laymen's understanding and acceptance, the laws which you apply and the courts in which you preside cannot continue to exist." What Mr. Burr omitted to observe was that the "Perry Mason" TV program,

in which he plays the lead, is typical of that intensely participational kind of TV experience that has altered our relation
to the laws and the courts.

The mode of the TV image has nothing in common with
film or photo, except that it offers also a nonverbal *gestalt*
or posture of forms. With TV, the viewer is the screen. He is
bombarded with light impulses that James Joyce called the
"Charge of the Light Brigade" that imbues his "soulskin with
sobconscious inklings." The TV image is visually low in data.
The TV image is not a *still* shot. It is not photo in any sense,
but a ceaselessly forming contour of things limned by the
scanning-finger. The resulting plastic contour appears by
light *through*, not light *on*, and the image so formed has the
quality of sculpture and icon, rather than of picture. The TV
image offers some three million dots per second to the receiver.
From these he accepts only a few dozen each instant, from
which to make an image.

The film image offers many more millions of data per
second, and the viewer does not have to make the same
drastic reduction of items to form his impression. He tends
instead to accept the full image as a package deal. In contrast,
the viewer of the TV mosaic, with technical control of the
image, unconsciously reconfigures the dots into an abstract
work of art on the pattern of a Seurat or Rouault. If anybody
were to ask whether all this would change if technology
stepped up the character of the TV image to movie data
level, one could only counter by inquiring, "Could we alter
a cartoon by adding details of perspective and light and
shade?" The answer is "Yes," only it would then no longer
be a cartoon. Nor would "improved" TV be television. The
TV image is *now* a mosaic mesh of light and dark spots which
a movie shot never is, even when the quality of the movie
image is very poor.

As in any other mosaic, the third dimension is alien to TV,
but it can be superimposed. In TV the illusion of the third

dimension is provided slightly by the stage sets in the studio; but the TV image itself is a flat two-dimensional mosaic. Most of the three-dimensional illusion is a carry-over of habitual viewing of film and photo. For the TV camera does not have a built-in angle of vision like the movie camera. Eastman Kodak now has a two-dimensional camera that can match the flat effects of the TV camera. Yet it is hard for literate people, with their habit of fixed points of view and three-dimensional vision, to understand the properties of two-dimensional vision. If it had been easy for them, they would have had no difficulties with abstract art, General Motors would not have made a mess of motorcar design, and the picture magazine would not be having difficulties now with the relationship between features and ads. The TV image requires each instant that we "close" the spaces in the mesh by a convulsive sensuous participation that is profoundly kinetic and tactile, because tactility is the interplay of the senses, rather than the isolated contact of skin and object.

To contrast it with the film shot, many directors refer to the TV image as one of "low definition," in the sense that it offers little detail and a low degree of information, much like the cartoon. A TV close-up provides only as much information as a small section of a long-shot on the movie-screen. For lack of observing so central an aspect of the TV image, the critics of program "content" have talked nonsense about "TV violence." The spokesmen of censorious views are typically semiliterate book-oriented individuals who have no competence in the grammars of newspaper, or radio, or of film, but who look askew and askance at all non-book media. The simplest question about any psychic aspect, even of the book medium, throws these people into a panic of uncertainty. Vehemence of projection of a single isolated attitude they mistake for moral vigilance. Once these censors became aware that in all cases "the medium is the message" or the basic source of effects, they would turn to suppression of media as such, instead of seeking "content" control. Their current assumption that content or programming is the factor that influences outlook and action is derived from the book medium, with its sharp cleavage between form and content.

Is it not strange that TV should have been as revolutionary a medium in America in the 1950s as radio in Europe in the 1930s? Radio, the medium that resuscitated the tribal and kinship webs of the European mind in the 1920s and 1930s, had no such effect in England or America. There, the erosion of tribal bonds by means of literacy and its industrial extensions had gone so far that our radio did not achieve any notable tribal reactions. Yet ten years of TV have Europeanized even the United States, as witness its changed feelings for space and personal relations. There is new sensitivity to the dance, plastic arts, and architecture, as well as the demand for the small car, the paperback, sculptural hairdos and molded dress effects — to say nothing of a new concern for complex effects in cuisine and in the use of wines. Notwithstanding, it would be misleading to say that TV will retribalize England and America. The action of radio on the world of resonant speech and memory was hysterical. But TV has certainly made England and America vulnerable to radio where previously they had immunity to a great degree. For good or ill, the TV image has exerted a unifying synesthetic force on the sense-life of these intensely literate populations, such as they have lacked for centuries. It is wise to withhold all value judgments when studying these media matters, since their effects are not capable of being isolated.

Synesthesia, or unified sense and imaginative life, had long seemed an unattainable dream to Western poets, painters, and artists in general. They had looked with sorrow and dismay on the fragmented and impoverished imaginative life of Western literate man in the eighteenth century and later. Such was the message of Blake and Pater, Yeats and D. H. Lawrence, and a host of other great figures. They were not prepared to have their dreams realized in everyday life by the esthetic action of radio and television. Yet these massive extensions of our central nervous systems have enveloped Western man in a daily session of synesthesia. The Western way of life attained centuries since by the rigorous separation and specialization of the senses, with the visual sense atop the hierarchy, is not able to withstand the radio and TV waves that wash about the great visual structure of abstract Individual Man. Those who, from political motives, would now add their force to the anti-individual action of our electric technology are puny subliminal automatons aping the patterns of the prevailing electric pressures. A century ago they would, with equal somnambulism, have faced in the opposite direction. German Romantic poets and philosophers had been chanting in tribal chorus for a return to the dark unconscious for over a century before radio and Hitler made such a return difficult to avoid. What is to be thought of people who wish such a return to preliterate ways, when they have no inkling of how the civilized visual way was ever substituted for tribal auditory magic?

At this hour, when Americans are discovering new passions for skin-diving and the wraparound space of small cars, thanks to the indomitable tactile promptings of the TV image, the same image is inspiring many English people with race feelings of tribal exclusiveness. Whereas highly literate Westerners have always idealized the condition of integration of races, it has been their literate culture that made impossible real uniformity among races. Literate man naturally dreams of visual solutions to the problems of human differences. At the end of the nineteenth century, this kind of dream suggested similar dress and education for both men and women. The failure of the sex-integration programs has provided the theme of much of the literature and psychoanalysis of the twentieth century. Race integration, undertaken on the basis of visual uniformity, is an extension of the same cultural strategy of literate man, for whom differences always seem to need eradication, both in sex and in race, and in space and in time. Electronic man, by becoming ever more deeply involved in the actualities of the human condition, cannot accept the literate cultural strategy. The Negro will reject a plan of visual uniformity as definitely as women did earlier, and for the same reasons. Women found that they had been robbed of their distinctive roles and turned into fragmented citizens in "a man's world." The entire approach to these problems in terms of uniformity and social homogenization is a final pressure of the mechanical and industrial technology. Without moralizing, it can be said that the electric age, by involving all men deeply in one another, will come to reject such mechanical solutions. It is more difficult to provide uniqueness and diversity than it is to impose the uniform patterns of mass education; but it is such uniqueness and diversity that can be fostered under electric conditions as never before.

Temporarily, all preliterate groups in the world have begun to feel the explosive and aggressive energies that are released by the onset of the new literacy and mechanization. These explosions come just at a time when the new electric technology combines to make us share them on a global scale.

The effect of TV, as the most recent and spectacular electric extension of our central nervous system, is hard to grasp for various reasons. Since it has affected the totality of our lives, personal and social and political, it would be quite unrealistic to attempt a "systematic" or visual presentation of such influence. Instead, it is more feasible to "present" TV as a complex *gestalt* of data gathered almost at random.

The TV image is of low intensity or definition, and therefore, unlike film, it does not afford detailed information about objects. The difference is akin to that between the old manuscripts and the printed word. Print gave intensity and

422

uniform precision, where before there had been a diffuse texture. Print brought in the taste for exact measurement and repeatability that we now associate with science and mathematics.

The TV producer will point out that speech on television must not have the careful precision necessary in the theater. The TV actor does not have to project either his voice or himself. Likewise, TV acting is so extremely intimate, because of the peculiar involvement of the viewer with the completion or "closing" of the TV image, that the actor must achieve a great degree of spontaneous casualness that would be irrelevant in movies and lost on stage. For the audience participates in the inner life of the TV actor as fully as in the outer life of the movie star. Technically, TV tends to be a close-up medium. The close-up that in the movie is used for shock is, on TV, a quite casual thing. And whereas a glossy photo the size of the TV screen would show a dozen faces in adequate detail, a dozen faces on the TV screen are only a blur.

The peculiar character of the TV image in its relation to the actor causes such familiar reactions as our not being able to recognize in real life a person whom we see every week on TV. Not many of us are as alert as the kindergartener who said to Garry Moore, "How did you get off TV?" Newscasters and actors alike report the frequency with which they are approached by people who feel they've met them before. Joanne Woodward in an interview was asked what was the difference between being a movie star and a TV actress. She replied: "When I was in the movies I heard people say, 'There goes Joanne Woodward.' Now they say, 'There goes someone I think I know.' "

The owner of a Hollywood hotel in an area where many movie and TV actors reside reported that tourists had switched their allegiance to TV stars. Moreover, most TV stars are men, that is, "cool characters," while most movie stars are women, since they can be presented as "hot" characters. Men and women movie stars alike, along with the entire star system, have tended to dwindle into a more moderate status since TV. The movie is a hot, high-definition medium. Perhaps the most interesting observation of the hotel proprietor was that the tourists wanted to see Perry Mason and Wyatt Earp. They did not want to see Raymond Burr and Hugh O'Brian. The old movie-fan tourists had wanted to see their favorites as they were in *real* life, not as they were in their film roles. The fans of the cool TV medium want to see their star in *role*, whereas the movie fans want the *real thing*.

A similar reversal of attitudes occurred with the printed book. There was little interest in the private lives of authors under manuscript or scribal culture. Today the comic strip is close to the preprint woodcut and manuscript form of expression. Walt Kelly's *Pogo* looks very much indeed like a gothic page. Yet in spite of great public interest in the comic-strip form, there is as little curiosity about the private lives of these artists as about the lives of popular-song writers. With print, the private life became of the utmost concern to readers. Print is a hot medium. It projects the author at the public as the movie did. The manuscript is a cool medium that does not project the author, so much as involve the reader. So with TV. The viewer is involved and participant. The *role* of the TV star, in this way, seems more fascinating than his private life. It is thus that the student of media, like the psychiatrist, gets more data from his informants than they themselves have perceived. Everybody experiences far more than he understands. Yet it is experience, rather than understanding, that influences behavior, especially in collective matters of media and technology, where the individual is almost inevitably unaware of their effect upon him.

Some may find it paradoxical that a cool medium like TV should be so much more compressed and condensed than a hot medium like film. But it is well-known that a half-minute of television is equal to three minutes of stage or vaudeville.

The same is true of manuscript in contrast to print. The "cool" manuscript tended toward compressed forms of statement, aphoristic and allegorical. The "hot" print medium expanded expression in the direction of simplification and the "spelling-out" of meanings. Print speeded up and "exploded" the compressed script into simpler fragments.

A cool medium, whether the spoken word or the manuscript or TV, leaves much more for the listener or user to do than a hot medium. If the medium is of high definition, participation is low. If the medium is of low intensity, the participation is high. Perhaps this is why lovers mumble so.

Because the low definition of TV insures a high degree of audience involvement, the most effective programs are those that present situations which consist of some process to be completed. Thus, to use TV to teach poetry would permit the teacher to concentrate on the poetic process of actual *making*, as it pertained to a particular poem. The book form is quite unsuited to this type of involved presentation. The same salience of process of do-it-yourself-ness and depth involvement in the TV image extends to the art of the TV actor. Under TV conditions, he must be alert to improvise and to embellish every phrase and verbal resonance with details of gesture and posture, sustaining that intimacy with the viewer which is not possible on the massive movie screen or on the stage.

There is the alleged remark of the Nigerian who, after seeing a TV western, said delightedly, "I did not realize you valued human life so little in the West." Offsetting this remark is the behavior of our children in watching TV westerns. When equipped with the new experimental head-cameras that follow their eye movements while watching the image, children keep their eyes on the faces of the TV actors. Even during physical violence their eyes remain concentrated on the facial *reactions*, rather than on the eruptive *action*. Guns, knives, fists, all are ignored in preference for the facial expression. TV is not so much an action, as a re-action, medium.

The yen of the TV medium for themes of process and complex reactions has enabled the documentary type of film to come to the fore. The movie *can* handle process superbly, but the movie viewer is more disposed to be a passive consumer of actions, rather than a participant in reactions. The movie western, like the movie documentary, has always been a lowly form. With TV, the western acquired new importance, since its theme is always: "Let's make a town." The audience participates in the shaping and processing of a community from meager and unpromising components. Moreover, the TV image takes kindly to the varied and rough textures of Western saddles, clothes, hides, and shoddy match-wood bars and hotel lobbies. The movie camera, by contrast, is at home in the slick chrome world of the night club and the luxury spots of a metropolis. Moreover, the contrasting camera preferences of the movies in the Twenties and Thirties, and of TV in the Fifties and Sixties spread to the entire population. In ten years the new tastes of America in clothes, in food, in housing, in entertainment, and in vehicles express the new pattern of interrelation of forms and do-it-yourself involvement fostered by the TV image.

It is no accident that such major movie stars as Rita Hayworth, Liz Taylor, and Marilyn Monroe ran into troubled waters in the new TV age. They ran into an age that questioned all the "hot" media values of the pre-TV consumer days. The TV image challenges the values of fame as much as the values of consumer goods. "Fame to me," said Marilyn Monroe, "certainly is only a temporary and a partial happiness. Fame is not really for a daily diet, that's not what fulfills you.... I think that when you are famous every weakness is exaggerated. This industry should behave to its stars like a mother whose child has just run out in front of a car. But instead of clasping the child to them they start punishing the child."

The movie community is now getting clobbered by TV, and lashes out at anybody in its bewildered petulance. These words of the great movie puppet who wed Mr. Baseball and Mr. Broadway are surely a portent. If many of the rich and successful figures in America were to question publicly the absolute value of money and success as means to happiness and human welfare, they would offer no more shattering a precedent than Marilyn Monroe. For nearly fifty years, Hollywood had offered "the fallen woman" a way to the top and a way to the hearts of all. Suddenly the love-goddess emits a horrible cry, screams that eating people is wrong, and utters denunciations of the whole way of life. This is exactly the mood of the suburban beatniks. They reject a fragmented and specialist consumer life for anything that offers humble involvement and deep commitment. It is the same mood that recently turned girls from specialist careers to early marriage and big families. They switch from jobs to roles.

The same new preference for depth participation has also prompted in the young a strong drive toward religious experience with rich liturgical overtones. The liturgical revival of the radio and TV age affects even the most austere Protestant sects. Choral chant and rich vestments have appeared in every quarter. The ecumenical movement is synonymous with electric technology.

Just as TV, the mosaic mesh, does not foster perspective in art, it does not foster lineality in living. Since TV, the assembly line has disappeared from industry. Staff and line structures have dissolved in management. Gone are the stag line, the party line, the receiving line, and the pencil line from the backs of nylons.

With TV came the end of bloc voting in politics, a form of specialism and fragmentation that won't work since TV. Instead of the voting bloc, we have the icon, the inclusive image. Instead of a political viewpoint or platform, the inclusive political posture or stance. Instead of the product, the process. In periods of new and rapid growth there is a blurring of outlines. In the TV image we have the supremacy of the blurred outline, itself the maximal incentive to growth and new "closure" or completion, especially for a consumer culture long related to the sharp visual values that had become separated from the other senses. So great is the change in American lives, resulting from the loss of loyalty to the consumer package in entertainment and commerce, that every enterprise, from Madison Avenue and General Motors to Hollywood and General Foods, has been shaken thoroughly and forced to seek new strategies of action. What electric implosion or contraction has done inter-personally and inter-nationally, the TV image does intra-personally or intra-sensuously.

It is not hard to explain this sensuous revolution to painters and sculptors, for they have been striving, ever since Cézanne abandoned perspective illusion in favor of structure in painting, to bring about the very change that TV has now effected on a fantastic scale. TV is the Bauhaus program of design and living, or the Montessori educational strategy, given total technological extension and commercial sponsorship. The aggressive lunge of artistic strategy for the remaking of Western man has, *via* TV, become a vulgar sprawl and an overwhelming splurge in American life.

It would be impossible to exaggerate the degree to which this image has disposed America to European modes of sense and sensibility. America is now Europeanizing as furiously as Europe is Americanizing. Europe, during the Second War, developed much of the industrial technology needed for its first mass consumer phase. It was, on the other hand, the First War that had readied America for the same consumer "take-off." It took the electronic *implosion* to dissolve the nationalist diversity of a splintered Europe, and to do for it what the industrial *explosion* had done for America. The industrial explosion that accompanies the

428

fragmenting expansion of literacy and industry was able to exert little unifying effect in the European world with its numerous tongues and cultures. The Napoleonic thrust had utilized the combined force of the new literacy and early industrialism. But Napoleon had had a less homogenized set of materials to work with than even the Russians have today. The homogenizing power of the literate process had gone further in America by 1800 than anywhere in Europe. From the first, America took to heart the print technology for its educational, industrial, and political life; and it was rewarded by an unprecedented pool of standardized workers and consumers, such as no culture had ever had before. That our cultural historians have been oblivious of the homogenizing power of typography, and of the irresistible strength of homogenized populations, is no credit to them. Political scientists have been quite unaware of the effects of media anywhere at any time, simply because nobody has been willing to study the personal and social effects of media apart from their "content."

America long ago achieved its Common Market by mechanical and literate homogenization of social organization. Europe is now getting a unity under the electric auspices of compression and interrelation. Just how much homogenization via literacy is needed to make an effective producer-consumer group in the post-mechanical age, in the age of automation, nobody has ever asked. For it has never been fully recognized that the role of literacy in shaping an industrial economy is basic and archetypal. Literacy is indispensable for habits of uniformity at all times and places. Above all, it is needed for the workability of price systems and markets. This factor has been ignored exactly as TV is now being ignored, for TV fosters many preferences that are quite at variance with literate uniformity and repeatability. It has sent Americans questing for every sort of oddment and quaintness in objects from out of their storied past. Many Americans will now spare no pains or expense to get to taste some new wine or food. The uniform and repeatable now must yield to the uniquely askew, a fact that is increasingly the despair and confusion of our entire standardized economy.

The power of the TV mosaic to transform American innocence into depth sophistication, independently of "content," is not mysterious if looked at directly. This mosaic TV image had already been adumbrated in the popular press that grew up with the telegraph. The commercial use of the telegraph began in 1844 in America, and earlier in England. The electric principle and its implications received much attention in Shelley's poetry. Artistic rule-of-thumb usually anticipates the science and technology in these matters by a full generation or more. The meaning of the telegraph mosaic in its *journalistic* manifestations was not lost to the mind of Edgar Allan Poe. He used it to establish two startlingly new inventions, the symbolist poem and the detective story. Both of these forms require do-it-yourself participation on the part of the reader. By offering an incomplete image or process, Poe *involved* his readers in the creative process in a way that Baudelaire, Valéry, T. S. Eliot, and many others have admired and followed. Poe had grasped at once the electric dynamic as one of public participation in creativity. Nevertheless, even today the homogenized consumer complains when asked to participate in creating or completing an abstract poem or painting or structure of any kind. Yet Poe knew even then that participation in depth followed at once from the telegraph mosaic. The more lineal and literal-minded of the literary brahmins "just couldn't see it." They still can't see it. They prefer not to participate in the creative process. They have accommodated themselves to the completed packages, in prose and verse and in the plastic arts. It is these people who must confront, in every classroom in the land, students who have accommodated themselves to the tactile and nonpictorial modes of symbolist and mythic structures, thanks to the TV image.

Life magazine for August 10, 1962, had a feature on how "Too Many Subteens Grow Up Too Soon and Too Fast." There was no observation of the fact that similar speed of growth and precociousness have always been the norm in tribal cultures and in nonliterate societies. England and America fostered the institution of prolonged adolescence by the negation of the tactile participation that is sex. In this, there was no conscious strategy, but rather a general acceptance of the consequences of prime stress on the printed word and visual values as a means of organizing personal and social life. This stress led to triumphs of industrial production and political conformity that were their own sufficient warrant.

Respectability, or the ability to sustain visual inspection of one's life, became dominant. No European country allowed print such precedence. Visually, Europe has always been shoddy in American eyes. American women, on the other hand, who have never been equaled in any culture for visual turnout, have always seemed abstract, mechanical dolls to Europeans. Tactility is a supreme value in European life. For that reason, on the Continent there is no adolescence, but only the leap from childhood to adult ways. Such is now the American state since TV, and this state of evasion of adolescence will continue. The introspective life of long, long thoughts and distant goals, to be pursued in lines of Siberian railroad kind, cannot coexist with the mosaic form of the TV image that commands immediate participation in *depth* and admits of no delays. The mandates of that image are so various yet so consistent that even to mention them is to describe the revolution of the past decade.

The phenomenon of the paperback, the book in "cool" version, can head this list of TV mandates, because the TV transformation of book culture into something else is manifested at that point. Europeans have had paperbacks from

431

the first. From the beginnings of the automobile, they have preferred the wraparound space of the small car. The pictorial value of "enclosed space" for book, car, or house has never appealed to them. The paperback, especially in its highbrow form, was tried in America in the 1920s and thirties and forties. It was not, however, until 1953 that it suddenly became acceptable. No publisher really knows why. Not only is the paperback a tactile, rather than a visual package; it can be as readily concerned with profound matters as with froth. The American since TV has lost his inhibitions and his innocence about depth culture. The paperback reader has discovered that he can enjoy Aristotle or Confucius by simply slowing down. The old literate habit of racing ahead on uniform lines of print yielded suddenly to depth reading. Reading in depth is, of course, not proper to the printed word as such. Depth probing of words and language is a normal feature of oral and manuscript cultures, rather than of print. Europeans have always felt that the English and Americans lacked depth in their culture. Since radio, and especially since TV, English and American literary critics have exceeded the performance of any European in depth and subtlety. The beatnik reaching out for Zen is only carrying the mandate of the TV mosaic out into the world of words and perception. The paperback itself has become a vast mosaic world in depth, expressive of the changed sense-life of Americans, for whom depth experience in words, as in physics, has become entirely acceptable, and even sought after.

Just where to begin to examine the transformation of American attitudes since TV is a most arbitrary affair, as can be seen in a change so great as the abrupt decline of baseball. The removal of the Brooklyn Dodgers to Los Angeles was a portent in itself. Baseball moved West in an attempt to retain an audience after TV struck. The characteristic mode of the baseball game is that it features one-thing-at-a-time. It is a lineal, expansive game which, like golf, is perfectly adapted to the outlook of an individualist and inner-directed society. Timing and waiting are of the essence, with the entire field in suspense waiting upon the performance of a single player. By contrast, football, basketball, and ice hockey are games in which many events occur simultaneously, with the entire team involved at the same time. With the advent of TV, such isolation of the individual performance as occurs in baseball became unacceptable. Interest in baseball declined, and its stars, quite as much as movie stars, found that fame had some very cramping dimensions. Baseball had been, like the movies, a hot medium featuring individual virtuosity and stellar performers. The real ball fan is a store of statistical information about previous explosions of batters and pitchers in numerous games. Nothing could indicate more clearly the peculiar satisfaction provided by a game that belonged to the industrial metropolis of ceaselessly exploding populations, stocks and bonds, and production and sales records. Baseball belonged to the age of the first onset of the hot press and the movie medium. It will always remain a symbol of the era of the hot mommas, jazz babies, of sheiks and shebas, of vamps and gold-diggers and the fast buck. Baseball, in a word, is a hot game that got cooled off in the new TV climate, as did most of the hot politicians and hot issues of the earlier decade.

There is no cooler medium or hotter issue at present than the small car. It is like a badly wired woofer in a hi-fi circuit that produces a tremendous flutter in the bottom. The small European car, like the European paperback and the European belle, for that matter, was no visual package job. Visually, the entire batch of European cars are so poor an affair that it is obvious their makers never thought of them as something to look at. They are something to put on, like pants or a pullover. Theirs is the kind of space sought by the skin-diver, the water-skier, and the dinghy sailor. In an immediate tactile sense, this new space is akin to that to which the picture-window fad had catered. In terms of "view," the picture window never made any sense. In terms of an attempt to discover a new dimension in the out-of-doors by pretending to be a goldfish, the picture window does make sense. So do the frantic efforts to roughen up the indoor walls and textures as if they were the outside of the house. Exactly the same impulse sends the indoor spaces and furniture out into the patios in an attempt to experience the outside as inside. The TV viewer is in just that role at all times. He is submarine. He is bombarded by atoms that reveal the outside as inside in an endless adventure amidst blurred images and mysterious contours.

However, the American car has been fashioned in accordance with the *visual* mandates of the typographic and the movie images. The American car was an enclosed space, not a tactile space. And an enclosed space, as was shown in the chapter on Print, is one in which all spatial qualities have been reduced to visual terms. So in the American car, as the French observed decades ago, "one is not on the road, one is in the car." By contrast, the European car aims to drag you along the road and to provide a great deal of vibration for the bottom. Brigitte Bardot got into the news when it was discovered that she liked to drive barefoot in order to get maximal vibration. Even English cars, weak on visual appearance as they are, have been guilty of advertising that "at sixty miles an hour all you can hear is the ticking of the clock." That would be a very poor ad, indeed, for a TV generation that has to be *with* everything and has to *dig* things in order to get at them. So avid is the TV viewer for rich tactile effects that he could be counted on to revert to skis. The wheel, so far as he is concerned, lacks the requisite abrasiveness.

Clothes in this first TV decade repeat the same story as vehicles. The revolution was heralded by bobby-soxers who dumped the whole cargo of visual effects for a set of tactile ones so extreme as to create a dead level of flat-footed deadpanism. Part of the cool dimension of TV is the cool, deadpan mug that came in with the teenager. Adolescence, in the age of hot media, of radio and movie, and of the ancient book, had been a time of fresh, eager, and expressive countenances. No elder statesman or senior executive of the 1940s would have ventured to wear so dead and sculptural a pan as the child of the TV age. The dances that came in with TV were to match—all the way to the Twist, which is merely a form of very unanimated dialogue, the gestures and grimaces of which indicate involvement in depth, but "nothing to say."

Clothing and styling in the past decade have gone so tactile and sculptural that they present a sort of exaggerated evidence of the new qualities of the TV mosaic. The TV extension of our nerves in hirsute pattern possesses the power to evoke a flood of related imagery in clothing, hairdo, walk, and gesture.

All this adds up to the compressional implosion — the return to nonspecialized forms of clothes and spaces, the seeking of multi-uses for rooms and things and objects, in a single word—the iconic. In music and poetry and painting, the tactile implosion means the insistence on qualities that are close to casual speech. Thus Schönberg and Stravinsky and Carl Orff and Bartok, far from being advanced seekers of esoteric effects, seem now to have brought music very close to the condition of ordinary human speech. It is this colloquial rhythm that once seemed so unmelodious about their work. Anyone who listens to the medieval works of Perotinus or Dufay will find them very close to Stravinsky and Bartok. The great explosion of the Renaissance that split musical instruments off from song and speech and gave them specialist functions is now being played backward in our age of electronic implosion.

One of the most vivid examples of the tactile quality of the TV image occurs in medical experience. In closedcircuit instruction in surgery, medical students from the first reported

435

a strange effect — that they seemed not to be watching an operation, but performing it. They felt that they were holding the scalpel. Thus the TV image, in fostering a passion for depth involvement in every aspect of experience, creates an obsession with bodily welfare. The sudden emergence of the TV medico and the hospital ward as a program to rival the western is perfectly natural. It would be possible to list a dozen untried kinds of programs that would prove immediately popular for the same reasons. Tom Dooley and his epic of Medicare for the backward society was a natural outgrowth of the first TV decade.

Now that we have considered the subliminal force of the TV image in a redundant scattering of samples, the question would seem to arise: "What possible *immunity* can there be from the subliminal operation of a new medium like television?" People have long supposed that bulldog opacity, backed by firm disapproval, is adequate enough protection against any new experience. It is the theme of this book that not even the most lucid understanding of the peculiar force of a medium can head off the ordinary "closure" of the senses that causes us to conform to the pattern of experience presented. The utmost purity of mind is no defense against bacteria, though the confreres of Louis Pasteur tossed him out of the medical profession for his base allegations about the invisible operation of bacteria. To resist TV, therefore, one must acquire the antidote of related media like print.

It is an especially touchy area that presents itself with the question: "What has been the effect of TV on our political life?" Here, at least, great traditions of critical awareness and vigilance testify to the safeguards we have posted against the dastardly uses of power.

When Theodore White's *The Making of the President: 1960* is opened at the section on "The Television Debates," the TV student will experience dismay. White offers statistics on the number of sets in American homes and the number of hours of daily use of these sets, but not one clue as to the nature of the TV image or its effects on candidates or viewers. White considers the "content" of the debates and the deportment of the debaters, but it never occurs to him to ask why TV would inevitably be a disaster for a sharp intense image like Nixon's, and a boon for the blurry, shaggy texture of Kennedy.

At the end of the debates, Philip Deane of the London *Observer* explained my idea of the coming TV impact on the election to the *Toronto Globe and Mail* under the headline of "The Sheriff and the Lawyer," October 15, 1960. It was that TV would prove so entirely in Kennedy's favor that he would win the election. Without TV, Nixon had it made. Deane, toward the end of his article, wrote:

> Now the press has tended to say that Mr. Nixon has been gaining in the last two debates and that he was bad in the first. Professor McLuhan thinks that Mr. Nixon has been sounding progressively more definite; regardless of the value of the Vice-President's views and principles, he has been defending them with too much flourish for the TV medium. Mr. Kennedy's rather sharp responses have been a mistake, but he still presents an image closer to the TV hero, Professor McLuhan says — something like the shy young Sheriff — while Mr. Nixon with his very dark eyes that tend to stare, with his slicker circumlocution, has resembled more the railway lawyer who signs leases that are not in the interests of the folks in the little town.
>
> In fact, by counterattacking and by claiming for himself, as he does in the TV debates, the same goals as the Democrats have, Mr. Nixon may be helping his opponent by blurring the Kennedy image, by confusing what exactly it is that Mr. Kennedy wants to change.
>
> Mr. Kennedy is thus not handicapped by clear-cut issues; he is visually a less well-defined image, and

appears more nonchalant. He seems less anxious to sell himself than does Mr. Nixon. So far, then, Professor McLuhan gives Mr. Kennedy the lead without underestimating Mr. Nixon's formidable appeal to the vast conservative forces of the United States.

Another way of explaining the acceptable, as opposed to the unacceptable, TV personality is to say that anybody whose *appearance* strongly declares his role and status in life is wrong for TV. Anybody who looks as if he might be a teacher, a doctor, a businessman, or any of a dozen other things all at the same time is right for TV. When the person presented *looks* classifiable, as Nixon did, the TV viewer has nothing to fill in. He feels uncomfortable with his TV image. He says uneasily, "There's something about the guy that isn't right." The viewer feels exactly the same about an exceedingly pretty girl on TV, or about any of the intense "high definition" images and messages from the sponsors. It is not accidental that advertising has become a vast new source of comic effects since the advent of TV. Mr. Khrushchev is a very filled-in or completed image that appears on TV as a comic cartoon. In wirephoto and on TV, Mr. Khrushchev is a jovial comic, an entirely disarming presence. Likewise, precisely the formula that recommends anybody for a movie role disqualifies that same person for TV acceptance. For the hot movie medium needs people who look very definitely a type of some kind. The cool TV medium cannot abide the typical because it leaves the viewer frustrated of his job of "closure" or completion of image. President Kennedy did not look like a rich man or like a politician. He could have been anything from a grocer or a professor to a football coach. He was not too precise or too ready of speech in such a way as to spoil his pleasantly tweedy blur of countenance and outline. He went from palace to log cabin, from wealth to the White House, in a pattern of TV reversal and upset.

The same components will be found in any popular TV

figure. Ed Sullivan, "the great stone face," as he was known from the first, has the much needed harshness of texture and general sculptural quality demanded for serious regard on TV. Jack Paar is quite otherwise—neither shaggy nor sculptural. But on the other hand, his presence is entirely acceptable on TV because of his utterly cool and casual verbal agility. The Jack Paar show revealed the inherent need of TV for spontaneous chat and dialogue. Jack discovered how to extend the TV mosaic image into the entire format of his show, seemingly snaffling up just anybody from anywhere at the drop of a hat. In fact, however, he understood very well how to create a mosaic from other media, from the world of journalism and politics, books, Broadway, and the arts in general, until he became a formidable rival to the press mosaic itself. As "Amos and Andy" had lowered church attendance on Sunday evenings in the old days of radio, so Jack Paar certainly cut night-club patronage with his late show.

How about Educational Television? When the three-year-old sits watching the President's press conference with Dad and Grandad, that illustrates the serious educational role of TV. If we ask what is the relation of TV to the learning process, the answer is surely that the TV image, by its stress on participation, dialogue, and depth, has brought to America new demand for crash-programming in education. Whether there ever will be TV in every classroom is a small matter. The revolution has already taken place at home. TV has changed our sense-lives and our mental processes. It has created a taste for all experience *in depth* that affects language teaching as much as car styles. Since TV, nobody is happy with a mere book knowledge of French or English poetry. The unanimous cry now is, "Let's *talk* French," and "Let the bard be *heard*." And oddly enough, with the demand for depth, goes the demand for crash-programming. Not only deeper, but further, into all knowledge has become the normal popular demand since TV. Perhaps enough has been said about the nature of the TV image to explain why this should be. How could it possibly pervade our lives any more than it does? Mere classroom use could not extend its influence. Of course, in the classroom its role compels a reshuffling of subjects, and approaches to subjects. Merely to put the present classroom on TV would be like putting movies on TV. The result would be a hybrid that is neither. The right approach is to ask, "What can TV do that the classroom cannot do for French, or for physics?" The answer is: "TV can illustrate the interplay of process and the growth of forms of all kinds as nothing else can."

The other side of the story concerns the fact that, in the visually organized educational and social world, the TV child is an underprivileged cripple. An oblique indication of this startling reversal has been given by William Golding's *Lord of the Flies*. On the one hand, it is very flattering for hordes of docile children to be told that, once out of the sight of their governesses, the seething savage passions within them would boil over and sweep away pram and playpen, alike. On the other hand, Mr. Golding's little pastoral parable does have some meaning in terms of the psychic changes in the TV child. This matter is so important for any future strategy of culture or politics that it demands a headline prominence, and capsulated summary:

WHY THE TV CHILD CANNOT SEE AHEAD

The plunge into depth experience via the TV image can only be explained in terms of the differences between visual and mosaic space. Ability to discriminate between these radically different forms is quite rare in our Western world. It has been pointed out that, in the country of the blind, the one-eyed man is not king. He is taken to be an hallucinated lunatic. In a highly visual culture, it is as difficult to communicate the nonvisual properties of spatial forms as to explain visuality to the blind. In *The ABC of Relativity* Bertrand Russell began by explaining that there is nothing difficult about Einstein's ideas, but that they do call for total reorganization of our imaginative lives. It is precisely this imaginative reorganization that has occurred via the TV image.

The ordinary inability to discriminate between the photographic and the TV image is not merely a crippling factor in the learning process today; it is symptomatic of an age-old failure in Western culture. The literate man, accustomed to an environment in which the visual sense is extended everywhere as a principle of organization, sometimes supposes that the mosaic world of primitive art, or even the world of Byzantine art, represents a mere difference in degree, a sort of failure to bring their visual portrayals up to the level of full visual effectiveness. Nothing could be further from the truth. This, in fact, is a misconception that has impaired understanding between East and West for many centuries. Today it impairs relations between colored and white societies.

Most technology produces an amplification that is quite explicit in its separation of the senses. Radio is an extension of the aural, high-fidelity photography of the visual. But TV is, above all, an extension of the sense of touch, which involves maximal interplay of all the senses. For Western man, however, the all-embracing extension had occurred by means of phonetic writing, which is a technology for extending the sense of sight. All non-phonetic forms of writing are, by contrast, artistic modes that retain much variety of sensuous orchestration. Phonetic writing, alone, has the power of separating and fragmenting the senses and of sloughing off the semantic complexities. The TV image reverses this literate process of analytic fragmentation of sensory life.

The visual stress on continuity, uniformity, and connectedness, as it derives from literacy, confronts us with the great technological means of implementing continuity and lineality by fragmented repetition. The ancient world found this means

441

in the brick, whether for wall or road. The repetitive, uniform brick, indispensable agent of road and wall, of cities and empires, is an extension, via letters, of the visual sense. *The brick wall is not a mosaic form*, and neither is the mosaic form a visual structure. The mosaic can be *seen* as dancing can, but is not *structured* visually; nor is it an extension of the visual power. For the mosaic is not uniform, continuous, or repetitive. It is discontinuous, skew, and nonlineal, like the tactual TV image. To the sense of touch, all things are sudden, counter, original, spare, strange. The "Pied Beauty" of G. M. Hopkins is a catalogue of the notes of the sense of touch. The poem is a manifesto of the nonvisual, and like Cézanne or Seurat, or Rouault it provides an indispensable approach to understanding TV. The nonvisual mosaic structures of modern art, like those of modern physics and electric-information patterns, permit little detachment. The mosaic form of the TV image demands participation and involvement in depth of the whole being, as does the sense of touch. Literacy, in contrast, had, by extending the visual power to the uniform organization of time and space, psychically and socially, conferred the power of detachment and noninvolvement.

The visual sense when extended by phonetic literacy fosters the analytic habit of perceiving the single facet in the life of forms. The visual power enables us to isolate the single incident in time and space, as in representational art. In visual representation of a person or an object, a single phase or moment or aspect is separated from the multitude of known and felt phases, moments and aspects of the person or object. By contrast, iconographic art uses the eye as we use our hand in seeking to create an inclusive image, made up of many moments, phases, and aspects of the person or thing. Thus the iconic mode is not visual representation, nor the specialization of visual stress as defined by viewing from a single position. The tactual mode of perceiving is sudden but not specialist. It is total, synesthetic, involving all the senses. Pervaded by the mosaic TV image, the TV child encounters the world in a spirit antithetic to literacy.

The TV image, that is to say, even more than the icon, is an extension of the sense of touch. Where it encounters a literate culture, it necessarily thickens the sense-mix, transforming fragmented and specialist extensions into a seamless web of experience. Such transformation is, of course, a "disaster" for a literate, specialist culture. It blurs many cherished attitudes and procedures. It dims the efficacy of the basic pedagogic techniques, and the relevance of the curriculum. If for no other reason, it would be well to understand the dynamic life of these forms as they intrude upon us and upon one another. TV makes for myopia.

The young people who have experienced a decade of TV have naturally imbibed an urge toward involvement in depth that makes all the remote visualized goals of usual culture seem not only unreal but irrelevant, and not only irrelevant but anemic. It is the total involvement in all-inclusive *nowness* that occurs in young lives via TV's mosaic image. This change of attitude has nothing to do with programming in any way, and would be the same if the programs consisted entirely of the highest cultural content. The change in attitude by means of relating themselves to the mosaic TV image would occur in any event. It is, of course, our job not only to understand this change but to exploit it for its pedagogical richness. The TV child expects involvement and doesn't want a specialist *job* in the future. He does want a *role* and a deep commitment to his society. Unbridled and misunderstood, this richly human need can manifest itself in the distorted forms portrayed in *West Side Story*.

The TV child cannot see ahead because he wants involvement, and he cannot accept a fragmentary and merely visualized goal or destiny in learning or in life.

Murder by Television

Jack Ruby shot Lee Oswald while tightly surrounded by guards who were paralyzed by television cameras. The fascinating and involving power of television scarcely needed this additional proof of its peculiar operation upon human perceptions. The Kennedy assassination gave people an immediate sense of the television power to create depth involvement, on the one hand, and a numbing effect as deep as grief, itself, on the other hand. Most people were amazed at the depth of meaning which the event communicated to them. Many more were surprised by the coolness and calm of the mass reaction. The same event, handled by press or radio (in the absence of television), would have provided a totally different experience. The national "lid" would have "blown off." Excitement would have been enormously greater and depth participation in a common awareness very much less.

As explained earlier, Kennedy was an excellent TV image. He had used the medium with the same effectiveness that Roosevelt had learned to achieve by radio. With TV, Kennedy found it natural to involve the nation in the office of the Presidency, both as an operation and as an image. TV reaches out for the corporate attributes of office. Potentially, it can transform the Presidency into a monarchic dynasty. A merely elective Presidency scarcely affords the depth of dedication and commitment demanded by the TV form. Even teachers on TV seem to be endowed by the student audiences with a charismatic or mystic character that much exceeds the feelings developed in the classroom or lecture hall. In the course of many studies of audience reactions to TV teaching, there recurs this puzzling fact. The viewers feel that the teacher has a dimension of almost sacredness. This feeling does not have its basis in concepts or ideas, but seems to creep in uninvited and unexplained. It baffles both the students and the analysts of their reactions. Surely, there could be no more telling touch to tip us off to the character of TV. This is not so much a visual as a tactual-auditory medium that involves all of our senses in depth interplay. For people long accustomed to the merely visual experience of the typographic and photo-graphic varieties, it would seem to be the *synesthesia*, or tactual depth of TV experience, that dislocates them from their usual attitudes of passivity and detachment.

The banal and ritual remark of the conventionally literate, that TV presents an experience for passive viewers, is wide of the mark. TV is above all a medium that demands a creatively participant response. The guards who failed to protect Lee Oswald were not passive. They were so involved by the mere sight of the TV cameras that they lost their sense of their merely practical and specialist task.

Perhaps it was the Kennedy funeral that most strongly impressed the audience with the power of TV to invest an occasion with the character of corporate participation. No national event except in sports has ever had such coverage or such an audience. It revealed the unrivaled power of TV to achieve the involvement of the audience in a complex *process*. The funeral as a corporate process caused even the image of sport to pale and dwindle into puny proportions. The Kennedy funeral, in short, manifested the power of TV to involve an entire population in a ritual process. By comparison, press, movie and even radio are mere packaging devices for consumers.

Most of all, the Kennedy event provides an opportunity for noting a paradoxical feature of the "cool" TV medium. It involves us in moving depth, but it does not excite, agitate or arouse. Presumably, this is a feature of all depth experi-ence.

32

Weapons:
War of the Icons

Retaining the emphasis from the preceding chapter on the primary role of perception in the evolution of media, McLuhan notes that it was "the lineal stress of perspective that had channeled perception in paths that led to the creation of gunfire." War is characterized as an instance of equilibrium-seeking. And to the reader who may be seeking an explanation of how weapons take their place among other media, McLuhan offers the observation that all technology can be regarded as weapons in the age of electric hegemony.

— *(Editor)*

When the Russian girl Valentina Tereshkova, quite without pilot training, went into orbit on June 16, 1963, her action, as reacted to in the press and other media, was a kind of defacing of the images of the male astronauts, especially the Americans. Shunning the expertise of American astronauts, all of whom were qualified test pilots, the Russians don't seem to feel that space travel is related enough to the airplane to require a pilot's "wings." Since our culture forbids the sending of a woman into orbit, our only repartee would have been to launch into orbit a group of space children, to indicate that it is, after all, child's play.

The first sputnik or "little fellow-traveler" was a witty taunting of the capitalist world by means of a new kind of technological image or icon, for which a group of children in orbit might yet be a telling retort. Plainly, the first lady astronaut is offered to the West as a little Valentine-or heart throb — suited to our sentimentality. In fact, the war of the icons, or the eroding of the collective countenance of one's rivals, has long been under way. Ink and photo are supplanting soldiery and tanks. The pen daily becomes mightier than the sword.

The French phrase "*guerre des nerfs*" of twenty-five years ago has since come to be referred to as "the cold war." It is really an electric battle of information and of images that goes far deeper and is more obsessional than the old hot wars of industrial hardware.

The "hot" wars of the past used weapons that knocked off the enemy, one by one. Even ideological warfare in the eighteenth and nineteenth centuries proceeded by persuading individuals to adopt new points of view, one at a time. Electric persuasion by photo and movie and TV works, instead, by dunking entire populations in new imagery. Full awareness of this technological change had dawned on Madison

Avenue ten years ago when it shifted its tactics from the promotion of the individual product to the collective involvement in the "corporate image," now altered to "corporate posture."

Parallel to the new cold war of information exchange is the situation commented on by James Reston in a *New York Times* release from Washington:

> Politics has gone international. The British Labor Leader is here campaigning for Prime Minister of Britain, and fairly soon John F. Kennedy will be over in Italy and Germany campaigning for reelection. Everybody's now whistle-stopping through somebody else's country, usually ours.
>
> Washington has still not adjusted to this third-man role. It keeps forgetting that anything said here may be used by one side or another in some election campaign, and that it may, by accident, be the decisive element in the final vote.

If the cold war in 1964 is being fought by informational technology, that is because all wars have been fought by the latest technology available in any culture. In one of his sermons John Donne commented thankfully on the blessing of heavy firearms:

> So by the benefit of this light of reason they have found out *Artillery*, by which warres come to quicker ends than heretofore …

The scientific knowledge needed for the use of gunpowder and the boring of cannon appeared to Donne as "the light of reason." He failed to notice another advance in the same technology that hastened and extended the scope of human slaughter. It is referred to by John U. Nef in *War and Human Progress*:

> The gradual abandonment or armor as a part of the
> equipment of soldiers during the seventeenth cen-
> tury freed some metal supplies for the manufacture
> of firearms and missiles.

It is easy to discover in this a seamless web of interwoven
events when we turn to look at the psychic and social consequences of the technological extensions of man.

Back in the 1920s King Amanullah seems to have put his
finger on this web when he said, after firing off a torpedo:

"I feel half an Englishman already."

The same sense of the relentlessly interwoven texture of
human fate was touched by the schoolboy who said:

"Dad, I hate war."

"Why, son?"

"Because war makes history, and I hate history."

The techniques developed over the centuries for drilling
gun-barrels provided the means that made possible the steam
engine. The piston shaft and the gun presented the same
problems in boring hard steel. Earlier, it had been the lineal
stress of perspective that had channeled perception in paths
that led to the creation of gunfire. Long before guns, gunpowder had been used explosively, dynamite style. The use
of gunpowder for the propelling of missiles in trajectories
waited for the coming of perspective in the arts. This liaison of
events between technology and the arts may explain a matter
that has long puzzled anthropologists. They have repeatedly
tried to explain the fact that nonliterates are generally poor
shots with rifles, on the grounds that, with the bow and arrow,
proximity to game was more important than distant accuracy,
which was almost impossible to achieve—hence, say some
anthropologists, their imitation of hunted beasts by dressing
in skins to get close to the herd. It is also pointed out that bows
are silent, and when an arrow missed, animals rarely fled.

If the arrow is an extension of the hand and the arm, the

rifle is an extension of the eye and teeth. It may be to the point to remark that it was the literate American colonists who were first to insist on a rifled barrel and improved gun-sights. They improved the old muskets, creating the Kentucky rifle. It was the highly literate Bostonians who outshot the British regulars. Marksmanship is not the gift of the native or the woodsman, but of the literate colonist. So runs this argument that links gunfire itself with the rise of perspective, and with the extension of the visual power in literacy. In the Marine Corps it has been found that there is a definite correlation between education and marksmanship. Not for the nonliterate is our easy selection of a separate, isolated target in space, with the rifle as an extension of the eye.

If gunpowder was known long before it was used for guns, the same is also true of the use of the lodestone or magnet. Its use in the compass for lineal navigation had, also, to wait for the discovery of lineal perspective in the arts. Navigators took a long time to accept the possibility of space as uniform, connected and continuous. Today in physics, as in painting and sculpture, progress consists in giving up the idea of space as either uniform, continuous, or connected. Visuality has lost its primacy.

In the Second World War the marksman was replaced by automatic weapons fired blindly in what were called "perimeters of fire" or "fire lanes." The old-timers fought to retain the bolt-action Springfield which encouraged single-shot accuracy and sighting. Spraying the air with lead in a kind of tactual embrace was found to be good by night, as well as by day, and sighting was unnecessary. At this stage of technology, the literate man is somewhat in the position of the old-timers who backed the Springfield rifle against perimeter fire. It is this same visual habit that deters and obstructs literate man in modern physics, as Milic Capek explains in *The Philosophical Impact of Modern Physics*. Men in the older oral societies of middle Europe are better able to conceive the nonvisual velocities and relations of the subatomic world.

Our highly literate societies are at a loss as they encounter the new structures of opinion and feeling that result from instant and global information. They are still in the grip of "points of view" and of habits of dealing with things one at a time. Such habits are quite crippling in any electric structure of information movement, yet they could be controlled if we recognized whence they had been acquired. But literate society thinks of its artificial visual bias as a thing natural and innate.

Literacy remains even now the base and model of all programs of industrial mechanization; but, at the same time, it locks the minds and senses of its users in the mechanical and fragmentary matrix that is so necessary to the maintenance of mechanized society. That is why the transition from mechanical to electric technology is so very traumatic and severe for us all. The mechanical techniques, with their limited powers, we have long used as weapons. The electric techniques cannot be used aggressively except to end all life at once, like the turning off of a light. To live with both of these technologies at the same time is the peculiar drama of the twentieth century.

In his *Education Automation*, R. Buckminster Fuller considers that weaponry has been a source of technological advance for mankind because it requires continually improved performance with ever smaller means. "As we went from the ships of the sea to the ships of the air, the performance per pound of the equipment and fuel became of even higher importance than on the sea."

It is this trend toward more and more power with less and less hardware that is characteristic of the electric age of information. Fuller has estimated that in the first half century of the airplane the nations of the world have invested two and a half trillion dollars by subsidy of the airplane as a weapon. He added that this amounts to sixty-two times the value of all the gold in the world. His approach to these problems is more technological than the approach of historians, who have often tended to find that war produces nothing new in the way of invention.

"This man will teach us how to beat him," Peter the Great is said to have remarked after his army had been beaten by Charles XII of Sweden. Today, the backward countries can learn from us how to beat us. In the new electric age of information, the backward countries enjoy some specific advantages over the highly literate and industrialized cultures. For backward countries have the habit and understanding of oral propaganda and persuasion that was eroded in industrial societies long ago. The Russians had only to adapt their traditions of Eastern icon and image-building to the new electric media in order to be aggressively effective in the modern world of information. The idea of the Image, that Madison Avenue has had to learn the hard way, was the only idea available to Russian propaganda. The Russians have not shown imagination or resourcefulness in their propaganda. They have merely done that which their religious and cultural traditions taught them; namely, to build images.

The city, itself, is traditionally a military weapon, and is a collective shield or plate armor, an extension of the castle of our very skins. Before the huddle of the city, there was the food-gathering phase of man the hunter, even as men have now in the electric age returned psychically and socially to the nomad state. Now, however, it is called information-gathering and data-processing. But it is global, and it ignores and replaces the form of the city which has, therefore, tended to become obsolete. With instant electric technology, the globe itself can never again be more than a village, and the very nature of city as a form of major dimensions must inevitably dissolve like a fading shot in a movie. The first circumnavigation of the globe in the Renaissance gave men a sense of embracing and possessing the earth that was quite new, even as the recent astronauts have again altered man's relation to the planet, reducing its scope to the extent of an evening's stroll.

The city, like a ship, is a collective extension of the castle of our skins, even as clothing is an extension of our individual skins. But weapons proper are extensions of hands, nails, and teeth, and come into existence as tools needed for accelerating the processing of matter. Today, when we live in a time of sudden transition from mechanical to electric technology, it is easier to see the character of all previous technologies, we being detached from all of them for the time being. Since our new electric technology is not an extension of our bodies but of our central nervous systems, we now see all technology, including language, as a means of processing experience, a means of storing and speeding information. And in such a situation all technology can plausibly be regarded as weapons. Previous wars can now be regarded as the processing of difficult and resistant materials by the latest technology, the speedy dumping of industrial products on an enemy market to the point of social saturation. War, in fact, can be seen as a process of achieving equilibrium among unequal technologies, a fact that explains Toynbee's puzzled observation that each invention of a new weapon is a disaster for society, and that militarism itself is the most common cause of the breaking of civilizations.

By militarism, Rome extended civilization or individualism, literacy, and lineality to many oral and backward tribes. Even today the mere existence of a literate and industrial West appears quite naturally as dire aggression to nonliterate societies; just as the mere existence of the atom bomb appears as a state of universal aggression to industrial and mechanized societies.

On the one hand, a new weapon or technology looms as a threat to all who lack it. On the other hand, when everybody has the same technological aids, there begins the competitive fury of the homogenized and the egalitarian pattern against

455

which the strategy of social class and caste has often been used in the past. For caste and class are techniques of social slow-down that tend to create the stasis of tribal societies. Today we appear to be poised between two ages—one of detribalization and one of retribalization.

> Between the acting of a dreadful thing,
> And the first motion, all the interim is
> Like a Phantasma, or a hideous Dream:
> The genius and the mortal instruments
> Are then in council; and the state of man,
> Like to a little Kingdom, suffers then
> The nature of an insurrection.

> (*Julius Caesar*, Brutus II, i)

If mechanical technology as extension of parts of the human body had exerted a fragmenting force, psychically and socially, this fact appears nowhere more vividly than in mechanical weaponry. With the extension of the central nervous system by electric technology, even weaponry makes more vivid the fact of the unity of the human family. The very inclusiveness of information as a weapon becomes a daily reminder that politics and history must be recast in the form of "the concretization of human fraternity."

This dilemma of weaponry appears very clearly to Leslie Dewart in his *Christianity and Revolution*, when he points to the obsolescence of the fragmented balance-of-power techniques. As an instrument of policy, modern war has come to mean "the existence and end of one society to the exclusion of another." At this point, weaponry is a self-liquidating fact.

33

Automation:
Learning a Living

Automation is characterized as electric logic, *the logic that dictates the abolition of many dichotomies, including that between culture and technology. Key topics from early chapters are reprised and integrated here: light (it is energy and information), electricity (it stores and moves perception and information), the reversal principle (it is a clue to understanding). Important qualifiers continue to appear (the* mass *of* mass media *refers not to to the massive size of communication empires but to involvement of the masses;* feedback *"means introducing an information loop or circuit, where before there had been merely a one-way flow or mechanical sequence"). McLuhan speaks of "the shock of unfamiliarity in the familiar that is necessary for the understanding of the life of forms." This is an invitation to relish the situation of the farmer of Irish legend, napping so long in a corner of his own field that the grass grows up and prevents him from knowing where he is. His plight is the opposite of* déjà vu. *But* déjà vu *brings the comfort of recognition without conferring the power of cognition. McLuhan urges that we prefer* déjà new *to* déjà vu, *because the life of forms is constantly renewed by new interactions, and meaning is meshing.*

— *(Editor)*

A newspaper headline recently read, "Little Red Schoolhouse Dies When Good Road Built." One-room schools, with all subjects being taught to all grades at the same time, simply dissolve when better transportation permits specialized spaces and specialized teaching. At the extreme of speeded-up movement, however, specialism of space and subject disappears once more. With automation, it is not only jobs that disappear, and complex roles that reappear. Centuries of specialist stress in pedagogy and in the arrangement of data now end with the instantaneous retrieval of information made possible by electricity. Automation is information and it not only ends jobs in the world of work, it ends subjects in the world of learning. It does not end the world of learning. The future of work consists of learning a living in the automation age. This is a familiar pattern in electric technology in general. It ends the old dichotomies between culture and technology, between art and commerce, and between work and leisure. Whereas in the mechanical age of fragmentation leisure had been the absence of work, or mere idleness, the reverse is true in the electric age. As the age of information demands the simultaneous use of all our faculties, we discover that we are most at leisure when we are most intensely involved, very much as with the artists in all ages.

In terms of the industrial age, it can be pointed out that the difference between the previous mechanical age and the new electric age appears in the different kinds of inventories. Since electricity, inventories are made up not so much of goods in storage as of materials in continuous process of transformation at spatially removed sites. For electricity not only gives primacy to *process*, whether in making or in learning, but it makes independent the source of energy from the location of the process. In entertainment media, we speak of this fact as "mass media" because the source of the program

459

and the process of experiencing it are independent in space, yet simultaneous in time. In industry this basic fact causes the scientific revolution that is called "automation" or "cybernation."

In education the conventional division of the curriculum into subjects is already as outdated as the medieval trivium and quadrivium after the Renaissance. Any subject taken in depth at once relates to others subjects. Arithmetic in grade three or nine, when taught in terms of number theory, symbolic logic, and cultural history, ceases to be mere practice in problems. Continued in their present patterns of fragmented unrelation, our school curricula will insure a citizenry unable to understand the cybernated world in which they live.

Most scientists are quite aware that since we have acquired some knowledge of electricity it is not possible to speak of atoms as pieces of matter. Again, as more is known about electrical "discharges" and energy, there is less and less tendency to speak of electricity as a thing that "flows" like water through a wire, or is "contained" in a battery. Rather, the tendency is to speak of electricity as painters speak of space; namely, that it is a variable condition that involves the special positions of two or more bodies. There is no longer any tendency to speak of electricity as "contained" in anything. Painters have long known that objects are not contained in space, but that they generate their own spaces. It was the dawning awareness of this in the mathematical world a century ago that enabled Lewis Carroll, the Oxford mathematician, to contrive *Alice in Wonderland*, in which times and spaces are neither uniform nor continuous, as they had seemed to be since the arrival of Renaissance perspective. As for the speed of light, that is merely the speed of total causality.

It is a principal aspect of the electric age that it establishes a global network that has much of the character of our central nervous system. Our central nervous system is not merely an electric network, but it constitutes a single unified field of experience. As biologists point out, the brain is the interacting place where all kinds of impressions and experiences can be exchanged and translated, enabling us to *react to the world as a whole*. Naturally, when electric technology comes into play, the upmost variety and extent of operations in industry and society quickly assume a unified posture. Yet this organic unity of interprocess that electromagnetism inspires in the most diverse and specialized areas and organs of action is quite the opposite of organization in a mechanized society. Mechanization of any process is achieved by fragmentation, beginning with the mechanization of writing by movable types, which has been called "the monofracture of manufacture."

The electric telegraph, when crossed with typography, created the strange new form of the modern newspaper. Any page of the telegraph press is a surrealistic mosaic of bits of "human interest" in vivid interaction. Such was the art form of Chaplin and the early silent movies. Here, too, an extreme speedup of mechanization, an assembly line of still shots on celluloid, led to a strange reversal. The movie mechanism, aided by the electric light, created the illusion of organic form and movement as much as a fixed position had created the illusion of perspective on a flat surface five hundred years before.

The same thing happens less superficially when the electric principle crosses the mechanical lines of industrial organization. Automation retains only as much of the mechanical character as the motorcar kept of the forms of the horse and the carriage. Yet people discuss automation as if we had not passed the oat barrier, and as if the horse-vote at the next poll would sweep away the automation regime.

Automation is not an extension of the mechanical principles of fragmentation and separation of operations. It is rather the invasion of the mechanical world by the instantaneous

461

character of electricity. That is why those involved in automation insist that it is a way of thinking, as much as it is a way of doing. Instant synchronization of numerous operations has ended the old mechanical pattern of setting up operations in lineal sequence. The assembly line has gone the way of the stag line. Nor is it just the lineal and sequential aspect of mechanical analysis that has been erased by the electric speedup and exact synchronizing of information that is automation.

Automation or cybernation deals with all the units and components of the industrial and marketing process exactly as radio or TV combine the individuals in the audience into new interprocess. The new kind of interrelation in both industry and entertainment is the result of the electric instant speed. Our new electric technology now extends the instant processing of knowledge by interrelation that has long occurred within our central nervous system. It is that same speed that constitutes "organic unity" and ends the mechanical age that had gone into high gear with Gutenberg. Automation brings in real "mass production," not in terms of size, but of an instant inclusive embrace. Such is also the character of "mass media." They are an indication, not of the size of their audiences, but of the fact that everybody becomes involved in them at the same time. Thus commodity industries under automation share the same structural character of the entertainment industries in the degree that both approximate the condition of instant information. Automation affects not just production, but every phase of consumption and marketing; for the consumer becomes producer in the automation circuit, quite as much as the reader of the mosaic telegraph press makes his own news, or just *is* his own news.

But there is a component in the automation story that is as basic as tactility to the TV image. It is the fact that, in any automatic machine, or galaxy of machines and functions, the generation and transmission of power is quite separate from the work operation that uses the power. The same is true in all servo-mechanist structures that involve feedback. The source of energy is separate from the process of translation of information, or the applying of knowledge. This is obvious in the telegraph, where the energy and channel are quite independent of whether the written code is French or German. The same separation of power and process obtains in automated industry, or in "cybernation." The electric energy can be applied indifferently and quickly to many kinds of tasks.

Such was never the case in the mechanical systems. The power and the work done were always in direct relation, whether it was hand and hammer, water and wheel, horse and cart, or steam and piston. Electricity brought a strange elasticity in this matter, much as light itself illuminates a total field and does not dictate what shall be done. The same light can make possible a multiplicity of tasks, just as with electric power. Light is a nonspecialist kind of energy or power that is identical with information and knowledge. Such is also the relation of electricity to automation, since both energy and information can be applied in a great variety of ways.

Grasp of this fact is indispensable to the understanding of the electronic age, and of automation in particular. Energy and production now tend to fuse with information and learning. Marketing and consumption tend to become one with learning, enlightenment, and the intake of information. This is all part of the electric *implosion* that now follows or succeeds the centuries of *explosion* and increasing specialism. The electronic age is literally one of illumination. Just as light is at once energy and information, so electric automation unites production, consumption, and learning in an inextricable process. For this reason, teachers are already the largest employee group in the U.S. economy, and may well become the *only* group.

The very same process of automation that causes a withdrawal of the present work force from industry causes learning itself to become the principal kind of production and consumption. Hence the folly of alarm about unemployment. Paid learning is already becoming both the dominant employment and the source of new wealth in our society. This is the new *role* for men in society, whereas the older mechanistic idea of "jobs," or fragmented tasks and specialist slots for "workers," becomes meaningless under automation.

It has often been said by engineers that, as information levels rise, almost any sort of material can be adapted to any sort of use. This principle is the key to the understanding of electric automation. In the case of electricity, as energy for production becomes independent of the work operation, there is not only the speed that makes for total and organic interplay, but there is, also, the fact that electricity is sheer information that, in actual practice, illuminates all it touches. Any process that approaches instant interrelation of a total field tends to raise itself to the level of conscious awareness, so that computers seem to "think." In fact, they are highly specialized at present, and quite lacking in the full process of interrelation that makes for consciousness. Obviously, they can be made to simulate the process of consciousness, just as our electric global networks now begin to simulate the condition of our central nervous system. But a conscious computer would still be one that was an extension of our consciousness, as a telescope is an extension of our eyes, or as a ventriloquist's dummy is an extension of the ventriloquist.

Automation certainly assumes the servomechanism and the computer. That is to say, it assumes electricity as store and expediter of information. These traits of store, or "memory," and accelerator are the basic features of any medium of communication whatever. In the case of electricity, it is not corporeal substance that is stored or moved, but perception and information. As for technological acceleration, it now approaches the speed of light. All nonelectric media had merely hastened things a bit. The wheel, the road, the ship, the airplane, and even the space rocket are utterly lacking in the character of instant movement. Is it strange, then, that electricity should confer on all previous human organization a completely new character? The very toil of man now becomes a kind of enlightenment. As unfallen Adam in the Garden of Eden was appointed the task of the contemplation and naming of creatures, so with automation. We have now only to name and program a process or a product in order for it to be accomplished. Is it not rather like the case of Al Capp's Schmoos? One had only to look at a Schmoo and think longingly of pork chops or caviar, and the Schmoo ecstatically transformed itself into the object of desire. Automation brings us into the world of the Schmoo. The custom-built supplants the mass-produced.

Let us, as the Chinese say, move our chairs closer to the fire and see what we are saying. The electric changes associated with automation have nothing to do with ideologies or social programs. If they had, they could be delayed or controlled. Instead, the technological extension of our central nervous system that we call the electric media began more than a century ago, subliminally. Subliminal have been the effects. Subliminal they remain. At no period in human culture have men understood the psychic mechanisms involved in invention and technology. Today it is the instant speed of electric information that, for the first time, permits easy recognition of the patterns and the formal contours of change and development. The entire world, past and present, now reveals itself to us like a growing plant in an enormously accelerated movie. Electric speed is synonymous with light and with the understanding of causes. So, with the use of electricity in previously mechanized situations, men easily discover causal connections and patterns that were quite unobservable at the slower rates of mechanical change. If we play backward the long development of literacy and printing and their effects on social experience and organization, we can easily see how these forms brought about that high degree of social uniformity and homogeneity of society that is indispensable for mechanical industry. Play them backward, and we get just that shock of unfamiliarity in the familiar that is necessary for the understanding of the life of forms. Electricity compels us to play our mechanical development backward, for it reverses much of that development. Mechanization depends on the breaking up of processes into homogenized but unrelated bits. Electricity unifies these fragments once more because its speed of operation requires a high degree of interdependence among all phases of any operation. It is this electric speedup and interdependence that has ended the assembly line in industry.

This same need for organic interrelation, brought in by the electric speed of synchronization, now requires us to perform, industry-by-industry, and country-by-country, exactly the same organic interrelating that was first effected in the individual automated unit. Electric speed requires organic structuring of the global economy quite as much as early mechanization by print and by road led to the acceptance of national unity. Let us not forget that nationalism was a mighty invention and revolution that, in the Renaissance, wiped out many of the local regions and loyalties. It was a revolution achieved almost entirely by the speedup of information by means of uniform movable types. Nationalism cut across most of the traditional power and cultural groupings that had slowly grown up in various regions. Multi-nationalisms had long deprived Europe of its economic unity. The Common Market came to it only with the Second War. War is accelerated social change, as an explosion is an accelerated chemical reaction and movement of matter. With electric speeds governing industry and social life, explosion in the sense of crash development becomes normal. On the other hand, the old-fashioned kind of "war" becomes as impracticable as playing hopscotch with bulldozers. Organic interdependence means that disruption of any part of the organism can prove fatal to the whole. Every industry has had to "rethink through" (the awkwardness of this phrase betrays the painfulness of the process), function by function, its place in the economy. But automation forces not only industry and town planners, but government and even education, to come into some relation to social facts.

The various military branches have had to come into line with automation very quickly. The unwieldy mechanical forms of military organization have gone. Small teams of experts have replaced the citizen armies of yesterday even faster than they have taken over the reorganization of industry. Uniformly trained and homogenized citizenry, so long in preparation and so necessary to a mechanized society, is becoming quite a burden and problem to an automated society, for automation and electricity require depth approaches in all fields and at all times. Hence the sudden rejection of standardized goods and scenery and living and education in America since the Second War. It is a switch imposed by electric technology in general, and by the TV image in particular.

Automation was first felt and seen on a large scale in the chemical industries of gas, coal, oil, and metallic ores. The large changes in these operations made possible by electric energy have now, by means of the computer, begun to invade every kind of white-collar and management area. Many people, in consequence, have begun to look on the whole of society as a single unified machine for creating wealth. Such has been the normal outlook of the stockbroker, manipulating shares and information with the cooperation of the electric media of press, radio, telephone, and teletype. But the peculiar and abstract manipulation of information as a means of creating wealth is no longer a monopoly of the stockbroker. It is now shared by every engineer and by the entire communications industries. With electricity as energizer and synchronizer, all aspects of production, consumption, and organization become incidental to communications. The very idea of communication as interplay is inherent in the electrical, which combines both energy and information in its intensive manifold.

Anybody who begins to examine the patterns of automation finds that perfecting the individual machine by making it automatic involves "feedback." That means introducing an information loop or circuit, where before there had been merely a one-way flow or mechanical sequence. Feedback is the end of the lineality that came into the Western world with the alphabet and the continuous forms of Euclidean space. Feedback or dialogue between the mechanism and its environment brings a further weaving of individual machines into a galaxy of such machines throughout the entire plant. There follows a still further weaving of individual plants and factories into the entire industrial matrix of materials and services of a culture. Naturally, this last stage encounters the entire world of policy, since to deal with the whole industrial complex as an organic system affects employment, security, education, and politics, demanding full understanding in advance of coming structural change. There is no room for witless assumptions and subliminal factors in such electrical and instant organizations.

As artists began a century ago to construct their works backward, *starting with the effect*, so now with industry and planning. In general, electric speedup requires complete knowledge of ultimate effects. Mechanical speedups, however radical in their reshaping of personal and social life, still were allowed to happen sequentially. Men could, for the most part, get through a normal life span on the basis of a single set of skills. That is not at all the case with electric speedup. The acquiring of new basic knowledge and skill by senior executives in middle age is one of the most common needs and harrowing facts of electric technology. The senior executives, or "big wheels," as they are archaically and ironically designated, are among the hardest pressed and most persistently harassed groups in human history. Electricity has not only demanded ever deeper knowledge and faster interplay, but has made the harmonizing of production schedules as rigorous as that demanded of the members of a large symphony orchestra. And the satisfactions are just as few for the big executives as for the symphonists, since a player in a big orchestra can hear nothing of the music that reaches the audience. He gets only noise.

The result of electric speedup in industry at large is the creation of intense sensitivity to the interrelation and interprocess of the whole, so as to call for ever-new types of organization and talent. Viewed from the old perspectives of the machine age, this electric network of plants and processes seems brittle and tight. In fact, it is not mechanical, and it does begin to develop the sensitivity and pliability of the human organism. But it also demands the same varied nutriment and nursing as the animal organism.

With the instant and complex interprocesses of the organic form, automated industry also acquires the power of adaptability to multiple uses. A machine set up for the automatic production of electric bulbs represents a combination of processes that were previously managed by several machines. With a single attendant, it can run as continuously as a tree in its intake and output. But, unlike the tree, it has a built-in system of jigs and fixtures that can be shifted to cause the machine to turn out a whole range of products from radio tubes and glass tumblers to Christmas-tree ornaments. Although an automated plant is almost like a tree in respect to the continuous intake and output, it is a tree that can change from oak to maple to walnut as required. It is part of the automation or electric logic that specialism is no longer limited to just one specialty. The automatic machine may work in a specialist way, but it is not limited to one line. As with our hands and fingers that are capable of many tasks, the automatic unit incorporates a power of adaptation that was quite lacking in the pre-electric and mechanical stage of technology. As *any*thing becomes more complex, it becomes less specialized. Man is more complex and less specialized than a dinosaur. The older mechanical operations were designed to be more efficient as they became larger and more specialized. The electric and automated unit, however, is quite otherwise. A new automatic machine for making automobile tailpipes is about the size of two or three office desks. The computer control panel is the size of a lectern. It has in it no dies, no fixtures, no settings of any kind, but rather certain general-purpose things like grippers, benders, and advancers. On this machine, starting with lengths of ordinary pipe, it is possible to make eighty different kinds of tailpipe in succession, as rapidly, as easily, and as cheaply as it is to make eighty of the same kind. And the characteristic of electric automation is all in this direction of return to the general-purpose handicraft flexibility that our own hands possess. The programming can now include endless changes of program. It is the electric feedback, or dialogue pattern, of the automatic and computer-programmed "machine" that marks it off from the older mechanical principle of one-way movement.

This computer offers a model that has the characteristics shared by all automation. From the point of intake of materials to the output of the finished product, the operations tend to be independently, as well as interdependently, automatic. The synchronized concert of operations is under the control of gauges and instruments that can be varied from the control-panel boards that are themselves electronic. The material of intake is relatively uniform in shape, size, and chemical properties, as likewise the material of the output. But the processing under these conditions permits use of the highest level of capacity for any needed period. It is, as compared with the older machines, the difference between an oboe in an orchestra and the same tone on an electronic music instrument. With the electronic music instrument, any tone can be made available in any intensity and for any length of time. Note that the older symphony orchestra was, by comparison, a machine of separate instruments that *gave the effect of organic unity*. With the electronic instrument, one *starts* with organic unity as an immediate fact of perfect synchronization. This makes the attempt to create the effect of organic unity quite pointless. Electronic music must seek other goals.

Such is also the harsh logic of industrial automation. All that we had previously achieved mechanically by great exertion and coordination can now be done electrically without effort. Hence the specter of joblessness and propertylessness in the electric age. Wealth and work become information factors, and totally new structures are needed to run a business or relate it to social needs and markets. With the electric technology, the new kinds of instant interdependence and interprocess that take over production also enter the market and social organizations. For this reason, markets and education designed to cope with the products of servile toil and mechanical production are no longer adequate. Our education has long ago acquired the fragmentary and piecemeal character of mechanism. It is now under increasing pressure to acquire the depth and interrelation that are indispensable in the all-at-once world of electric organization.

Paradoxically, automation makes liberal education mandatory. The electric age of servomechanisms suddenly releases men from the mechanical and specialist servitude of the preceding machine age. As the machine and the motorcar released the horse and projected it onto the plane of entertainment, so does automation with men. We are suddenly threatened with a liberation that taxes our inner resources of self-employment and imaginative participation in society. This would seem to be a fate that calls men to the role of artist in society. It has the effect of making most people realize how much they had come to depend on the fragmentalized and repetitive routines of the mechanical era. Thousands of years ago man, the nomadic food-gatherer, had taken up positional, or relatively sedentary, tasks. He began to specialize. The development of writing and printing were major stages of that process. They were supremely specialist in separating the roles of knowledge from the roles of action, even though at times it could appear that "the pen is mightier than the sword." But with electricity and automation, the technology of fragmented processes suddenly fused with the human dialogue and the need for over-all consideration of human unity. Men are suddenly nomadic gatherers of knowledge, nomadic as never before, informed as never before, free from fragmentary specialism as never before—but also involved in the total social process as never before; since with electricity we extend our central nervous system globally, instantly inter-relating every human experience. Long accustomed to such a state in stock-market news or front-page sensations, we can grasp the meaning of this new dimension more readily when it is pointed out that it is possible to "fly" unbuilt airplanes on computers. The specifications of a plane can be programmed and the plane tested under a variety of extreme conditions before it has left the drafting board. So with new products and new organizations of many kinds. We can now, by computer, deal with complex social needs with the same architectural certainty that we previously attempted in private housing. Industry as a whole has become the unit of reckoning, and so with society, politics, and education as wholes.

Electric means of storing and moving information with speed and precision make the largest units quite as manage-able as small ones. Thus the automation of a plant or of an entire industry offers a small model of the changes that must occur in society from the same electric technology. Total interdependence is the starting fact. Nevertheless, the range of choice in design, stress, and goal within that total field of electromagnetic interprocess is very much greater than it ever could have been under mechanization.

Since electric energy is independent of the place or kind of work-operation, it creates patterns of decentralism and diversity in the work to be done. This is a logic that appears plainly enough in the difference between firelight and electric light, for example. Persons grouped around a fire or candle for warmth or light are less able to pursue independent thoughts, or even tasks, than people supplied with electric light. In the same way, the social and educational patterns latent in automation are those of self-employment and artistic autonomy. Panic about automation as a threat of uniformity on a world scale is the projection into the future of mechanical standardization and specialism, which are now past.

Appendix

Marshall McLuhan:
Report on Project in Understanding New Media

Preface: W. Terrence Gordon

Soon after assuming the presidency of the University of Toronto in 1958, Claude Bissell became aware that an English professor toiling in a remote corner of the campus was beginning to enjoy a reputation for his work on media. Speaking at convocation at Wayne State University in Detroit, Bissell invited questions from his audience and was asked if he could summarize McLuhan's ideas. There is no record of how Bissell rose to this challenge, but in replying to the same question in years to come, McLuhan himself emphasized that he had neither a theory nor a point of view — only percepts and probes.

As Bissell spoke in Detroit, McLuhan was in Toronto, gearing up for sabbatical. He was pondering various projects, when good news arrived and settled the choice for him. The U.S.-based National Association of Educational Broadcasters (NAEB) had approved McLuhan's proposal for a study of media. One evaluator had declared McLuhan a genius in his vision and a poet by temperament. And so, NAEB project #69 was launched; McLuhan referred to the project as "Vat 69," a cauldron where he tossed in as many ideas as he could seize, hoping eventually to turn his study into a book. Later, his enthusiastic but frustrated editors for the original edition of *Understanding Media* would innocently believe that he also intended to give his subject a more orderly treatment.

The NAEB set McLuhan to designing a teaching method and syllabus for introducing secondary school students to the nature and effects of media. With sponsorship from both the NAEB and the United States Office of Education, McLuhan's research focused on interviews with educators and a program of media-testing conducted with the help of the Ryerson Polytechnic Institute, Toronto (see text following). He began his work as soon as the project was approved, traveling to Philadelphia to confer on project design with Harry Skornia, then professor of radio and television broadcasting at the University of Illinois at Urbana and president of the NAEB, and with author Gilbert Seldes.

It was clear that the project would absorb McLuhan fully, a final report scheduled for submission at the end of his sabbatical year in June 1960. But he had a sample syllabus ready (see text following) before the end of 1959. As an objective for prospective courses, he gives priority to examining the interaction of media, the nature of print, and new electronic technologies. He points out that achieving increased awareness of the forms of media is not an end in itself but provides a means of anticipation and control over powerful new environments. This was to become a key notion in *Understanding Media*.

Like the book into which it would grow, McLuhan's model syllabus aims at overview and integration, while attempting to focus on particular media. One reader evaluating an early draft commented that "media" was not in the average teacher's vocabulary and needed introduction and explanation. Where McLuhan writes about print learnt in childhood as a potential obstacle to learning in subjects other than music the same evaluator cautioned him that a backlash from teachers was all but inevitable and urged him to ease up on that point. But McLuhan was not about to soft-pedal the very insight that had launched him on investigating media effects.

McLuhan's correspondence during his 1959-60 sabbatical reveals his preoccupation with the ideas that his media project was generating. Many of his correspondents during this period were persons he interviewed in connection with the project, including his longtime friend Bernie Muller-Thym, with whom McLuhan shared new ideas as soon as they hatched. Among these was the high-definition/low-definition contrast as a basic defining feature of media (see Chapter 2), presented to Muller-Thym through an illuminating example tracing the evolution of a single medium, the road (see Chapter 10) from country path to city scar. McLuhan's principle of reversal, as a feature of both the operation and effect of a medium, a linchpin of his thinking throughout his writings, is linked here, for the first time, to the high-definition/low-definition contrast. With Muller-Thym and Harry Skornia, McLuhan shared and explored what he considered to be the crucial breakthroughs carrying him toward an overview of media. In correspondence with poet and English professor Wilfred Watson, who would be his collaborator for *From Cliché to Archetype*, he emphasized the engagement of artists with technology (see *inter alia* references to Charles Baudelaire, T. S. Eliot, James Joyce, Wyndham Lewis, Pablo Picasso, and Georges Seurat in Name Index) and opened questions relating to the scholarship of Northrop Frye, questions he and Watson would examine in their book.

As a sop to impatient readers, McLuhan stated in the final version of *Report on Project in Understanding New Media* that all his recommendations could be reduced to the imperative to scrutinize modes of media, in order to reveal the assumptions they impose and the perceptual habits they foster. Such scrutiny is not far removed from the essential technique that McLuhan learned at Cambridge from F. R. Leavis and I. A. Richards. But McLuhan evoked his Cambridge mentors in only the subtlest of ways in a section of his NAEB report titled "The New Criticism and the New Media."

He linked work in American linguistics to his original media probe, noting that Robert A. Hall's notion of the organizing pattern imposed on experience by language was, in effect, the same fundamental idea encapsulated in 'the medium is the message." This is a long way from what "the medium is the message" meant when McLuhan said it to reassure worried radio broadcasters that their medium was not threatened with extinction by television. The new interpretation is also distinct from the protean formulations the same probe would undergo throughout the 1960s and '70s. It is much closer to the thesis of Chapter 4 in Ogden and Richards's *The Meaning of Meaning*, "Signs in Perception"— the notion that all perception demands interpretation — an idea related to McLuhan's observations on the transformations of sensory input into sensory closure. (see Glossary and Subject Index)

Report on Project in Understanding New Media gave much prominence to television and the discoveries that McLuhan considered to be among the most important results of his sabbatical research. His findings about the medium itself are crucially tied to those that emerged from the Ryerson experiment with respect to the educational value of the medium.

McLuhan noted that throughout his year-long project he had found it easier to talk with business people than with educators. Not content to make this an incidental remark, he took the occasion to unsettle his audience, castigating educators for insulating themselves against cultural change by taking shelter in bureaucratic structures. He was the first to benefit from his own research, as he made plain in a section of his report entitled "What I Learned" (see text below). It is also clear that McLuhan believed he had made discoveries about media effects that compelled a program of further research and writing that would go beyond tackling hidebound educators. And so, within four years, *Report on Project in Understanding New Media* became *Understanding Media*.

The first extract from *Report on Project in Understanding New Media* that follows is the complete text describing the Ryerson experiment in comparative evaluation of media effectiveness in teaching. The second is the complete sample syllabus for the integration of media studies into secondary school curricula. Those portions of *Report* which constitute first drafts of material revised as the chapters of *Understanding Media* have been omitted.

— *W. Terrence Gordon*

The Ryerson Media Experiment

The Ryerson Media Experiment in the maximized testing of the media was made possible by the following people:

> *A. Roy Low, Department of Physics*
> *Carl Williams, Department of Psychology*
> *Isabel Macbeth, School of Radio & Television*
> *James Peters, Department of English*
> *Gerald Kane, Depart of Radio*
> *William Sokira, Department of Radio*
> *Geofrey Jamieson, Department of Television*

Mass Media and Learning — an Experiment

Introduction

A seminar on culture and communication has frequent cause to concern itself with the mass media. The experiment here reported was the culmination of our first year effort. While in a very real sense an interdisciplinary product, the responsibility for the design, analysis and presentation of results fell to the psychologists in the seminar as being most familiar with the techniques involved.

Most research on mass media is concerned with either of two objectives: studies of the influence of one medium on attitude changes, and consumer research designed ultimately to help sell soap or whatnot. Little if any work has been done on the degree to which various media facilitate or impede learning, if indeed they have any influence at all. The question does not occur readily because the mass media themselves are seldom seen as educational devices. The silent assumption that mass media exist primarily for entertainment and propaganda, which underlies most such research, automatically excludes research with an educational bias.

Problem

In its most general form, the problem investigated can be stated thus: Is learning affected by the channel over which information comes? If so, how and to what extent? While we usually assume that television, for instance, is more compelling than radio in securing our attention, we also assume that we can easily compensate psychologically for this differential advantage. Whenever our attention is really aroused, we can and do attend to the radio address, news or weather report with the firm conviction that we will end up with all the information we require. An extra effort of attention, we assume, will easily make up for the fact that we could have gleaned the same information with less effort over television.

With these considerations in mind, the experiment was designed to provide the "same" information in the identical wording, to four similar audiences, each of which had the "same" motivation to seek out and remember the information presented. Given the same objective examination on that information, would the only systematic difference remaining, namely the different media used, make a statistically significant difference to the average scores of those audiences? Television and radio were obvious choices for an experiment on mass communication. Since they are often contrasted with "real" situations, a "live" lecture audience was added. The fourth medium chosen was the printed page since it is widely regarded as the essential carrier of Culture

—with a capital C—and is most often thought of as being threatened by the newer media in terms of its continued existence.

Design

From the standpoint of design, all that was required was that the factual content be clearly transmitted without undue distortion over each of the four media and that it be cast in such a way that no one medium was favoured over others. The method employed was the method of constant stimuli whereby the lecturer himself provides the stimulus without reliance on the peculiarities of particular medium "props". The fact that his gestures, intonations, etc., are differently transmitted by the different media is precisely the point of investigation. That is, since each medium carries the information *in its own way*, do these differences affect the learning process of the audience?

The subjects were 108[1] undergraduates in the General Course in Arts at the University of Toronto, all of whom were studying anthropology as one of five courses comprising their year's work. The lecture topic "Thinking Through Language", was unfamiliar to them, and from their point of view, both difficult and stimulating. The class list was arranged in descending order of academic grades, based on first year results, and then arbitrarily divided into four groups or audiences on a stratified sampling basis, such that each audience contained an equal number of high, average and low students. For this purpose, 'high' means grades of A and B+ , 'average' means grades of B and B-, 'low' means grades of C+ and below. After the four audiences had been selected in this way, another

1. Actually the number was larger, but to make the groups as equal as possible and to make the classification on previous academic standing clear, the final number was reduced to 108.

person arbitrarily assigned each audience to a medium. These were announced to the students on arrival at the CBC studios. Each group went to a separate room in the CBC buildings where they were supervised by members of the seminar. The lecture was delivered before the studio audience and simultaneously relayed to the television audience and the radio audience. At the same time, mimeographed copies of the lecture were distributed to the reading group, who read at their own speed and for the same length of time as it took to deliver the lecture. Immediately thereafter, each group wrote a thirty minute examination on the lecture. This consisted of twenty multiple-choice questions (four alternatives each), plus one broad essay type question to be answered in 200-300 words. Most students finished before the nominal time limit. The test should therefore be regarded as a 'power' rather than a 'speed' test. The papers were graded by the anthropology section of the seminar and turned over to the psychology section for analysis.

Here is a section from the lecture and its covering question:

> I recall one experience I had several years ago while living with the Eskimos. I was riding along on a dog sled one bitterly cold day — the wind hit me in the back and seemed to come out the other side — when I turned to a hunter with me and said, as best I could in Eskimo, 'The wind is cold.' He roared with laughter. 'How,' he asked, 'can the *wind* be cold? *You're* cold, *you're* unhappy. But the *wind* isn't cold or unhappy!' Now this involves more than just another way of speaking; it involves another way of *seeing* things. Consider how different human action must appear when seen through the filter of the Eskimo language where, owing to the lack of transitive or action verbs, it is likely to be perceived as a sort of happening without an active element in it. In Eskimo one cannot say: 'I *kill* him' or 'I *shoot* the arrow', but

only 'He dies to me', 'The arrow is flying away from me', just as 'I *hear*' is 'me-sound-is'. Similarly where we say, 'The lightning *flashed*' as if the lightning *did* something, as if it involved something more than just being lightning the Eskimo merely says 'Flash'. Eskimo philosophers, if there were any, would be likely to say that what we call action is really a pattern of succeeding impressions.

When we say, 'The lightning flashed' we:

    a   read action into the event
    b   use an intransitive verb
    c   describe the event as being without action
    d   describe the event in the only possible way

The essay question called for an understanding of the whole lecture: 'The lecturer described two native philosophies, but at the same time said that the Eskimos, for example, had no philosophers. How would you interpret these two statements in terms of the lecture?'

Controls

It is a truism that whereas the 19th century public sought to learn the *results* of science, the 20th century public is, more realistically, interested in the *methods* whereby results are achieved. For this reason, if for no other, some discussion of the controls used in this study forms an essential part of this report.

The term 'control' itself is a highly ambiguous term, as our seminar quickly learned. As used here, it means only those measures which were taken to hold constant all factors, other than the four media themselves, which might be expected to bias the results.

It does not mean that experiments of this type are totalitarian, that social scientists are dictators at heart, that science scorns understanding and seeks only prediction and control, that our subjects were humiliated or 'pushed around' without their consent, that we laboured under the illusion of playing God with other people's lives, that the study was undertaken to fool, bully, delude, hoax or otherwise cajole an innocent group of students.

In terms of controls, the lecturer was his own control. His choice of topic and his organization was his own. The controls were first, that his information be basically accessible via each channel and second, that it should not rely on external 'props' of any kind. Finally, and most difficult of all, the lecture had to be memorized so that the reading group would receive exactly the same content as the other audiences. In order to compensate, as far as possible, for the fact that the reading audience was deprived of both the sound and sight of the lecturer, certain key words in the mimeographed material were capitalized to give something of the same emphasis they received as delivered.

The subjects were selected to be as homogeneous as possible, i.e., same course, same class and age range, and sharing a common subject matter. Academic ability was controlled by the method of stratified sampling described above, since it was a fair assumption that good students in general learn more than poor ones, even in lectures.

Motivation was controlled by an arrangement with the class instructor who agreed to incorporate performance on our examination into the course term mark. In order to avoid undue anxiety, the arrangement was that those who did well would get a term mark bonus, while those who did indifferently or poorly would suffer no penalty. These factors also operated to produce a good attendance at the studio, and to offset, if not entirely eliminate, factors of personal preference for one or another medium. In addition, the students were fully informed about the experiment and its objectives, and afterwards, were the first group to hear an analysis of the results.

No attempt was made to equate groups for age, sex, socio-economic status, familiarity with television, radio, etc. These were assumed to be roughly controlled (i.e., equated) by random assignment to each group.

The examination was controlled by the use of the objectives, multiple-choice type of question, which permits of easy quantification. The score on this section was simply the number right. It should be noted that since each question contained a best answer among the four alternatives presented, the measure yielded is a measure of immediate *recognition*, not recall.

No note-taking was allowed during the lecture, in an attempt to stimulate normal conditions of television and radio listening. Whereas the lecturer automatically 'paced' the studio, television and radio audiences, thereby conferring a precise degree of control on them, it was not possible to duplicate this pacing for the reading audience. In this sense, this group was not as well controlled as the others.

Results

The results given here are confined to an analysis of the multiple-choice section of the examination. The statistics used were the analysis of variance and the 't' test of significance[2] of differences between the means, i.e., averages.

The analysis of variance showed that media in general *do* make a significant difference in the amount learned as measured by the multiple-choice test. It also showed, as we suspected, that academic ability makes a significant difference in the amount learned. Having established the fact that the

---

2. Statistically speaking, a difference between two averages is called significant if it could not have occurred by chance more often than 5 times in 100 occurrences. Therefore, the betting in this study is that we have 95 chances in 100 of being sure that the differences obtained are 'real' differences and not due to chance. In some cases we have 99 chances in 100 of being right.

four media *per se* were significant to the learning process, it was then possible to test the audience averages for significance of difference in order to rank them in effectiveness. This analysis showed that the television average was superior to the radio average—at the 1% level of confidence (i.e., there are 99 chances in 100 that this is a true difference). It also showed radio to be significantly above both the reading and studio performances—significant at the 5% level of confidence (i.e., there are 95 chances in 100 that this is a true difference).

The graph shows the examination results by audiences and by academic ability, shown here at three levels. This display is more revealing than the averages for each medium, since it shows how the media affect each level of academic ability. The clearest indications come from the television, radio, and reading comparisons, where it can be seen that the media exert their effects at all three academic levels. Note for example that the low students on television do exactly as well as the middle students on radio, a clear instance of medium effect. Note too, that the greatest single discrepancy on the graph occurs between the good students on television and radio. Apparently television has its greatest effect on the best students!

The studio results are puzzling. The 'lows' and 'highs' reflect presumably the distractions and excitement of the studio itself, but if they do, why were the middles unaffected by this to the point that they did as well as the television middle group? Originally the studio group was proposed as the equivalent of a lecture audience. One glance at the confusion of the television studio convinced us, *before* the statistical analysis, that whatever this group was, it was *not* a lecture group. We retained it in the study but with the new name 'studio' group.

The table beneath the chart below shows the number of cases (N), and the averages for each audience together with the confidence level at which the differences can be accepted as significant.

Conclusions and Comments

One experiment does not establish a generalization, but it is plain that under these conditions at least, certain of the mass media, and notably television, are very effective channels for conveying information. The astonishing feature of the study is the relatively poor performance of the reading group. Many members of the seminar predicted it to be the best of the four! One feature involved in these standings became clear from an examination of the results of a single question. In one portion of the talk the lecturer stressed gestures and delivery to accent his words; the question covering this passage was accurately answered by most of the television audience, half the radio audience and few of the reading audience.

It is then fair to conclude that media do make a difference in immediate recognition, using undergraduates as subjects. It is also fair to rank the media from television through radio to reading in terms of their effectiveness *under these conditions*. No conclusion is drawn on the studio group.

At this stage of research, generalization is dangerous. The study does not prove that television is 'better' than radio or that either is preferable to books, or that 'live' audiences learn little. Would one get similar results with housewives, with engineers or even with these same students presented with a totally unfamiliar topic, say, the devolution of estates in Athenian law? Would persons of average or below average intelligence react in the same way? Would children? These and a host of similar questions suggest that at least an interesting and important area of research has been tapped by this exploration.

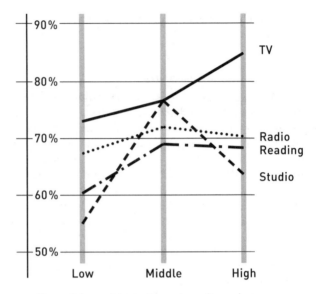

N = 108 (27 subjects in each audience)
each 'x' represents the average of 9 subjects

Academic Levels of Students

Audience	N	Mean	Difference significant at
TV	27	77.2%	1% level
Radio	27	69.2%	5% level
Reading	27	65.1%	not
Studio	27	64.9%	significant

Table showing average scores by media

492

Mass Media and Retention

We have now described in detail the experiment conducted by a Communications Seminar at the University of Toronto in February, 1954, to test learning via various media. The experiment has since been summarized in a number of journals, but unfortunately this interest has not always been accompanied by understanding.

In October, 1954, the original test was re-administrated to the 74 students available of the 108 who took part in the first experiment. They were unaware beforehand that a re-examination was intended.

The multiple choice questionnaire used had 20 items, each with four alternative answers. It could be expected that a group knowing nothing of the subject would get 25% of the answers correct by guessing. Our subjects, however, were university Arts students who had taken courses in the social sciences and who could be assumed to do better than chance, even if they had neither seen nor heard the lecture. To check this, a control group was used: the questionnaire was given to 15 Second Year Honour psychology students, selected because, though their general training in social sciences was similar to the experimental groups', the lecture was unknown to them and they had received no instruction from the lecturer.

During the eight-month interval some students had heard the lecture a second time when its kinescope was shown on television to the general public and some may have discussed it with friends. It is assumed here, however, that such reinforcement was random. To determine whether or not the 74 students re-examined were representative of the original 108, we compared the performance means of the two groups (Figure 1). Although in each case the retest means were slightly lower, these differences were fairly uniform, not great, and therefore, not fatal.

It was anticipated that on the average the re-examination marks would be significantly lower than the ones obtained on the first test. It was further assumed that if the media did not continue to influence the retention of learning over time, there would be no 'real' differences on the re-test among the four groups who received this lecture through different media. If significant differences were found on re-examination among the groups, this could be fairly attributed to differential effects of the media through which the information was originally obtained. The results of the two tests for the 74 students are shown in Figure 1, broken down into the four groups, each of which was exposed to one medium. For comparison, the mean percentages of the original four groups are also given.

Two questions were considered: Were the media differences demonstrated by the first experiment still demonstrable after eight months? Did the media have a differential effect on forgetting during this period?

An affirmative answer to the first question was obtained by an analysis of variance of the means of the groups on the second test. It was found that there were still significant differences (i.e., could have occurred by chance only once in 100 times) between them. Unfortunately, it was not possible to analyze these results in more detail since there were unequal losses in subjects in the four groups, so that further comparisons were not statistically justifiable. The answer to the first question is, however, clear: after eight months significant differences exist between the groups exposed to the different media. The results showed there had been one change in the order of ranking the four media: the studio group moved from last to second place. The results from this group were regarded with doubt in the original experiment and were not included in the conclusions; no interpretation is now made of this change in rank.

In order to answer the second question, it is necessary to compare the differences between the first and second tests for each group. The losses here when tested by analysis of variance could not be considered to differ significantly from each other. This implies that the amount retained after a period of time is proportional to the amount originally learned. In other words, the rate of forgetting information is independent of the medium by means of which it was acquired. This was graphically demonstrated, in that the original ranking of media in order of effectiveness—television, radio, print—held after eight months.

Since it was found that for every group the mean percentage for the second test was significantly lower than for the first test, a third question was asked. If, after eight months, the students have in general lower scores, how much better than intelligent guessing are their second test results? This is answered in Figure 1 in a comparison of their results with those of the control group of psychology students. Their scores are better than random guessing, but significantly lower than the lowest of the four media groups.

In this particular experiment media made a difference in learning, not only in immediate recall, but after eight months. The original order of effectiveness—television, radio, print— held after this interval. In this experiment, different media influenced retention by influencing the amount of original learning.

The qualifications given in "Mass Media and Learning — An Experiment" about misinterpretations of the original findings apply to these later findings as well.

	Re-Test Group			Original Group	
	Test % First	Test % Second		Test % First	
Television	75.4	61.5	*14*	77.2	*27*
Radio	65.5	52.6	*21*	69.2	*27*
Studio	62.9	56.0	*21*	64.9	*27*
Reading	63.9	47.5	*18*	65.1	*27*
			N = 74 *objects*	*N = 108* *objects*	

In this repeat performance, pains were taken to allow
each medium full play of its possibilities with reference to
the subjects, just as in the earlier experiment each medium
was neutralized as much as possible. Only the mimeograph
form remained the same in each experiment. Here we added
a printed form in which an imaginative typographical layout

was followed. The lecturer used the blackboard and permitted discussion. Radio and TV employed dramatization, sound effects and graphics. In the examination, radio easily topped TV (see Chart 2). Yet, as in the first experiment, both radio and TV manifested a decisive advantage over the lecture and written forms. As a conveyor both of ideas and information, TV was, in this sound experiment, apparently enfeebled by the deployment of its dramatic resources, whereas radio benefited from such lavishness. 'Technology is explicitness', writes Lyman Bryson. Are both radio and TV more explicit than writing or lecture? Would a greater explicitness, if inherent in these media, account for the ease with which they top other modes of performance?

Announcement of the results of the first experiment evoked considerable interest. Advertising agencies circulated the results with the comment that here, at last, was scientific proof of the superiority of TV. This was unfortunate and missed the main point, for the results didn't indicate the superiority of one medium over others. They merely directed attention toward differences between them, difference so great as to be of kind rather than degree. Some CBC officials were furious, not because TV won, but because print lost. Scratch most and you find Student Christian-types who understand little of literature and contribute less, but, like publishers, have a vested interest in book culture. At heart they hate radio and TV, which they employ merely to disseminate the values of book culture. They feel they should dedicate themselves to *serious* culture. This is why they can't use radio and TV with conviction and are afraid to use it comically, and so they end up with wishy-washy. They are like 16th century scholars who saw the book revolution as simply a means of propagating old ideas and failed to realize it was a monumental change in sensibility, in thinking and feeling.

Official culture still strives to force the new languages to do the work of the old. But the horseless carriage didn't do the work of the horse; it abolished the horse and did what the horse could never do. Horses are fine. So are books.

Nobody yet knows the languages inherent in the new technological culture; we are all deaf-blind mutes in terms of the new situation. Our most impressive words and thoughts betray us by referring to the previously existent, not to the present.

The problem has been falsely seen as democracy vs. the mass media. But the mass media are democracy. The book itself was the first mechanical mass medium. What is really being asked, of course, is: can books' monopoly of knowledge survive the challenge of the new languages? The answer is, no. What should be asked is: What can print do better than any other medium and is that worth doing?

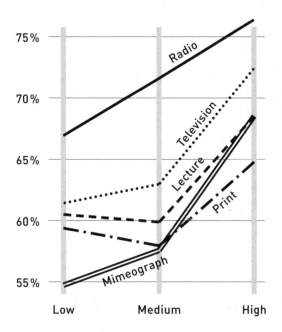

Thinking through Language

I recall one experience I had several years ago while living with the Eskimos. I was riding along on a dog sled one bitterly cold day—the wind hit me in the back and seemed to come out the other side—when I turned to a hunter with me and said, as best I could in Eskimo, "The wind is cold." He roared with laughter. "How," he said, "can the *wind* be cold? *You're* cold, *you're* unhappy, but the *wind* isn't cold or unhappy!"

Now this involves more than just another way of speaking, it involves another way of *seeing* things. Consider how different human actions must appear when seen through the filter of the Eskimo language where, owing to the lack of transitive or action verbs, it is likely to be perceived as a sort of happening without an active element in it. In Eskimo one cannot say: "I *kill* him or I *shoot* the arrow," but only "He dies to me," "The arrow is flying away from me," just as "I *hear*" is "me-sound-is". Similarly, where we say, "The lightning *flashed*," as if the lightning *did* something, as if it involved something more than just being lightning, the Eskimo merely says "flash". Eskimo philosophers, if there were any, would be likely to say that what we call action is really a pattern of succeeding impressions. Such differences between languages—I don't mean Indo-European ones like French and German but native ones—are really tremendous. Some languages lack tenses. Of course, I really shouldn't say "lack" because this implies a deficiency, and there is nothing deficient about these languages. In fact some can express ideas that English cannot.

Take metaphors: in English when we want to express an emotional or philosophical experience, we have no choice but to use words which refer to real objects or real actions. For example, I might say: I grasp the thread of your argument, but if its level is over my head my imagination may wander. How can my imagination wander? Most native languages, on the other hand, distinguish between inner psychological experiences and those that belong to the world of matter.

Let's take the language of the Trobrianders, a group of Pacific Islanders who live not far from New Guinea. Two famous anthropologists, Bronislaw Malinowski and Dorothy Lee, studied these people, and we probably know more about them than any other native group.

Now the Trobrianders are concerned with being, and only with being (nature, essence, existence). Change and becoming are foreign to their thinking. An object or event is grasped and evaluated in terms of itself alone, that is, irrespective of other beings.

The Trobriander usually refers to it by a word and one word only. To describe it would be redundant. We define an object in terms of what it is like or what it is unlike or what it can do. The Trobriander is interested only in what it is. To Trobrianders each object is usually grasped timelessly. We must describe it in terms of past, present, future, but, for the Trobriander, these distinctions are non-existent. A word for an object implies the existence of all the qualities it incorporates. If I were to go with a Trobriander to a garden where the *taytu* (a type of yam) had just been harvested, I would come back and tell you: "There are good taytu there, just the right degree of ripeness, large and perfectly shaped; not a blight to be seen, not one rotten spot; nicely rounded at the tips, with no spiky points; all first-run harvesting; no second gleanings." The Trobriander would come back and say TAYTU, and he would have said all that I said and more. Even the phrase "there are taytu" would be repetitious since existence is implied in being.

And all the attributes, even if he could find words for them in his own language, would have been repetitious since the concept of *taytu* contains them all. In fact, if one of these were absent, the object would not have been a *taytu*. If it is unripe, *bwanawa*, if overripe, spent, it is not a spent *taytu*, but a *yowana*. If blighted, *nukunokuna*. If it has a rotten patch, *taboula*. If misshapen, *ususu*. If perfect in shape, but small, *yagogu*. If tuber, whatever shape or condition, is a post-harvest gleaning, *ulumadala*. When the spent tuber, the *yowana*, sends its shoots underground, as we would put it, it is not a *yowana* with shoots, but a *silisata*. When new tubers have formed on these shoots, it is not a *silisata*, but *gadena*. In short, an object is; it cannot change an attribute and retain its identity. As soon as it changes, it ceases to be itself. As being is identical with the object, there is no verb *to be*. As being is changeless, there is no verb to *become*. Becoming involves time, but being to a Trobriander has no reference to time.

With us, however, time is all important. We cannot respond with approval or disapproval unless we know that a thing is getting bigger and better. If I am told that John Smith earns $4000 I cannot respond to this adequately unless I compare it with another salary. But If I am told that John Smith has been promoted to $4000 I will say 'good,' or if I hear that he has been demoted to $4000 I will say 'pity.' But simply John Smith at $4000 is something I cannot respond to. Our language is full of terms like *demote* and *promote* where value is attached to change.

Our language requires that we express nearly all views in terms of time. The Trobriand language however, has no tenses. Verbs are timeless. Being is apprehended as a whole, not in terms of attributes. This is very difficult for you and me to do. We rarely value sheer being in itself, except perhaps when we are "blindly in love." When you are in love the girl can have large feet, a small mustache, an I.Q. of 6 and a father in the penitentiary, but you love her for herself alone. Even mothers in our society are often incapable of valuing children in this way, demanding instead attributes and achievements before they can respond with love. Am I overstating the case when I say that most mothers love the successful child more than the unsuccessful one? The inability to react to being itself sometimes creates embarrassing predicaments for us. Several years ago a friend, visiting our house, held in his arms our youngest child, age several weeks. "Can he talk?" he asked. "No of course not," I said. "Can he walk?" "No," I laughed. He was a philosopher, you understand. He just stood there, holding a mass of protoplasm, not knowing what to say. We would say "How bright" or "How clever". The Trobriander would say "How baby" and he would respond emotionally in a situation where we cannot.

In our society, the tendency both in love and friendship is to be attracted by qualities rather than persons. We like people not for what they are in themselves, but because they are beautiful or rich or amusing, so if they lose their looks or their money or their wit, we lose our interest. But for the Trobriander, being is evaluated in terms of itself alone, not in comparison with other things. Again, this is foreign to our thinking, except perhaps in the sphere of art, and even here, we are entering a "twilight of the absolute." Nowadays we can respond to a work of art only if we know how much it it worth, or who painted it; we are incapable of judging it for itself alone.

Generally for us, to be good, being has to be as good as, or preferably better than, something else. Our language makes it easy to compare beings at every turn. Our vocabulary provides us with a large number of comparatives. The Trobriander has no such means. Where we use similes, where we say I am your JUNIOR, where we stress self-improvement and competition, they employ metaphors. The Trobriander says young man I. The Trobriander emphasizes the status quo and co-operation. Phrases which we commonly hear— "My toy is bigger than Johnnie's"—"Your dress is prettier than Mabel's"—have no equivalent in Trobriand. To be, an object must be true to itself, not in terms of its relationship to other beings. To be good, it must be the same always. In Trobriand, nothing — not even the world — ever came into existence; it has always been, exactly as now. Their mythology contains no concept of creation from nothingness. No supreme being ever acted as creator, artificer or transformer. The Trobrianders don't find themselves in the awkward position of trying to answer that unanswerable question: 'Who created the creator?'

The Trobriander simply isn't interested in chronological sequence. For example, he gives his autobiography in complete disregard for chronology, an effect achieved only deliberately by our sophisticated writers. He begins with the crisis, so to speak, and weaves backwards and forwards in time with many omissions and repetitions on the tacit assumption that your mind is moving in the same groove as his, and that no explanation is needed.

But for us chronological sequence is of vital importance, largely because we are interested not so much in the event itself, but rather in its place within a related series of events. To the Trobriander, events do not fall of themselves into a pattern of cause and effect as they do for us. We in our culture automatically see and seek relationships, not essence. We express relationship mainly in terms of cause or purpose. The maddeningly persistent question of our young children is WHY, because this is the question implicit in most of our ordinary statements and behavior. Every aspect of our lives teaches us to ask WHY, WHY, to seek causes. The Trobriander parent, however, does not entirely escape this questioning, for their children ask WHAT. I might add, our children do too, up until the age of about two or two and a half; then, when they are unconsciously beginning to learn the implicit philosophy in English grammar, then they ask why, why?

Yet the Trobriander has no term for why, nor for because, so as to, cause, reason, effect, purpose, to this end, so that,

Malinowski's frequent why evoked from the Trobrianders either confused or self-contradictory answers, or the usual 'It was ordained of old.' Just as the conservative Torontonian says, "This is the way we do it here." I might add that the Eskimos always reply to questions of this type, "Happy people don't ask questions."

Now being the value only in a specific setting. Let me illustrate: The early pearl traders offered the natives money and trade goods as an inducement to get them to dive for pearls. But the natives refuse, having no use for money and little interest in foreign trinkets. The traders noticed, however, that the Trobrianders set great store by certain large stone blades. First they imitated them carelessly, but the natives didn't want them. Next they had them made of slate in Europe and shipped half-way around the world. But the natives didn't want them. Finally they had the native stone quarried and sent to Parisian craftsmen but these beautiful blades were also rejected. And indeed, why would the natives want them? For the blades had meaning only within a patterned activity. Let me give an analogy. Let's say you're walking down a street in Toronto, you glance in a window and see a girl who has just received a Valentine. She is excited and so is the family. There's hope! Obviously, to her, that Valentine is the most valuable thing in the world. You ring the door bell and offer her a job. When she asks the salary you say, 'One Valentine a week.' It is a ridiculous situation, or course, because the value of the Valentine lies in the fact it's February 14, she's young and in love. In any other setting but its own the Valentine is worthless and so is the Trobriander's stone blade.

Now the Trobrianders are not blind to causality; they are quite capable of perceiving events in a lineal pattern. But when a pattern assumes lineality, it is utterly despicable. For example, dating a girl includes giving her a gift. But if a boy gives a girl a gift so as to win her favour, he is despised. Or if she accepts him as a sweetheart just to obtain the gift, not because she loves him, she is regarded as being callous. Similarly, in activity in which the men exchange necklaces and arm bands with one another, some men are accused of giving gifts as an inducement to their partner to give them in turn an especially good gift. Such men are labeled with a vile phrase; he *barters*. For the receipt of a gift should not cause the Trobriander to do something to give a gift in return; it is understood that he was going to do it anyway. In other words, the Trobriander can experience events lineally, even causally, but when he does, value is either absent or destroyed. In a sense, the gift exchange is not unlike our Christmas giving. I recall that at Christmas time my mother always kept in an upstairs room several presents—the kind suitable for any age, any sex—wrapped but unlabelled, just in case someone she had forgotten brought her a present. She would then thank them, say 'Just a minute while I run up and get your present,' hurry upstairs, quickly label the present, and then present it. For Christmas presents and cards are spoiled, in a way, if we think the other person has forgotten us or perhaps didn't plan to give us anything until forced to by the receipt of our gift.

The Christmas pattern is really an exception in our society: it is perhaps significant that children enjoy it most whereas in Trobriand life, it is the rule. Trobriand behaviour is not motivated by a sequence of events, or by any line of activities leading up to something. On the contrary, they do their best to ignore, to refute such a sequence or line.

But is this line really present? Perhaps it is; maybe it isn't. But we feel happier when we think it is there. Then the situation has meaning and we can respond to it. If I tell you that Sally married a millionaire, that she's selling notions in Woolworth's, that she once worked for Vogue, went to Vassar and was poor—it's mere jumble. But if I say Sally was poor, worked at Woolworth's, saved her money, went to Vassar, worked for *Vogue*, and then married a millionaire—now it all falls in line and makes sense. Our idea of happiness is bound with this motion along an envisioned line leading to a desired end. Our conception of freedom rests on the principle of non-interference with this moving line which leads to a desired climax. As we see our history climactically, so we plan future experiences climactically, leading up to future satisfaction or meaning. Who but a very young child would think of starting a meal with strawberry shortcake, and ending up with spinach? The Trobriander meal has no dessert, no line, no climax. The special bit, the relish, is eaten with the staple food; it is not something to look forward to while disposing of a meaningless staple.

Now for members of our culture, value lies ideally in change, in moving away from the established pattern. We hopefully expect next year to be better, brighter, different; we hope it brings change. Our advertisers thrive on this value of the different. It's new, it's brighter! Our industries have long depended on our love for new models: 1956, just out! New! Our writers cannot plagiarize; our inventors must invent.

The Trobriander on the other hand expects and wants next year to be the same as this year. The new is not good; the old is known and valued. In repetition of experience he finds, not boredom, but satisfaction and safety. Members of our culture go into uncharted seas fearlessly. We explore new lands eagerly—mountain peaks, sea floors, inaccessible jungles and polar wastes. The Trobriander goes into known seas. He explores nothing. New lands, new thoughts, new ways, hold no interest for him. He was born into a culture that was operating long before he was born and will continue to operate long after he is dead.

Test Paper

NAME:

ADDRESS:

PHONE:

CIRCLE THE LETTER BEFORE THE BEST ANSWER.
DO ALL OF THEM IF YOU CAN BUT DO NOT GUESS.

1. The Trobriand Islands are in:
   a the Pacific
   b the Caribbean
   c near New Guinea
   d Polynesia

2. The Trobriander values:
   a creativeness
   b the old and traditional
   c the new and the different
   d that which is useful

3. An art form in our society is valued for:
   a its position within a patterned activity
   b itself alone
   c its financial value
   d for many attributes, not all of which concern art

4. When we say, "The lightning flashed" we:
   a read action into the event
   b use an intransitive verb
   c describe the event as being without action
   d describe the event in the only possible way

5. The stone blades were:
   a used as inducement
   b equivalent to money
   c part of a patterned activity
   d gifts in payment for other gifts

6. English does not:
    a stress causality
    b differentiate between psychological and external experiences
    c emphasize time
    d employ a variety of adjectives

7. By lineality we mean:
    a patterned activities
    b emphasis upon being
    c connections, usually sequential, between things
    d the use of gestures

8. Our language and our culture structure experience so that it:
    a leads or should lead to a desired climax
    b emphasizes repetition and sameness
    c can be responded to
    d stands by itself, without reference to other experiences

9. In our language much value is attached to:
    a changes in temporal sequences
    b essence of being
    c ability of emotional expression
    d change and becoming

10. We stress causality and lineality because:
    a we are interested in relationships
    b we are interested in being
    c our language and culture teach us to value them
    d this is the most accurate way to describe reality

11. The Eskimo language is characterized by:
    a lack of nouns
    b lack of tenses
    c lack of transitive or action verbs
    d lack of categories differentiating living from non-living

12. The English language makes continual use of:
    a several tenses (simultaneously)
    b spatial metaphors
    c words describing subjective, psychological experiences
    d categories differentiating internal and external experiences

13. The Trobriand language emphasizes:
    a change and becoming
    b varied use of adjectives
    c temporal aspects of objects
    d being and existence

14. A noun in the Trobriand language refers to:
    a a highly unique object
    b being as a whole
    c an object at a particular stage of growth
    d useful objects only

15. Native languages differ from English because:
    a they cannot express causality
    b they are not fully evolved
    c they do not deal with temporality
    d they contain other metaphysical systems

16. In our culture we tend to judge things in terms of:
    a intrinsic value
    b qualities and attributes
    c aesthetic satisfaction
    d relation to other beings

17. Disregard of chronological sequence is characteristic of:
    a all primitive languages
    b English language
    c the thinking of small children
    d Trobriand language

18. The implicit philosophy of English grammar makes us:
    a seek essence
    b ask "why"
    c value money
    d see lineality and being

19. In the lecture an analogy was drawn between:
    a money and a Valentine
    b foreign trinkets and necklaces
    c stone blades and Valentines
    d arm bands and Valentines

20. Gift exchange among the Trobrianders is significant because it:
   a stimulates barter
   b symbolizes kinship and friendship
   c involves gifts which influence behaviour
   d is similar to Christmas-giving

The lecturer described two native philosophies, but at the same time said that the Eskimo, for example, had no philosophers. How would you interpret these two statements in terms of the lecture? Write 200–300 words. Use back of paper if necessary.

On the scale below indicate with a check mark
how you feel about the lecture

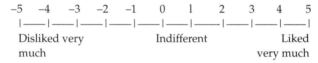

| –5 | –4 | –3 | –2 | –1 | 0 | 1 | 2 | 3 | 4 | 5 |

Disliked very   Indifferent   Liked
much   very much

What I Learned on the Project

Correction for Lasswell formula—not who is speaking to whom, but what is speaking to whom. Lasswell ignores the media; but obviously if a person is speaking into a P.A. system or into a radio microphone, etc., the who and the what are profoundly transformed.

That staples are media and media are staples. When iron ore and oil and lumber and fish are available to the population of a particular area, their patterns of association are much modified by this fact. Dorson in his recent volume *American Folklore* draws attention to the power of cotton in the United States in creating homogeneous culture capable of creating a spontaneous folklore. The same homogenizing power over human institutions is exercised by any economic staple like wheat, or lumber in Canada, but this serves to draw attention to the same power which resides in the media of communication. The media are, in fact, themselves staples or new natural resources. Media are extensions of the human senses. They modify the patterns of human association while remaining rooted in this or that sense, and these staples are not limited to any geographical area, but are co-extensive with the human family itself.

Another peculiarity of media as staples or natural resources is this: as they step up the speed of human transactions the information levels of the community rise. As information levels rise, one commodity becomes substitutable for another. No staple becomes indispensable as information levels rise. The tendency is for information to move, rather than commodities. The stress shifts in human study and attention from subject matters to the learning process itself. The Universities which have long tended to be processing plants become models of the learning process in action instead. Let us illustrate the implications of this change from that most popular of all subject matters — the ever present threat of war.

To-day, civil defence would seem to consist in protection against media fallout. In the past, war has consisted in the movement of commodities back and forth across frontiers. To-day, when the largest commodity of all is information itself, war means no longer the movement of hardware, but of information. What had previously been "a peace time" activity within our own boundaries now becomes the major "cold-war" activity across frontiers. Instead of competing for the franchise and dollars of our own citizens, we are now engaged in trying to win the favourable attention of Asian and African millions for the star turn or top show. Our own conceptions of education and of warfare are so completely tied to 3,500 years of literacy that the meaning of the electronic revolution is much less obvious to us in the West than to a Japanese or a Chinese or to an African.

Another basic aspect of the electronic is this: it telescopes centuries of development and evolution into weeks or months.

In speeding up actual change, it makes the understanding of change much more feasible just as a movie of an organic process may reveal years of growth in seconds. But such acceleration of growth in no way prepares the human community to adapt to it. Suddenly there is a nine foot redwood where in the morning you had experienced a bedroom.

Our educational, political and legal establishments are scarcely contrived to cope with such change. There is no mercy for culture-lag in our new technology. There is no possibility of human adaptation. Yet in all these situations we confront only ourselves and extensions of our own senses. There is always the possibility of escape into understanding. We can live around these new situations, even if we cannot live with them.

The New Criticism and the New Media

The so-called new criticism which followed after the new poetry which followed after the new developments in our Western world has most typically been engaged in explaining why works of art have no content and no subject matter. It was the new media themselves, from the telegraph (1830) onward which created the situation which the poets and painters tried to explain to us by "prophetic" new art forms. Is it not ludicrous that the very scientists who expected the radical changes should stand around with yammering and incoherent gesture while complaining of their inability to understand modern art?

For the past century, the artist has been our only navigator in social and political terms. The models which he makes are not wishful dreams that money can buy, but urgent factual instructions of the means of avoiding disaster. Top industry understands this a little bit to-day. Artists move onward and upward in the commercial field through the departments of industrial and package design. In the field of operations research, the artist accepts the priority which is not his reward, but his responsibility.

To suppose that the teaching of media in our schools should be a peripheral feature of an august and a well-tested curriculum could be a disastrous supposition indeed.

In purely realistic terms, I feel that the associated power of specialist and vested interest of many kinds definitely insures that we shall fail to meet any and every challenge that is offered to us in the electronic age. Why should we understand new media when no generation of the Western past has understood all media? However, now that we have begun to understand all media for the first time (see H. A. Innis, *Empire and Communications*) there is the outside possibility that we might decide to consider them as fit objects of study and control.

— END —

Marshall McLuhan
Toronto, Canada, June 30th, 1960.

Project in Understanding New Media

Sample of Syllabus
H. M. McLuhan

Teachers today face a quite different environment from the one they themselves grew up in. The new problems in school resulting from new procedures, new aids, etc., are minor. The big changes are outside in new relations between classroom and community, in the new relevance of higher education to business of production (assembly line now obsolete), in the new relation of education to politics and to world responsibilities, in the multi-language situation and in the multi-media world in which print is no longer monarch.

Unfortunately, the problems for teachers and students, created by our changing environment, do not disappear when they are ignored. In the same way, the unnoticed and non-verbalized aspects and effects of print, movies, of TV, do not cease to work upon us because neglected. Media do not cease bothering about us, nor about one another, just because we do not bother about them.

Education has become everybody's business in our society. The globe has become a community of learning. The communications industries are many times larger even in capitalization than heavy industry. With learning and teaching becoming the business of everybody, round-the-clock, and round-the-globe, what becomes of the older roles and relations of teacher and student? Must both teacher and student become more aware of the inter-relations of specialist areas? Must they now study the action of media upon our habits of perception and judgment, in order to remain reasonable and autonomous? Can we escape into understanding, as well as into success, as it were?

Objectives

In starting a syllabus for media instruction in high-schools, we might assign as long-term objectives for such a course of instruction:

(a) Identification of major media developments in our environment, and their relation to new science and technology.
(b) Familiarity with the various and often contradictory qualities and effects of media.
(c) Greater awareness of ways in which words, concepts, and data may be used, or misused, in the handing of media matters.

One reason for the greater awareness of all media matters, which we now take for granted, is quite simply the co-existence and constant interaction of many media in our ordinary environment. What we might call the climate of education has undergone radical change because of the co-presence of several major media. Their daily operation upon one another, as in the mere use of newspaper or the use of the telephone, in movies and TV shows and on the stage, of itself and without verbalization, creates a new kind of increased awareness which, in turn, has become a major factor in changing the climate or over-all situation within which our educational establishment functions.

What the physicists have developed in our century as a basic means of research is the bombarding of nucleii of unknown structure with nucleii of known structure. In effect, this is what happens when one medium crashes into another.

A sort of x-ray occurs in which the skeleton and the organic structure, as it were, of a medium of communication is revealed. For example, when printing intrudes into a non-literate area, the structure of the print form with its whole sub-structure of assumptions and its superstructure of achievements is as fully revealed as is the corresponding structure of the archaic or pre-literate culture receiving the impact of print.

Print is, of course, quite different from writing in its nature and in its effects. To-day in parts of Russia, China, India and Africa, large sections of the world population are experiencing what the West did 500 years ago. Meantime, the West has, via electronics, entered upon a phase of what may prove to be "post-literacy". At any rate, print is no longer the prime nor prior experience in our society, nor does it condition production in industry as it did. The assembly-line as Peter Drucker shows in *Landmarks of To-morrow* (Harpers, 1959) is as obsolete in production as delegated authority is in management. And *Parkinson's Law* makes hay with the written and typed word, in the worlds of bureaucracy and business, showing how the letter and the memorandum have inflated themselves into a kind of Marx Brothers world of surrealist nonsense.

Yet, from the time of Gutenberg in the fifteenth century, when printing from movable types taught the West how to organize its procedures, in culture, commerce and science, on the one-thing-at-a-time principle, the West has extended the Gutenberg principle of organization to many phases of living and learning.

Our new electric technology seems to have made all aspects of print-culture somewhat obsolete. Are we to stand by while the new technology brainwashes the assumptions and values of print-culture out of our world? Or is there a valid way of translating the achieved values of literate society into the new languages of film, of radio and television? Can we, by studying the non-verbal assumptions and forms of print, create an effective osmosis between print values and the new forms? For these new forms have unexplored and unknown levels of power and meaning which they are just beginning to release themselves into our living and learning situation. It took writing centuries to re-structure the forms of human association in the ancient world. It took print until the Industrial Revolution in England, and the French Revolution in France, to express itself as new power and new order. How long will it take our electronic technology to replace the industrial, political and educational forms we built on the printed word?

Are we able to escape from these dilemmas into understanding? Writing on "The Social Role of the Confession Magazine" (*Social Problems*, Vol. VI, No. 1, 1958) George Gerbner says of the typical heroine: (pp 35–36) "The price for human dignity and compassion is the basic irrelevance of the narrator's desperate and confused protest. Buffeted by events she cannot understand and is not permitted to "wrest about to suit herself", the heroine's headlong flight "down the line of least resistance" leads to her inevitable "sin". As she has no conscious relatedness to the larger social context with which she must, in fact, contend, her act becomes irrelevant as social protest. It only brings calamity to her and to those she loves. Her suffering is a spine-tingling object lesson in bearing up under relentless blows of half-understood events. Through her agony comes not insight into the circumstances of her act, transcending the immediate causes of her misfortune, but, if anything, a remote glimpse of such "happiness" as might be had in coming to terms with an unbending, punitive and invisible code of justice. Responding to hidden authority rather than being permitted to be self-directing, her "problem-solving" becomes an un-reflecting drift towards "adjustment." The heroine's plight here has been the role of the educator in other times than ours when faced by misunderstood consequences of technological change. We would seem to have no alternative but to face the subliminal levels of media, old and new, in order to maintain autonomy and relevance of response to new situations to-day.

But increased awareness of the forms of the media, as they operate upon our modes of perception and judgment, is not merely a means of understanding, but of prediction and control. Awareness of media is itself a change and a cause of change. In understanding media, some teachers will experience uncertainty and doubt and insecurity. They will sense that areas of knowledge which they have acquired with difficulty are being undermined, as it were, by their own new insights into the structure and assumptions of that knowledge. They will find difficulty in talking about media, in which they themselves have a deep personal and also a vested interest, to adolescents who have not as yet any awareness of having a stake in such knowledge.

The mental discipline necessary for transposing the realities of our lives into new spheres, and new media, turns us all into mental D.Ps. to-day. But we are mental D.Ps. confronted with undeveloped countries of the mind. For merely to translate the ordinary procedures of teaching English into the audio-tape medium is to be released into the previously unknown world of structural linguistics.

The scholar-teacher, who has lived in the class-room and the study all his life, is suddenly released into the enterprise and domain of publishing by radio, or television teaching. For, whereas learning and teaching have since Gutenberg been based on the publisher of print, educational TV makes of the teacher and researcher himself a publisher on a very large scale.

Laurence Burns, the President of R.C.A., in announcing the imminence of video-tapes for ordinary home uses, argued that the day is at hand when teachers will make his kind of salary, $170,000 a year. Such a view of the new educational dimensions seems to occur naturally within industry if not in education to-day. For teachers, of course, it serves merely to indicate new patterns and landmarks set up before the old scene has disappeared.

Such a prospect serves also to point out that a mere personal *point-of-view* is no longer a way of coming to grips with such a fast-changing situation. In the midst of accelerated change, the only relevant strategy is what Wells Foshay has described as "swarming all over the situation". James Joyce called his very 20th century *Finnegans Wake* an everyway roundabout with intrusions from above and below. The inclusive "field" has succeeded the specialist or single "point-of-view" approach. Peter Drucker, in his *Landmarks of Tomorrow*, refers to this over-all non-point-of-view approach as "organized ignorance". And it permits the ordinary teacher in an ordinary class to engage in first-hand research.

It was during World War II that Operations Research hit upon the strategy of pulling specialists out of their fields. A weapons problem was handed right off to biologists and psychologists, instead of to engineers and physicists. Because it was found that specialists inevitably directed their acquired knowledge at a problem. The non-specialist knowing nothing of the difficulties involved, could only ask: "What would I have to know in order to make sense of this situation?" In a word, he organized his ignorance not his knowledge. The result was many break-throughs and solutions that otherwise would not have happened. In retrospect the greatest discoveries seem quite obvious. Is it not because the beclouding assumptions and unconscious bias from earlier and irrelevant training have disappeared?

J. Lewis Powell, speaking at the Armed Forces Communications Electronics Association's National Convention, last June, made a basic point for educators to-day. He said:

When I was a child, people used to go around and say how we were unprepared when World War I happened. And later on I heard that we were unprepared when World War II happened. This is a lot of folklore and it just isn't so! We have never been unprepared! When World War I came along we were thoroughly prepared to fight the Spanish-American War. And when World War II came along, we could have fought World War I brilliantly ... Ours is a tendency to fight the last one over, to be handicapped by experiences!... Now, in these times things are exploding! If you are 20 or 30 months behind the times, you are further behind than your dad was when he was 20 or 30 years behind.

The problem is that you and I were born, and educated, and lived during the age of plodding progress. We are almost incapable of realizing that progress is now made on purpose by inspiration rather than made by accident with perspiration. Brainpower has replaced manpower as a national resource. This fact dictates that you need to educate everyone according to his aptitude. We need better brains across the board in all fields of human endeavour.

Earlier, Mr. Powell pointed out that: "when we were making progress very slowly, we weren't too bright in keeping up with it". Today, when there has been much more technological change in a single lifetime than in all the previous history of the world, it is sheer necessity for educators to *understand* the implications of these changes for school and society alike. Because there is no opportunity for teacher or student to adjust to these changes by the older habits of social osmosis and gradual adjustment. We have now no choice but to understand and to control the new technology, since its power is too sudden and total to be absorbed by casual osmotic or subliminal means.

For the first time in human history, progress is on purpose. Inspiration has supplanted perspiration. Education must, therefore, assume the same power of prediction and control of overall situations and change. To eliminate the subliminal levels from all media, new and old, is a minimal step in educational strategy under such conditions.

To achieve this end, to-day, is not as difficult as it would have been a few years ago, precisely because the multimedia world is luminous from within. The x-ray action of one medium upon another reveals the structure of the bones and organs of society. We have long known structure of external machinery. But we have known little about the organic structure of society. Yet the organs of society are, quite naturally, networks of human association. And just as the dynamic structure of our vernacular tongues was hidden until the tape-recorder gave us ready access to them, so the structure and operation of the languages, as it were, of the media of writing and gesture and print and photography, remained concealed beneath the vesture of social accidents and overt uses.

Just as educational broadcasting adds new dimensions to teaching and transforms the teacher into a large-scale publisher, so the daily operation of the media, under the disguise of "entertainment" has extended the walls of the class-room to include the entire globe. The global movement of information by technological means is by far the largest business in the world. Culture has become our business and has swallowed the business man as well as the educator.

Rabelais had a vision of printing as "the world in Pantagruel's mouth". It is a vision whose meaning is easy to discern now that rival mouths, and the maws of newer media await us. We can, as it were, take our pick among the giants who will chomp upon us. Or, we can assume the role of Jack the Giant Killer and bring these vast projections or extensions of our own ingenuity under the control of our conscious purposes. The educator is now called to bring the alchemy of history within our management.

In business to-day, the budget item which exceeds both production and marketing costs is research. In a situation in which change is so fast as to approximate the condition of an explosion, big business has to outstrip change by anticipation and research. The goal of this research is to be far ahead of change in order to control it.

Educators have a much higher social obligation to anticipate and to control technological change as it affects the learning process. Faced with multiple media of codified information, we must learn which ones have a natural priority for invading and structuring the sensuous and perceptual lives of the young. For the first invaders, be it the English language or French or print, or pictures, leaves an imprint on sensibility which affects all subsequent learning.

Carl Orff will not admit to his school of music training in Vienna a child who has learned to read and write. He insists that when the eye has been given such a powerful training before the ear and other senses, real music training is impossible.

We have yet to learn which of the media may safely be given precedence in education, consistent with the fullest use of our faculties over the whole extent of our adult lives. For we can no longer think of education as the training of the young, because even the very young are constantly subjected to adult fare via press, magazine and television to-day. And the adult must frequently embark on new programs of learning in the normal course of his business and social life. The successful business executive of 35-45 is regularly sent to big management centers of education for "brain-washing" and re-instruction.

If our minds have been blocked by types of perceptual training, which make future learning difficult, we are as much casualties of technological change as the actor who can't make the shift from movies to television.

The basic question of which media of instruction constitute the least ultimate blockage of future learning is thus quite distinct from the question of which medium provides the quickest result in learning for the young. Print learnt in childhood may prove to have been a major block to learning in many subjects besides music.

Related to this aspect of media, in shaping the over-all learning process, is the question of how far it is desirable to subject either the individual or whole societies to all media at once. This is a social question which arises to-day in Africa, and India, and China. Because, whether they receive all the media at once, or whether they get one-at-a-time may have quite drastic effects on their development and on their relation to one another.

Print has great power to detribalize a society, but it also creates intense individualism and nationalism. Radio, on the contrary, in Africa intensifies the existing tribal patterns. It is only the print bias of our own training which leads to suppose that the program "content" of radio broadcasts in Africa are the cause of the ensuing emotions.

What can be said then is that no medium has its meaning alone to-day. During the centuries when writing and printing have had a monopoly of educational influence, the assumption grew that these forms were basic means only. Their inherent and non-verbalized patterns and powers were quite unnoticed until photography, and film, and telegraph, and other rival forms of codifying experience revealed the hidden educational powers of writing and printing.

Since however several rival forms are part of the ordinary childhood experience to-day, it is quite unrealistic to expect children to be impressed by the ancient prestige of the printed word. Faced with a training procedure based on the traditional assumptions of the role of print in the school and in industry, the child registers an immediate loss of motivation.

This loss of motivation is his natural response to a confusion which exists in the minds of the entire community. In Russia, print has all the challenge of novelty and of new technology related to their first conquest of industrial and social organization. It is unlikely that the Russians understand this, any more than they are prepared for the growth of individual initiative and national feeling that will follow.

Print has spread in Russia the age of radio, and television, and of nuclear physics. Therefore, its pattern of growth will not be the same as it was in the Western world when print had no such rivals, or coexistence with other media and knowledge.

Likewise, the course of the development of print in the Chinese culture will be quite unlike its path in Russia. For the power of the ideogrammic writing of the Chinese is much greater than any opposition met by print in the West, or in Russia on the other hand. The ideogram is closer to the modes of nuclear phenomena and investigation than anything known in Russia or the West. The Chinese will, eventually, be more at home with non-Euclidean mathematics and physics than anybody else.

It is no accident that Western poets and artists have, for more than a century, been drawn into a fascinated study of Chinese art and writing. The poets and artists have always been the *avant-garde* of coming change in media technology. Their work provides the earliest clues to the meaning of new technology for altered patterns of human living and learning. This role of the artist, as herald and model-maker, affords a major resource in understanding media. In fact, it indicates a natural means of integration of the curriculum whether in history, mathematics, or language and media study.

This syllabus will attempt, in some degree, to provide a means of over-all integration, while also coming directly to grips with particular media problems.

SESSION I
Writing

It may be desirable to omit this Session from the Syllabus, since it involves discussion of the character of pre-literate societies. The argument in favor of starting at this point is owing to the character of "post-literacy" that, in a techno-logical sense, seems to belong to electronic man. The gist of a Session on the origin of writing is as follows:

(a) Writing is an achievement of sedentary societies. Nomadic peoples do not develop any form of writing.

(b) Writing is preceded by sculpture.

(c) Sculpture is what André Malraux calls the "voice of silence". It is the visible modelling of sound or "auditory space".

(d) Writing is the translation of "auditory space" or of verbal sound into a pictorial or visual space. It is enclosed space.

(e) A sedentary society develops a division of labor which fosters a partial dissociation of sensory life as well.

(f) A non-nomadic people, accustomed to specialized tasks, learns how to give effect to one sense in terms of another sense. It learns how to visualize sound in a variety of ways.

(g) The phonetic alphabet was a further technological step. It divorced the sounds and visual forms of let-ters from "meaning". It created a means of having "average sounds" for a wide range of actual speech patterns. This happens again to-day, with words themselves, when we try to build a translating "ma-chine" or language computer. The multiple meanings and shades of words must be averaged as we learn in Information Theory.

(i) The phonetic alphabet, with its structure of averages, gave the West the power to translate into its sign system any language at all. It became a means of cultural and political conquest and control. (See H. A. Innis, *Empire and Communications*, Oxford University Press, 1950).

(j) Why is similar conquest not possible for other kinds of writing such as the Chinese? Because hieroglyphs, pictograph and ideograms do not break up complex situations into analytical bits and fragments.

(k) The phonetic alphabet not only arrests verbal sound visually, but enables man to analyse and to capture the movements of his mind by writing them down as static propositions and enunciations.

In our electronic age we have discovered deep impatience with this form of recording and representing thought. The symbolic logicians, and others, are aware of the distortions of thought so recorded, and find little good in traditional logic.

In the same way, anthropologists, seeking to communicate a full account of the archaic cultures, find the literacy means of description inadequate compared to the movie which can capture multiple facets and relations of situations in a single shot.

(l) Writing first gave man the power to consider one-thing-at-a-time.

(m) Oral or pre-literate man did not have the means to do this, but took whole situations in their togetherness as we find in "primitive" languages.

(n) What we call "myth" is such all-at-onceness, and in the electronic age we not only have great sympathy and understanding for myth, but we ourselves create myths all the time. A pictorial advertisement has much of the formal character of a myth in providing an inclusive image of production know-how,

527

and of consumer satisfaction in a single moment or situation.

(o) In its one-at-a-timeness, writing fostered the habit of subliminal life. That is, when man attends to one-thing-at-a-time, most of the situation he sees or hears or experiences is thrust under the level of attention.

Literate man develops a very large subliminal life, while primitive man, paradoxically, is much more aware of over-all situations. Electronic man finds he has to develop this kind of awareness, again, since the instant character of information movement, today, makes it necessary to understand the total field of impact and effect. Our field, in fact, is the globe which, as it were, is reduced to village size by electronics.

(p) Again paradoxically, pre-literate and electronic man face situations of similar kind, in living and learning.

Will their strategies of culture also be the same?

Will we too seek fixity, and family as the pattern of all organization?

SESSION II
Manuscript Culture

Session I was concerned with the cycle from pre-literate to post-literate living. This was the shift from audile, verbal, tribal, organization through the literate stages of detribalization and new social structure to "post-literacy", or electronics.

The whole Session may be dispensable, as also this one on the manuscript. I say this because, should these two Sessions be thought necessary, they would naturally have to be *expanded* a great deal. They would also be supplied with actual selections from various works, and not just reference reading.

Manuscript culture was handicraft culture. Print was the first true mechanization of an ancient handicraft. And printing may have owed less to the handicrafts than to the scholastic philosophers, with their insistence upon the universality and, in a sense, the uniformity of all truth and knowledge. For print gave material expression to this universalism, uniting, as it has often been claimed, science and democracy.

The fortunes of manuscript culture in the West fluctuated with the supply of writing materials. David Diringer's *The Hand-Produced Book* (Hutchinson, 1953) is the best source for data on all aspects of pre-print writing and publishing. But H. J. Chaytor's *From Script to Print* (Dufour, Philadelphia) is the major work for studying the effects of the manuscripts on habits of composing poetry in the Middle Ages. Moses Hadas, *Ancilla to Classical Reading* (Columbia University Press, N.Y., 1954) reports on some of the relations of reader to writer in antiquity. Studies of the effect and uses of writing, and of the habits of readers, are almost totally lacking for all periods, including the modern.

The reason seems to be inherent in the nature of printing in particular. For print not only fosters the illusion of the neutrality of all media (via its pattern of static shots) but makes a theory of change difficult if not impossible.

(a) The manuscript was read aloud both in the ancient world, and in the Middle Ages. Silent reading was regarded as an almost impossible stunt until printing. The consequences of such reading for living and learning, writing and reading, were very great.

(b) The manuscript was written slowly and read slowly. In the Middle Ages, parchment was scarcer than time. Copies of works were few.

(c) Because reading was slow, and works were scarce and hard to refer to, students naturally tended to memorize all they read.

(d) Writers and scribes tended to condense and to epitomize works of others, as well as their own, for these reasons. Manuscript culture works by compendia, on one hand, and by oral teaching, on the other.

(e) The elementary classes were often engaged in helping students to make their own copies of the poets, as well as their own grammars, lexicons, and common place books of condensed wisdom.

(f) Manuscript conditions foster great self-reliance in textual matters and great readiness of oral disputation in dialogue.

(g) The reader of a manuscript has no illusion of moving freely over the contours of an author's mind. For the most part, an author, whether ancient or medieval, had no notion of a close personal relation with his reader and, therefore, no ambition of self -expression or self-revelation.

(h) The habits of memorization of compendia also fostered the aim of encyclopedism and overall coverage.

(i) When learned men had all their texts and data in memory, the habit of disputation by the bombarding of text with text was natural. When memorization stopped, so did oral disputation.

(j) When texts were memorized and exchange was oral, why was it also natural to insist upon minute precision, and constant exactitude of definition of terms?

Why should printing have changed all this?

SESSION III
Printing

The reason why the previous two Sessions are entirely tentative, to my thinking, is that they concern cultural climates remote from the experience of students to-day.

Since print is not remote I shall try to set up this Session in a series of Sections that will approximate to my *present* idea of how to teach the medium of print in Grade XI. That means that around this Session I shall hope for a variety of practical criticism and suggestion from many sources.

It seems to be inadvisable to approach print in a merely chronological fashion.

SECTION A: The Message of Repeatability

The Chinese had achieved printing from blocks, around the 8th century A.D. Just as the cave-painters did not make their pictures for human spectators, but for the gods, as it were, so Chinese printing was a kind of visual prayer-wheel by which invocations to the gods could be multiplied in exactly repeated ritual form. Such was not the intent of Gutenberg, but we can ask how far this Chinese magic was imbedded in the process and the consequences of printing from movable types.

(a) What were some of the first consequences of printing for education?

Print meant that everybody could have the same text. How would this change class-room teaching from manuscript conditions?

Is our return to-day to the direct oral teaching of languages a repudiation of print?

What are the new conditions to-day which favor the change of teaching methods in languages and other subjects, as compared to the methods developed from heavy stress on the printed text?

(b) In his *Prints and Visual Communication* (Routledge and Kegan Paul, London, 1953) William Ivins tells how science and mathematics could scarcely exist in our sense until printing.

What has this to do with the factor of exact repeatability?

(c) How far is repeatability a verbal or non-verbal aspect of printing?

Is this the same as conscious and unconscious, or subliminal?

(d) Why would the printed book have seemed at first to be only a cheaper and more accessible manuscript?

(e) Print could be read at high speed compared to manuscripts. It could also be read silently and in private. Why could the manuscript book not be read silently?

Why could it not be readily taken home?

How would private, silent reading at high speed change methods and outlook in study?

Why was the manuscript of small use as a work of reference?

SECTION B: *Print and perspective*

(a) The private point of view of the silent, solitary reader of print coincides with the rise of perspective in painting and writing and in politics. Does this seem a natural, or necessary, development?

(b) How do the new media of our day, such as newspaper, photography, film, radio and TV, foster a private point-of-view in living and learning?

(c) Does the private point-of-view of the print reader seem consistent with the uniformity and repeatability of the printed form?

(d) Why should the reading of even lines of continuous print, by a private, solitary reader, foster habits of perspective?

Why does perspective require a single, fixed position?

Is perspective natural for the eye?

Does it exist outside print cultures?

Why has it disappeared since Cézanne?

(e) In *The Sacred and the Profane* (Harcourt Brace, N.Y., 1959) Mircea Eliade compares and contrasts the views of space and time of pre-literate man and literate man.

He shows that, whereas we take for granted that space is "homogeneous and continuous", these aspects of space and time are unknown to pre-literate or archaic man.

To-day, in physics, we now take for granted that space and time are neither homogeneous nor continuous. That each space is a unique space-time event, and that each time is likewise unique. Does this mean that we, in the electronic age, have begun to share the outlook of the cave-man?

Have we, during the recent centuries of print, of uniformity and repeatability, developed a way of life that is incompatible with our electrical technology?

Can understanding the effect of our technologies upon our habits and outlook help to avoid confusion and undue stress during period of accelerated change?

(f)   During the very decades when printing from movable types first came into use, "another innovation was gaining a foothold in Italy". (See pp 13 ff, of Erik Barnouw's *Mass Communication*, Rinehart, N.Y., 1956)

This was the *Camera Obscura* described by Leonardo da Vinci in his unpublished notes:

"If on a sunny day you sit in a darkened room with only a pinhole open on one side, you see on an opposite wall, or other surface, images of the outside world—a tree, a man, a passing carriage". (Barnouw, p 13).

The discovery and its uses was described in detail by Giovanni Battista della Porta in his *Natural Magic* in 1558.

Soon a lens was inserted to sharpen the images pouring through the pinhole. But the images were upside down. So the lens was put in one side of a box, instead of in the wall of a room, and the image was turned right side up by mirrors. But the box, considered as a small room, was called a *Camera Obscura*. And it "could be aimed at a landscape, a street, or a garden party. A group of people looking in amazement at the moving images in the box may well have resembled a group watching television". (Barnouw, p 14). But the earlier form, before the lens phase, when people merely sat in a dark room, had also much resemblance to movies.

The question arises then whether there is any resemblance or correlation between the magic of the *Camera Obscura*, and the magic of exactly repeatable and uniform pages and volumes that rolled off the lines of movable types of Johannes Gutenberg and the other printers?

(g)   Is there anything in the quickly-read lines of type that corresponds to the fast-moving sequences of still shots that make up a movie?

Do the black printed words on the white page form a sequence of "shots" of the mind in motion?

Does the reader "travel", in mind and fancy, over the contours of another mind, as it were, in the act of reading print? (See Keats's sonnet "On First Reading Chapman's Homer") "Much have I travelled in the realms of gold. And many goodly states and kingdoms seen".

(h) Does the possibility of fast reading open up vistas of history and of remote ages and cultural climates inaccessible to the slow-moving manuscript reader?

(i) Does the first age of print coincide with the age of exploration of the world, by accident?

Medieval maps show that, until the sixteenth century, men thought of the spaces of towns and countries and oceans as not continuously inter-related by lines, as it were. The idea of simply keeping moving ahead in a line, whether geographically, or in terms of social and intellectual mobility, was new and revolutionary.

(j) The first age of print was also the first age of individualism and nationalism. It is easy to see why print readers might have developed a private point-of-view and habits of individual initiative and of "inner direction" towards self-appointed goals. (See *The Lonely Crowd* by David Riesman, Anchor Paperback).

But why should print have fostered a new and intense nationalism unknown to medieval communities? Is it because the seeing of one's vernacular in new technological dress provides a new image of one's community?

(These questions seem not ever to have been asked before, so that the student can feel free to speculate and probe). See Carleton Hayes, *The Historical Evolution of Modern Nationalism* (Richard R. Smith, N.Y., 1931).

(k) Does print present an image of organized and mobilized power as it were?

(l) Does the assembly-line of movable types present a non-verbal image of the means organizing military and industrial life, by tackling one-thing-at-a-time, in uniform segments?

(m) Are the assembly-lines of the first Industrial Revolution in England dependent on a print-reading community?

(n) Are the citizen armies that Napoleon sent through Europe dependent on the uniformity and repeatability of lines of movable types?
        Was Napoleon's failure with his new kind of army, in pre-industrial Russia, a clash of old and new technologies and strategies?

SECTION C: *Print and Industry*

At first, the printed book was regarded as a cheap manuscript. Such a small fact as the printers not putting in page numbers for the reader is indicative. Because a manuscript was read very slowly, it was not used as a work of reference. A printed book could be read very fast. It was no longer necessary, or convenient, to memorize when reading quickly. But it became necessary to refer to particular passages in a book. Yet it was a century before such means of reference were provided by printers.

It took less time for creative writers, like Rabelais, to grasp the new art-dimensions of printing. And Aretino quickly discovered that the printed pamphlet could be "the scourge of princes". But it is not till Montaigne and Cervantes that print comes into its own as a new kind of art form. The effect of printing on industry, and commerce, was speedy and extensive.

(a) In what sense was an exactly repeatable commodity, such as printed book, a new event in commerce?

(b) In the East to-day at a bazaar, does the procedure of bargaining have anything to do with the absence of exactly repeatable commodities?

(c) How could a price structure be devised or accepted in a community unaccustomed to print?

[d] If engineers introduce mass production into India, or Africa, or Russia, before printing and reading are taken for granted, how can commodities be sold in a non-existent market?

Is communism the inevitable consequence of introducing machine production before printing, reading, price systems and uniform commodity markets?

(e) Does the reader of print present the aspect of "consumer"?

Marshall McLuhan: Report on Project in Understanding New Media

Afterword: W. Terrence Gordon
Transforming the Report

With his Canadian Governor-General's Award in hand for *The Gutenberg Galaxy* (1962), McLuhan was ready to stir "Vat 69." His friend and colleague, anthropologist Ted Carpenter, advised him against revising his NAEB report under the title of "Understanding Media," preferring the phrase "extensions of man," from Emerson's *Works and Days*. In the end, McLuhan retained "Understanding Media" for its echo of Cleanth Brooks's and Robert Penn Warren's *Understanding Poetry* and relegated Emerson's phrase to the subtitle.

McLuhan's collaborator on *Understanding Media* was to have been Harry Skornia, who had read the draft and the proofs of *The Gutenberg Galaxy* and wrote the publicity, but their collaboration never progressed beyond a fond wish on the part of both men. To Skornia, McLuhan had spoken of his terror at the disaster civilization faced if mankind could not learn to use new media wisely, adding that this required understanding them in order to control their consequences.

McLuhan transformed his *Report on Project in Understanding New Media* into the manuscript of *Understanding Media* by raiding his own library. References to scores of authors sprang up throughout the new text, though some of these would eventually be obscured by editorial cuts. McLuhan had sent a proposal and a sketchy first draft to McGraw-Hill in New York in June 1961. By the time associate editor Leon Wilson responded with an evaluation, McLuhan had sent him an expanded second draft. With it came a second proposal, for a project titled "Child of the Mechanical Bride," a study of advertising intended to update McLuhan's first book, *The Mechanical Bride* (1951).

Wilson found the manuscript brilliant but deemed its writing style to be sloppy, offering McLuhan criticisms by conventional editorial standards. Changes would be needed to make the work intelligible. Wilson wrote to McLuhan in October 1962, without acknowledging — perhaps without understanding — McLuhan's definition of "message" as effect on the reader. The manuscript read like a plug for *Finnegans Wake* to Wilson's editorial eyes. And though he grudgingly admitted to taking some pleasure in reading McLuhan's volcanic prose, he was committed to containing the eruptions. McGraw-Hill were committed to publishing the work, but asked McLuhan for smoother copy.

He was incensed. Having been stung by the critics who detected a moral point of view in *The Mechanical Bride*, he feared the same reaction to *Understanding Media*, if Wilson were to have his way in forcing revisions. In any case, he had already rewritten parts of his manuscript many times, keeping at it with a concern for provoking his readers, satisfied only when he could find fresh twists bold enough to satisfy himself that his prose would banish complacency.

When Leon Wilson left McGraw-Hill in 1963, the *Understanding Media* file devolved on David Segal. Having prepared himself by reading the correspondence between Wilson and McLuhan, his curiosity was more than mildly stimulated. McLuhan seemed to believe that *Understanding Media* was scheduled for release in the fall of 1963. Segal quickly disabused him of this notion but despaired, as Wilson had, about the state of the manuscript. Segal sensed that it might be pointless to attempt foisting standard editorial guidelines on McLuhan and even conceded that there could be some symbiosis of style and subject matter in the manuscript. In this respect, he came close to divining the purpose of *Understanding Media*, but with no more advantage than future readers who would only discover its theme explicitly announced in Chapter 31: "It is the theme of this book that not even the most lucid understanding of the peculiar force of a medium can head off the ordinary 'closure' of the senses that causes us to conform to the pattern of experience presented."

Segal finally agreed to back down entirely on his editorial demands, but it was McLuhan who believed that he had been compelled to back down on giving his mosaic style full throttle. By comparison with both his earlier output, e.g. *The Gutenberg Galaxy*, and later works, such as *From Cliché to Archetype*, *Understanding Media* would give the appearance of a conventional book. When asked about the stylistic transition between *The Gutenberg Galaxy* and *Understanding Media*, McLuhan replied with something of a *non sequitur* about a McGraw-Hill house rule requiring authors to quote only sources with which they disagree. As for McLuhan's original idea for "Child of the Mechanical Bride," it was apparently dropped in a mutual conspiracy of silence born of the antagonism between author and publisher.

In the third and final draft of *Understanding Media*, submitted in the late fall of 1963, "The Medium is the Message," previously Chapter 2, became Chapter 1. However, it no longer opened with "The 'content' of any medium is always another medium," though this idea remained the focus of the chapter. The striking metaphor of a cloak of invisibility cast over any medium *from within*, i.e., by its content, was dropped from the second draft, along with the original statement of media effects: the effect of any medium is overwhelmingly from the unnoticed medium itself, and the effect is always the greater when its source is ignored.

McLuhan made major revisions chiefly in the first part of the book; in part two, the chapter heading "Clocks: Time with an American Accent" became "Clocks: The Scent of Time," "The Credit Card: The Tinkling Symbol" became "Money: The Poor Man's Credit Card" (with the tinkling symbol moving to the telephone chapter). A long section on advertising, planned originally with illustrations in the style of *The Mechanical Bride*, was produced in drastically shortened final form and without a single visual (a source of complaint on the part of one reviewer), becoming "Ads: Keeping Upset with the Joneses."

Where McLuhan discusses the wheel as an obsolete form in the twentieth century, his exhausted editor, too obstinate or incapable of making a leap with his author, even with a safety net of logic securely in place, complained about talk of clocks in the same breath. In his second draft, McLuhan had written: All manner of utensils are a yielding to this bodily pressure [to extend storage, mobility functions, and portability] … as in vases, jars, and 'slow matches' (stored fire). The cave or the hole in the ground precedes the house as storage place, because it is not enclosed space but an extension of the body. His editor protested that "because it is not enclosed space" was a baffling phrase; McLuhan responded by scratching it out. The final typescript of *Understanding Media*, much revised and expanded during its gestation, was shortened when the book went into production. But the chapters on "Hybrid Energy" and "Media as Translators," added by McLuhan to his final draft, were retained.

The book was released in the spring of 1964; by January of 1965, McLuhan had decided to tackle a second edition. By late summer, perhaps recalling his earlier, tongue-in-cheek offer to clarify *Understanding Media* by redrafting it in the 850-word system of Basic English devised by C. K. Ogden (championed by Ogden's collaborator and McLuhan's Cambridge mentor, I. A. Richards), he was proposing a children's edition to McGraw-Hill. McLuhan's enthusiasm was inspired by what he regarded as new insights in his own thinking about the medium as message, about new technologies as creators of new human environments. These insights he developed and refined to the end of his career; the children's edition of *Understanding Media* would never see the light of day.

— *W. Terrence Gordon*

Critical Reception of *Understanding Media*

As reviews began to appear, *The New Statesman* proclaimed: "Marshall McLuhan is now a power in more than one land."[1] It was no compliment. The reviewer alluded to "Mr. McLuhan's clutch of crystal balls" and found that the book's themes of electric speed, media as extensions, and media effects "cohabit not very fruitfully." He sensed the importance of those themes but remarked that "they are altogether drowned by the style." (A wry smile must have crossed the faces of McLuhan's editors.) The probes of *Understanding Media*, intended to goad readers into thinking, prevented the same reviewer from recognizing that McLuhan redefines content in terms of media. Just as reviews a decade earlier had managed to mistake whimsical style for moral stance in *The Mechanical Bride*, *The New Statesman* ignored the explicit deferral of moral judgment in the opening paragraph of *Understanding Media* and cast McLuhan somewhere between evangelist and showman.[2]

Commonweal Review called the book infuriating, brilliant, and incoherent, dismissing McLuhan's view that print

---

1. Christopher Ricks, "Electronic Man," *New Statesman*, 11 December 1964.

2. Ibid. "And there is, too, a heady market for prophecies, especially those which skilfully and at the last moment substitute a sermon for a prophecy. Like Jacques Barzun, Mr. McLuhan has the suspenseful air of being about to lift the veil."

technology gave rise to nationalism as "grotesquely inadequate."[3] *The Tamarack Review* detected "an air of almost smug knowingness."[4] The *Toronto Star* pronounced *Understanding Media* a rich, sprawling book, but found the author guilty of the exaggeration that blemishes Pasteur, Freud, Darwin, and Marx.[5] Reviewers skirmished with each other: the CBC'S Lister Sinclair fumed that *Understanding Media* was full of words about pictures but offered not so much as a single illustration.[6] Alan Thomas chastised Sinclair for his "fit of pique," noting that he had chosen "to deal with McLuhan's work by the easiest method ever known. If you don't understand something, but can't afford to ignore it, condemn it ..."[7] Thomas offered a strategy for approaching the book: "If you have any skill at reading poetry use it."[8]

A reader reproached *Time* for its review that said not one word about the theme that *Understanding Media* encapsulates in its title.[9] But there was worse to come: a popular misreading of McLuhan's thought was not long in taking hold. In a review obscurely titled "Reverse Canadian," the *Riverside Press-Enterprise* of Riverside, California, branded *Understanding Media* "a book against books."[10]

---

3 C. J. Fox, "Our Mass Communications," *Commonweal Review*, 16 October 1964.

4 Ronald Bates, "The Medium Is the Message," *The Tamarack Review* 33 (1964), pp 79-86.

5 Arnold Rockman, "A rich, sprawling book," *Toronto Star*, 13 June 1964.

6 Lister Sinclair, "*Understanding Media* hard to understand," *The Globe Magazine*, 11 July 1964.

7 Alan Thomas, "Misunderstanding Media," *Toronto Telegram*, 22 August 1964.

8 Ibid.

9 Joseph B. Ford, "Letters to the Editor," *Time,* 10 July 1964.

10 Douglas Parker, "Reverse Canadian," *Riverside (California) Press-Enterprise*, 24 May 1964. This view arose out of passages from *Understanding Media* such as the following: "The achievements of the Western world, it is obvious, are testimony to the tremendous values of literacy.

Misunderstandings of McLuhan in the work of high-profile commentators moved him toward the development of a key idea. Charles Reich's *The Greening of America* contained what McLuhan called a "prize misinterpretation"[11] of the percept that the medium is the message. Unlike many of McLuhan's critics, Reich actually said what he thought the phrase meant. McLuhan was grateful for this and the unintended stimulant it provided. "He assumes that I simply mean that the medium has no content, as with the electric light. It will serve as a useful quote in the revision, which I am gradually working at."[12]

Reich's illusion became McLuhan's illumination: "[Reich's] statement is actually one of the few useful remarks that has ever come to my attention about anything I have written. It enables me to see that the user of the electric light, or a hammer, or a language, or a book, is the content."[13] McLuhan wrote Reich a letter of thanks and attempted to disabuse him of his blunder, but Reich insisted that "what you say in your letter is very relevant to the point I was attempting to make in my book—the point that people 'ignore' but are shaped by the media they use in the form of the daily life

---

(*cont.*) But many people are also disposed to object that we have purchased our structure of specialist technology and values at too high a price" (p84). In his public pronouncements and in correspondence, McLuhan reworked this theme in various forms such as the following: "In so far as print bias renders us helpless and ineffectual in the new electronic age, I am strongly inclined to cultivate the kinds of perception that are relevant to our state" (*Letters of Marshall McLuhan*, p3). From such statements commentators construed his stance as "anti-book." Yet in the same context, McLuhan noted: "Not that I am an anti-book person, or an anti-lineal thinker. If I have any normal and natural preferences, they are for the values of the literate world" (ibid.).

11 Marshall McLuhan to Eric McLuhan, *Letters*, p418.

12 Ibid.

13 Marshall McLuhan to Edward T. Hall, *Letters*, p422.

routines." [14]

McLuhan shared his new insight with Peter Newman, then editor of the *Toronto Star*: "The user is always the content of any medium. He creates the *meaning* by gradually discovering the potential of the medium of which he is the content. This is just as true of language as of a house or a car. The *meaning* is the interplay. We make sense by gradually focussing and exploring the various media that surround us. But in making sense, we also make new service / disservice environments which in turn become new media. If the meaning is the process of interplay between us and a technology, the effect or message results from the projection of this interplay between us and the media. To say, therefore, that 'the medium is the message' is to coalesce these stages a bit." [15]

Since writing to Reich, McLuhan had moved from thinking about one medium as the *content* of another to that of a medium as *meaning*, because it is the content of another. Or, to emphasize the structural dynamics in play, because of *its relation* to another medium. In *Take Today* (with Barrington Nevitt), McLuhan gave this idea pride of place on the first page of the text as "the meaning of meaning is relationship."

Here is an echo of the title of the book co-authored by C. K. Ogden and McLuhan's Cambridge mentor I. A. Richards: *The Meaning of Meaning*. If McLuhan was looking back to the ideas he had absorbed from Richards, he was also moving toward the idea of all technologies as linguistic in structure. This move may have been prompted at first by an observation on his original media probe from one of his correspondents [16]

14 National Archives of Canada (NAC), Charles Reich to Marshall McLuhan, 22 January 1971.
15 NAC, Marshall McLuhan to Peter Newman, 5 February 1971.
16 Max Nanny: "By the way, I have found out that the difficulties concerning your insight 'The Medium is the Message' can elegantly be solved by a recourse to [Ferdinand] de Saussure's distinction between 'langue' [language system] and 'parole' [speech]: the medium as 'langue'

and was certainly stimulated by criticisms of McLuhan's work from Jonathan Miller and Umberto Eco.

McLuhan took notice of Jonathan Miller's criticisms of his work in early 1965: "Jonathan Miller is a medical man whose interest in my stuff derives from his concern with the sensory modalities. His interest in sensory modalities stems directly from his interest in neurology."[17] At first, McLuhan had earned high praise from Miller, who said that McLuhan was "doing for the fact of visual space what Freud did for sex, namely to reveal its pervasiveness in the structuring of human affairs."[18] In his correspondence with McLuhan, Miller declared himself a McLuhanite but had suggestions to make and added his own observations on the television medium: "I am one of your more vehement disciples and keep wishing that you would cool down the jargon a little bit ... The other thing which has always worried me is the slightly promiscuous way in which you use 'cool' and 'hot'"[19] Miller noted four distinct meanings he had discovered for these terms in McLuhan's writings, a polysemic permissiveness of which he disapproved. If Miller was a disciple, he was quickly becoming a doubting Thomas about points McLuhan had set out in *Understanding Media*; he was particularly keen to challenge him on his description of television as the cool medium *par excellence*.

The points McLuhan makes in explaining television in *Understanding Media* looked as if they could provide easily scored debating points to Jonathan Miller. He differed with McLuhan above all on the nature of the image on the television

---

(*cont.*) is the message; the medium as 'parole' is rather content oriented, or as you said, the user!" NAC, volume 32, file 45. Max Nanny to Marshall McLuhan, 3 June 1971.

17 NAC, Marshall McLuhan to Lou Forsdale, 28 January 1965.

18 NAC, Marshall McLuhan to Peter Drucker, 2 February 1965.

19 NAC, Jonathan Miller to Marshall McLuhan, 28 April 1965

screen, and noted: "The filling in which occurs in watching TV ... really comes down to the simple matter of completing perspectives and not giving some spatial meaning to the different patches of grey tones. In this respect, therefore, I do not think that you can really put too much weight on the participation which an audience brings to TV. It is *important certainly in the neuro-psychological argument*." [20]

Here Miller was allowing a point that would be demonstrated by psychologist Herbert Krugman in support of McLuhan's view,[21] but Miller pursued in his argument with a notion radically different from McLuhan's: "In fact TV does work by a series of frame by frame presentations and although within each frame the picture has a grain which is due to the scanning line on the camera, the picture which the audience gets is a compilation of *frames* and the scanning process which makes up this picture does *not* impose scanning on the *observer*. In a sense, both movies and TV are particulate presentations ... The difference between the two is purely quantitative and not, in fact, qualitative." [22]

Miller eventually dropped this argument against McLuhan's view, but in 1965 he was still insisting that the iconic quality of television was pure metaphor, ascribing it to "the smallness of the screen which makes it like a tiny devotional object in the drawing room." [23] McLuhan ignored this absurd claim, responding simply: "Don't allow my terminology to 'put you off.' I use language as probe, not as package. Even when I seem to be making very dogmatic statements, I am exploring contours." [24]

---

20 NAC, Jonathan Miller to Marshall McLuhan, 28 April 1965. (emphasis added)
21 Marshall McLuhan to Hugo McPherson, *Letters*, p409.
22 NAC, Jonathan Miller to Marshall McLuhan, 28 April 1965. McLuhan never maintained that the scanning process is imposed on the TV viewer.
23 Ibid.
24 NAC, Marshall McLuhan to Jonathan Miller, 4 May 1965.

McLuhan was delighted when he learned that Frank Kermode had commissioned Miller to write *McLuhan* for the prestigious Modem Masters series. [25] He admitted that he was "enormously flattered that Jonathan Miller should take the trouble to even look at my stuff." [26] The trouble would prove to be McLuhan's: Miller's book was a scathing attack. McLuhan read it only when he realized how much attention it was drawing. His first response was to write to Kermode: "[Miller] is not inquiring nor discussing along the lines I have opened up. He assumes that our sensory order is not violatable by new technologies. This is a universal assumption of our entire establishments, humanist and scientific alike. Merely to challenge it creates panic, for it means that we have polluted not just the physical but the psychic and perceptual order of our societies without questioning our procedures. To argue whether there is any quantitative proof of this is part of the panic. Nobody *wants* any proof. Most people desperately don't want it." [27]

Later, McLuhan wrote to the BBC's Martin Esslin: "May I ask you whether you think it reasonable that I should write a letter to *Encounter* concerning Jonathan Miller? In his book he refers to my work as a pack of lies, without referring to any of my percepts whatever. In what sense do you think he means 'lies'? Does he imagine that I have consciously and deliberately falsified evidence? Surely Miller is a clown with the habits of a sixth form debator?" [28]

The little book stayed in the public eye, annoying McLuhan, who began referring to Miller's "anti-McLuhan

---

25 "Alan Williams was telling me that you reacted very favourably to the news that Jonathan Miller was writing a book about you for my series" (NAC, Frank Kermode to Marshall McLuhan, 25 June 1969).
26 *Letters*, p375.
27 NAC, Marshall McLuhan to Frank Kermode, 4 March 1971.
28 *Letters*, p440.

crusade." [29] He eased his irritation by deciding to treat Miller's work as a spoof: "Apropos your review of Miller's *McLuhan*. Dr. Miller, remember, is the cabaret artist and satirist who did 'Beyond the Fringe,' in part at least. His book is a spoof, of course. Quite deliberately, he ignored my Cambridge work on Thomas Nashe and the history of rhetoric, of the trivium and quadrivium from the fifth century to the present. He ignored all my symbolist poetry studies in order to create the nostalgic image of a prairie boy yearning for the fleshpots of culture." [30]

McLuhan's major foray into linguistics came in the final decade of his career. His eclectic reading in modern linguistic theory began in the very year that Jonathan Miller published *McLuhan*, with its final (and longest) chapter, "Language, Literacy and the Media," couching criticism of McLuhan in terms of linguistic theory. Miller charges that McLuhan's work lacks the scientific basis that linguistics offers. McLuhan turned to linguistics to arm himself for a meeting on the turf Miller had chosen. This is evident from the subtitle of his *Laws of Media: The New Science*, with its deliberate echo of both Francis Bacon and Giambattista Vico.

Miller might not have moved McLuhan toward linguistics, if he had not been adding his voice to that of Umberto Eco. Here was a critic of McLuhan's thought sniping from the lofty intellectual ridge above linguistics—semiotic theory. Unlike linguistics, semiotics does not confine itself to the

---

29 "Jonathan Miller has been continuing his anti-McLuhan crusade in the July 15th *Listener*. Having at last taken time to read his *McLuhan* book, I can honestly say I am amazed that he would take so much trouble to accomplish such a pitiable objective. He doesn't know what I am talking about because he has never taken a step outside the boundaries of visual space so dear to the positivist and the quantifier in these recent centuries. It all ended with Lewis Carroll and the new science of nuclear physics, to say nothing of poetry and the arts in this century. These things Miller has no inkling about, except as mysterious, irrational mythologies that he

study of human language but investigates any and all forms of communication. Its wide scope could reasonably be expected to welcome free-ranging thinkers such as McLuhan, but Eco will admit no such accommodation. Understanding the criticisms of McLuhan put forward by Eco and Miller, understanding their misguided thrust, provides a key to understanding McLuhan's work, and to appreciating his difficulties in getting his message through — even to an intellectually sophisticated audience.

The medium is not the message. The starting point for this observation is Eco's reflection on a cartoon showing a cannibal chieftain wearing an alarm clock as a necklace. (Cf. pages 199, 210, 287) Eco disagrees with McLuhan's claim that the invention of the clock universally fostered a concept of time as space divided into uniform chunks. He will concede that this happened for some persons but believes that the message of the clock could have a different meaning for others (as it did for the cannibal). Eco refers to this as the residual freedom of the individual to interpret in different ways. If we are willing to grant this point, Eco says, "it is still equally untrue that acting on the form and contents of the message convert the person receiving it."[31]

As a criticism of the idea that the medium is the message, Eco's comment strays amazingly far from McLuhan: (1) it does not deal with the unperceived effects of a technology at the sensory level, but conscious reflection on the technology; (2)

(*cont.*) finds extremely repugnant. In the name of 19th century rationalism he mounts an anti-Catholic crusade against McLuhan in the spirit of the Rev. Paisley of Belfast…He belongs to the same vintage as Lewis Mumford who, likewise, cannot understand us at all" (NAC, Marshall McLuhan to Dave Dooley, 12 August 1971).

30 NAC, volume 23. file 79, Marshall McLuhan to Marshall Fishwick, 30, November 1972.

31 *Travels in Hyperreality*, p138.

it puts form and content together as if this were part of McLuhan's view, when in fact he always separates them.

McLuhan uses the term "media" broadly. This part of Eco's discussion sets out techniques of semiotics and then charges that McLuhan does not respect the distinctions required by those techniques: "To say that the alphabet and the street are 'media' is lumping a code together with a channel" [32] This is no different than viewing the cannibal (with the residual freedom Eco gave him) as a threat to professional standards of clock-making. Allowing McLuhan the freedom to define "media" broadly does not undermine the work of the semiotician.

But Eco continues, noting that electric light can be signal, message, or channel, whereas McLuhan is concerned only with the third of these when he says that electric light is a medium without a message. Now Eco's objection is to too narrow a focus rather than too broad a definition. His examples of the cases where light is a signal (using light to flash a message in Morse code) or a message in itself (light left on in a window as an all-clear to a lover) do not have any effect on the scale, speed, or patterns of organization in society as a whole (McLuhan's definition of message) and do not, therefore, damage McLuhan's view. McLuhan is concerned with the effect of light, produced whether light is signal, message (Eco's sense), or channel.

Just as McLuhan's broad definition of "media" does not subvert semiotics, the semiotician's distinctions do not detract from the relevance of McLuhan's observations to his own purpose. Eco concludes his comments on McLuhan's sense of "media" by stating that "it is the code used that gives the light-signal its specific content." [33] This neither undermines nor is undermined by any of

---

32 Ibid.
33 Ibid., p139.

McLuhan's pronouncements on the subject. In fact, it has nothing to do with them.

All media are not active metaphors. Here Eco's argument is still that McLuhan ignores the semiotician's all-important concept of code: that although languages translate the form of an experience because they *are* codes, a metaphor is just a replacement *within* a code. On the one hand, Eco argues that the sense in which print is a medium should be distinguished from the sense in which language is a medium, and that this makes McLuhan's talk of media as metaphors too all-encompassing. On the other hand, he argues that McLuhan's analysis would be improved by replacing the notion of media as metaphors with that of code, as if there could be a unified code of media. He admits that "the press does not change the coding of experience with respect to the written language."[34] It would be of no use to McLuhan, therefore, to analyze in terms of codes, when it is precisely the changes brought about by new media that he wishes to study.

"The medium is the message" has three possible meanings. This is a far cry from saying, as Eco did earlier, that the medium is *not* the message. Now Eco is concerned that the potential meanings of McLuhan's phrase are contradictory. He gives them as (1) the form of the message is the real content of the message; (2) the code is the message; (3) the channel is the message. And while these are only potential meanings, they are somehow proof for Eco that "it is not true, as McLuhan states, that scholars of information have considered only the content of information without bothering about formal problems."[35]

Language is not a medium. This criticism comes not from Eco but from Jonathan Miller, who takes the view that McLuhan makes a "spurious assumption that one can consider lan-

34 Ibid., pp233-34.
35 Ibid, p235.

guage as a technical medium which exists independently of the mind which uses it." [36] The alleged evidence for this view comes from a long ride on which Miller takes the reader, through the territory of linguistics to a destination that McLuhan never set for himself. At that point Miller no longer speaks of McLuhan's "spurious assumption" but "difficulties which arise when language is regarded as a medium." [37] Just as Eco condemns the absence of the semiotician's specialist viewpoint in McLuhan work, Miller condemns the absence there of the distinction linguists make between knowing language and using it.

Miller writes as if making that distinction causes the collapse of McLuhan's claim that language is a medium. This simply could not be. The distinction made in linguistic theory between knowing language and using it does not even deal with words as expression of thought. But Miller continues to draw heavy artillery from the linguistic arsenal, concluding: "Any theory of human communication which does not take its implied differences into consideration has very little right to be taken seriously." [38]

In fact, McLuhan's objective is not to offer a theory of human communication, but to probe the effects of anything and everything we use in dealing with the world around us, including language. This purpose is not served by the specialized focus of linguistics, nor does linguistics discredit McLuhan's approach.

McLuhan takes metaphors literally. Jonathan Miller quotes the following passage from *Understanding Media*: "The TV image is not a *still* shot. It is not photo in any sense, but a ceaselessly forming contour of things limned by the scanning-finger. The resulting plastic contour appears by light

---

36 *McLuhan*, p108.
37 Ibid., p111.
38 Ibid., p110.

through, not light *on*, and the image so formed has the quality of sculpture and icon, rather than of picture." (page 418) Here Miller finds "a vivid example of a metaphor illicitly conjured into a concrete reality." [39] This charge is based on misinterpretation. McLuhan does not refer to the TV image as tactile because of the metaphorical finger scanning the screen, as Miller believes, but because the TV image requires of the eye a degree of involvement as intense as that of touch. McLuhan takes involvement and makes it tactile metaphorically; Miller takes a metaphorical term involving tactility and makes it into a mistake. It is his own, not McLuhan's. The proof that McLuhan does not give a concrete sense to the metaphors of tactility, sculpture, and iconicity is to be found in the same chapter that Miller quotes: "[I]conographic art *uses the eye as we use our hand* to create an inclusive image, made up of many moments, phases, and aspects of the person or thing." (page 442, emphasis added)

Speech is as linear as print. With this claim, Miller presumes to undo a McLuhan percept. The crux of the argument is that sounds can be uttered only one at a time. Miller, believing himself to be clinching a point with the observation that speech can be recorded on a length of magnetic tape, asks, "How linear can one get?" [40] This rather lame line of thinking ignores the very different qualities of the marginal linearity Miller describes and the much more powerful linearity of print, a linearity constantly forcing the eye to move from left to right, from top to bottom, over visible figures against visible ground.

Facing such irrelevant criticisms from Eco and Miller, McLuhan could have countered them easily by arguing from his own position. And so he did, in the case of Miller, in both

---

39 *McLuhan*, p121.
40 *McLuhan*, p113.

private correspondence and the popular press. [41] But he also chose to delve into linguistics, and in so doing he moved into a final and little-known phase of his intellectual journey. [42]

— W. Terrence Gordon

---

41 McLuhan's reply to Miller's piece on McLuhan in *The Listener*, 15 July 1971, was published in the magazine on 26 August 1971 and is reprinted in *Letters of Marshall McLuhan*, pp435-38.
42 For a discussion, see W. Terrence Gordon, *Marshall McLuhan: Escape into Understanding* (Gingko Press, 2003).

Glossary

alphabet

> (phonetic): a writing system that shows how its units
> are pronounced. Compare *and* (phonetic) with *&*
> (non-phonetic). The study of the phonetic alphabet
> is crucial in identifying the profoundly different
> qualities and effects between it and other writing
> systems such as hieroglyphics and ideograms. (See
> also *lineal structure*)

autoamputation

> a concept borrowed from medical research into
> stress. It refers to the human body reacting to an
> unidentifiable source of irritation by shutting down
> the affected area. The concept is extended as an
> explanation for the development of human technol-
> ogy.

automation

> the term is not to be understood in its commonest
> sense of complex industrial assembly by a succession
> of machines but rather as the technique of convert-
> ing a mechanical process to automatic operation by
> electronic control.

center-margin dynamics

> the dynamics of any system that operates by one-way
> expansion from its center to its peripheral elements.
> Center-margin dynamics are associated with a host
> of technological extensions and social structures
> from the phonetic alphabet and the Roman Empire

to assembly lines and the separation of the individual from the state.

closure

not in whatever sense(s) the term may have had as the buzzword of the mid-1990s, but in connection with *autoamputation* (see above), to refer to the human body's attempt to regain equilibrium among its channels of sensory input, whenever this equilibrium is disturbed by new media.

common sense

a direct and literal translation of the Latin phrase *sensus communis*, referring to the conscious elucidation of experience by systematically transferring or transposing it from one bodily sense to another or all others. Touch, the "meeting-place of the senses" is the common sense.

content

in its various conventional senses of conceptual material conveyed, information, meaning, or messages, distracts attention from the technology or medium conveying the content. Any medium is more powerful in itself than whatever content it might be used to convey, and it is in this sense that *the medium is the message*.

cool medium

one that demands the user's participation (See also *medium*)

counter-irritant

When technologies produce stress and pressure through speed-up or overload, new technologies develop to offset those effects. Counter-irritants can be benign (games) or as destructive as the original irritant (a drug habit).

decentralization

of political power, of urban structures, of work environments, etc., is the chief effect of electricity

detachment

or the capacity to act without reacting, goes hand in hand with the fragmenting nature of the phonetic alphabet and is a powerful, primary effect of literacy. (See also *literacy*)

detribalization

or collapse of tribal structure and pattern is a consequence of the introduction of literacy and mechanical technologies to societies and cultures.

environment

understood as total situation, may refer variously to physical surroundings or social contexts but more frequently to features of these that warrant examination for the effects created in them by new technologies. *Environment* is another term for *medium*. (See also *medium*)

explosion

a key term for describing the fragmenting and specializing effect of mechanical technologies (see also *fragmentation, cf. implosion*)

field

the inclusive or unified properties of a subject or of a method of analysis. The concept of the field occurs in linguistic studies *(semantic field theory)* and physics *(unified field theory)*

fragmentation

> the principal effect of mechanical technologies that operate on the repeatability principle (see below). It characterizes technologies from the phonetic alphabet and moveable type to specialist knowledge and the assembly line.

global village

> a metaphor for our planet reduced in all aspects of its functioning and social organization to the size of a village by the effect of electricity since the advent of the telegraph.

high definition

> describes media that convey sharply defined forms and sharply-separated forms—*e.g.* printed letters on page, images fixed on film. Synonymous with "hot," non-involving (*cf. low definition*)

hot medium

> one that inhibits the user's involvement. (See also *medium*)

hybridization

> compounding or interaction of media, which proves particularly revealing of their structural features and effects

icon

> in the broad sense of any inclusive form; related to the properties of whatever functions by inclusiveness. Iconic qualities are associated with the senses of smell and touch and with the interplay of the senses, as with television.

implosion

> the principal effect of electricity, which has permitted the most powerful technological extension of all, that of the central nervous system, with the effective elimination of space and time and the reversal of media effects associated with mechanical technology (*cf. explosion*)

information

> occasionally in the sense of *facts* or *data*, but more frequently in the sense of a medium and its property of being *in formation* or in relation to another medium in its operation

lineal structure

> a key feature of the phonetic alphabet and all its technological derivatives characterized by isolating and fragmenting parts of a whole and wedding that fragmentation to the repeatability principle, whether for scientific method, industrial production, or social organization

literacy

> generally, the ability to read. Specifically, the use of letters and the phonetic alphabet. (See also *alphabet*)

low definition

> describes media that convey weakly defined forms —*e.g.* speech sounds, images on a television screen. Synonymous with "cool," involving. (*cf. high definition*)

medium

> not limited to the stereotypical medium of mass communication such as radio, a medium is any extension of the human body (wheel as extension of foot, com-

puter as extension of central nervous system) or form of social organization and interaction (language, roads, money). Specifically, a medium is a side-effect of a technology, generally invisible; it consists of all the psychic and social adjustments that its users and their society undergo when they adopt the new form. It is the "message" sent by the new technology; so "the medium is the message."

message

exceptionally, in the sense of content (see above); usually in the sense of change in scale, pace or pattern of human action and interaction created by new media, thus *the* [new] *medium is the message* [of any innovation].

mosaic

integral form and integrating process, closely related to *field* (see above)

myth

free of the negative connotation it can carry in contemporary usage (*e.g. urban myth*), refers to instant recognition or comprehension of a complex process (*cf. icon, organic, pattern*)

Narcissus trance

an interpretation of the Greek myth of Narcissus as the inability to recognize technologies as extensions of the human being and the failure to detect the message, or new environment, created by new technologies. See also *numbing, somnambulism.*

numbing

see *Narcissus trance, somnambulism*

obsolescence

>the typical effect of a medium to render obsolete older media with the same, or similar, function

organic

>a quality of interrelated wholes. See also *myth, structure.*

participation

>used not in reference to *intellectual* or reflective participation but to mean involvement and degree of involvement of the *physical* senses of the user of a medium. Hot media (see above) are low in participation or sensory involvement; cool media (see above) require high participation.

pattern

>defined early in the text as unity of form and function

repeatability

>the principle of indefinitely repeating a process or reusing specialized, fragmented units of a mechanical, linear technology

retribalization

>the recovery of ancient tribal structures and patterns in a society previously under the dominant influence of mechanical technology and newly under electric technology

reversal

>a typical effect produced when a technology is pushed to its extreme form and acquires its opposite characteristics

role(s)

> the form of social behavior in the work place, education, etc., imposed on Western society by electric technology, in contrast to individual goals under mechanical technology and literacy

sense ratios

> the degree of balance among the physical senses in relation to the technologies that extend them and that dominate society

simultaneity

> a principal culural side effect of electricity.

specialism

> see *fragmentation*

structure

> an integrated whole or configuration, particularly of the type imposed by electric technology. See also *field, mosaic, myth*.

symbolism

> refers principally to the aesthetic manifestations of structural awareness. *Cf. field, mosaic, myth, structure.*

synesthesia

> transfer of perceptions from one sense to another and / or integration of the senses, in the arts, for aesthetic or educational purposes

tactility

> used not only in reference to the sense of touch but to describe the quality of a medium requiring a high degree of involvement of one or more of the other senses

technology

 in the broad sense, the same as *medium* above

translation

 the process of transfering knowledge from one mode to another

tribal structure

 social interaction characterized by the intense mutual involvement of the members of the community and the opposite of the detached individualism and private identities characteristic of a literate culture under mechanical technology

uniformity

 all manifestations of the homogeneous quality resulting from mechanical technologies and the repeatability principle (see above)

Works Cited

This is a complete list of works cited in *Understanding Media* but not a representative bibliography of media analysis in the sense qualified by its subtitle: *The Extensions of Man*. McLuhan was the first to develop such an analysis in the pages of this book, and his references are, therefore, largely the primary sources from which he fashioned that development. But these references are not confined to his sources of inspiration (some of which—most notably the writings of I. A. Richards and Harold Innis—are absent), for many of the cited works serve McLuhan as approaches and models to avoid. Frequently, a title will be introduced by such comments as "But Mumford takes no account of…," "Eliade is unaware…," "Bergson seems not to have noticed…" What emerges from even a casual inspection of the following pages is that McLuhan's approach to understanding media is informed in some instances by academic studies in domains as diverse as psychology, history, philosophy, economics, and psychology, but more often inspired by the insights of playwrights, poets, and novelists.

References as given in the text have been left in their original form but corrected and/or given in fuller form below, including date of original publication, whenever possible. Exceptionally, dates for Shakespearean plays have been omitted; any inadvertent attention to controversies surrounding those dates would only serve to debase the McLuhan dictum that the medium is the message.

Anonymous. *The Book of Kells.* c. 800.

Baudelaire, Charles. *Les Fleurs du mal,* 1857.

Benda, Julien. *The Great Betrayal.* London, 1928. (Translation of *La Trahison des clercs,* 1927)

Bergson, Henri. *Creative Evolution,* 1911.

Bernard, Claude. *An Introduction to the Study of Experimental Medicine,* 1927.

Betjeman, John. "Slick but Not Streamlined." *Poems and Short Pieces,* 1947.

Blake, William. *Jerusalem: The Emancipation of the Great Albion,* 1804–20.

Boorstin, Daniel J. *The Image, or What Happened to the American Dream,* 1961.

Bosanquet, Theodora. *Henry James at Work,* 1927.

Boulding, Kenneth. *The Image: Knowledge in Life and Society,* 1956.

Burns, Robert. "The Cotter's Saturday Night." In *Poems Chiefly in the Scottish Dialect,* 1786.

Burroughs, William. *Naked Lunch,* 1959.

Butler, Samuel. *Erewhon,* 1872.

Byron, George Gordon. *Childe Harold's Pilgrimage,* 1812.

Canetti, Elias. *Crowds and Power,* 1962.

Capek, Milic. *The Philosophical Impact of Contemporary Physics,* 1961.

Carnegie, Dale. *How to Win Friends and Influence People,* 1936.

Carothers, J. C. *The African Mind in Health and Disease,* 1953.

Carroll, Lewis. *Alice's Adventures in Wonderland,* 1865.

Cater, Douglass. *The Fourth Branch of Government,* 1959.

Cervantes, Miguel de. *Don Quixote,* 1605.

Cheyney, Peter. *You Can't Keep the Change*, 1940.

Christie, Agatha. *The Labours of Hercules*, 1947.

Dantzig, Tobias. *Number: The Language of Science*, 1930.

Deane, Philip. *I Was a Captive in Korea*, 1953.

De Grazia, Sebastian. *Of Time, Work and Leisure*, 1962.

Dewart, Leslie. *Christianity and Revolution: The Lesson of Cuba*, 1963.

Dickens, Charles. *The Cricket on the Hearth*, 1846.

____. *David Copperfield*, 1849-50.

____. *Great Expectations*, 1860-61.

____. *Pickwick Papers*, 1837-39.

Donne, John. "The Sun Rising." In *Poems*, 1633.

Doob, Leonard. *Communication in Africa: A Search for Boundaries*, 1961.

Eisenstein, Sergei. *Notes of a Film Director*, 1959.

Eliade, Mircea. *The Sacred and the Profane: The Nature of Religion*, 1959.

Eliot, T. S. *Four Quartets*, 1944.

____. "The Love Song of J. Alfred Prufrock." In *Prufrock and Other Observations*, 1917.

____. *Murder in the Cathedral*, 1935.

____. *Sweeney Agonistes: An Aristophanic Fragment*, 1932.

____. *The Waste Land*, 1922.

Fitzgerald, F. Scott. *The Great Gatsby*, 1925.

Flaubert, Gustave. *Sentimental Education*, 1941.

Forster, E. M. *A Passage to India*, 1924.

Fuller, R. Buckminster. *Education Automation: Freeing the Scholar to Return to his Studies*, 1962.

Genet, Jean. *The Balcony*, 1957.

Gibbon, Edward. *The History of the Decline and Fall of the Roman Empire*, 1776-81.

Gogol, Nikolai. *The Overcoat*, 1842.

Gombrich, E. H. *Art and Illusion: A Study in the Psychology of Pictorial Representation*, 1960.

Gray, Thomas. *Elegy*, 1751.

Hall, Edward T. *The Silent Language*, 1959.

Handlin, Oscar. *Boston's Immigrants 1790-1865: A Study in Acculturation*, 1941.

Heisenberg, Werner. *The Physicist's Conception of Nature*, 1962.

Hussey, Christopher. *The Picturesque*, 1927.

Joyce, James. *Finnegans Wake*, 1939.

_____. *Ulysses*, 1922.

Jung, C. J. *Contributions to Analytical Psychology*, 1928.

Keats, John. *The Insolent Chariots*, 1958.

Keynes, J. M. *A Treatise on Money*, 1930.

Kierkegaard, Søren. *The Concept of Dread*, 1944.

Lam, Bernard. *The Art of Speaking*, 1696.

Lederer, William J. *The Ugly American*, 1958.

Lewis, Wyndham. *The Childermass*, 1928.

Liebling, A. J. *The Press*, 1961.

Marvell, Andrew. "To his Coy Mistress." In *Miscellaneous Poems*, 1681.

McLuhan, Marshall. *The Gutenberg Galaxy*, 1962.

Miller, Arthur. *Death of a Salesman*, 1949.

Milton, John. *Paradise Lost*, 1667.

Meiss, Millard. *Andrea Mantegna as Illuminator*, 1957.

Modupe, Prince. *I Was a Savage*, 1957.

Mumford, Lewis. *The City in History: Its Origins, Its Transformations, Its Prospects*, 1961.

____. *Technics and Civilization*. 1934.

Nef, John U. *War and Human Progress*, 1950.

Packard, Vance. *The Hidden Persuaders*, 1957.

Pope, Alexander. *The Dunciad*, 1726.

Potter, Stephen. *The Theory and Practice of Gamesmanship*, 1947.

Ross, Lillian. *Picture*, 1962.

Rourke, Constance. *American Humor: A Study of the National Character*, 1931.

Rowse, A. L. *Appeasement: A Study in Political Decline 1933–1939,* 1961.

Russell, Bertrand. *The ABC of Relativity*, 1925.

Sachs, Curt. *World History of the Dance*, 1963.

Sansom, G. B. *Japan*, 1931.

Schramm, Wilbur. *Television in the Lives of Our Children*, 1961.

Selye, Hans. *The Stress of Life*, 1956.

Shakespeare, William. *As You Like It*.

____. *Henry V*.

____. *King Lear*.

____. *Macbeth*.

____. *Richard III*.

____. *Romeo and Juliet*.

____. *Troilus and Cressida*.

Spengler, Oswald. *The Decline of the West*, 1932.

Synge, J. M. *Playboy of the Western World*, 1907.

Theobald, Robert. *The Rich and the Poor: A Study of the Economics of Rising Expectations*, 1960.

Toynbee, Arnold. *A Study of History*, 1948.

Volturius. *Art of War*, 1472.

White, Lynn. *Medieval Technology and Social Change*, 1962.

White, Theodore. *The Making of the President: 1960*, 1961.

Whyte, William H. *The Organization Man*, 1956.

Woodham-Smith, Cecil. *Lonely Crusader*, 1951.

Wordsworth, William. "Michael." In *Lyrical Ballads*, 1801.

Young, John Z. *Doubt and Certainty in Science: A Biologist's Reflections on the Brain*, 1951.

Publications
of Marshall McLuhan

Books

1951

The Mechanical Bride: Folklore of Industrial Man. New York: Vanguard Press, 1951; London: Routledge & Kegan Paul, 1967; Corte Madera, CA: Gingko Press, 2002.

1954

Selected Poetry of Tennyson. Edited by Marshall McLuhan. New York: Rinehart, 1954.

1960

Explorations in Communication: An Anthology. Edited by Edmund Carpenter and Marshall McLuhan. Boston: Beacon Press, 1960.

Report on Project in Understanding New Media. Washington, D.C.: U.S. Office of Education, 1960.

1962

The Gutenberg Galaxy: The Making of Typographic Man. Toronto: University of Toronto Press, 1962. Translated into French by Jean Paré and published as *La Galaxie Gutenberg: la genèse de l'homme typographique.* Montreal: Hurtubise HMH, 1967; Paris: Gallimard, 1977, 2 vols.

1964

Understanding Media: The Extensions of Man. New York: McGraw-Hill, 1964; second edition 1965; MIT Press edition, Cambridge, Mass.: MIT Press, 1994, with an introduction by Lewis Lapham. 1964 edition translated into French by Jean Paré and published as *Pour comprendre les média: les prolongements technologiques de l'homme.* Montreal: Hurtubise HMH, 1968. Reissued in a new edition (Bibliotheque Québécoise, #36) in 1993. Translations of *Understanding Media* have appeared in more than twenty languages.

Voices of Literature. (2 vols.) Edited by Marshall McLuhan and Richard J. Schoeck. New York: Holt, Rinehart and Winston, 1964, 1965.

1967

McLuhan: Hot & Cool. A Primer for the Understanding of and a Critical Symposium with a Rebuttal by McLuhan. Edited by Gerald Emanuel Stern. New York: Dial Press, 1967. New York: The New American Library, 1969. Thirty-one selections include reprinted essays (in whole or in part) from Howard Luck Gossage, Tom Wolfe, John Culkin, Walter Ong, Dell Hymes, Frank Kermode, George Steiner, Susan Sontag, and five selections from McLuhan's writings. The book concludes with the transcript of a dialogue between the editor and McLuhan, originally published in *Encounter* in June 1967, wherein McLuhan responds to commentaries on his work from some of the other contributors to the volume.

The Medium Is the Massage: An Inventory of Effects. With Quentin Fiore and Jerome Agel. New York: Bantam, 1967. Reprinted, Gingko Press, 2001. Translated into French and published as Message et massage. Montreal: Hurtubise HMH, 1968.

Verbi-Voco-Visual Explorations. New York: Something Else Press, 1967. (Reprint of Explorations, no. 8.)

1968

Through the Vanishing Point: Space in Poetry and Painting. With Harley Parker. New York: Harper & Row, 1968.

War and Peace in the Global Village: an inventory of some of the current spastic situations that could be eliminated by more feedforward. With Quentin Fiore and Jerome Agel. New York: Bantam, 1968. Reprinted, New York: Touchstone Books, 1989; Gingko Press, 2001. Translated into French as *Guerre et paix dans le village plané-taire: un inventaire de quelques situations spasmodiques courantes qui pourraient être supprimées par le feedforward.* Montreal: Hurtubise HMH, 1970; Paris: Laffont, 1970.

1969

Counterblast. With Harley Parker. New York: Harcourt, Brace and World, 1969. Translated into French by Jean Paré. (Montreal: Hurtubise, 1972; Paris: Mame, 1972.)

The Interior Landscape: The Literary Criticism of Marshall McLuhan 1943–1962. Edited by Eugene McNamara. New York: McGraw-Hill, 1969.

Mutations 1990. Translation by François Chesneau of *The Future of Sex* (see Other Works below). Montreal: Hurtubise HMH, 1969.

1970

Culture Is Our Business. New York: McGraw-Hill, 1970.

From Cliché to Archetype. With Wilfred Watson. New York: Viking, 1970. Translated into French by Derrick de Kerckhove and published as *Du cliché à l'archétype: la foire du sens.* Montreal: Hurtubise HMH, 1973; Paris: Mame, 1973. Translated into Italian by Francesca Valente and Carla Pezzini and published as *Dal cliché all'archetipo: l'uomo tecnologico nel villaggio globale.*

1972

Take Today. With Barrington Nevitt. Toronto: Longman, 1972.

1977

Autre homme autre chrétien à l'âge électronique. With Pierre Babin. Lyon: Editions du Chalet, 1977.

City as Classroom: Understanding Language and Media. With Eric McLuhan and Kathryn Hutchon. Toronto: Book Society of Canada Limited, 1977.

D'Oeil à oreille. Translation by Derrick de Kerckhove of articles by and interviews with McLuhan. Montreal: Hurtubise, 1977.

1978

The Possum and the Midwife. [text of McLuhan lecture on Ezra Pound.] Moscow: University of Idaho Press, 1978.

1987

Images from the Film Spiral. Selected by Sorel Etrog with text by Marshall McLuhan. Toronto: Exile Editions, 1987.

Letters of Marshall McLuhan. Selected and edited by Matie Molinaro, Corinne McLuhan, and William Toye. Toronto: Oxford University Press, 1987.

1988

Laws of Media: *The New Science*. With Eric McLuhan. Toronto: University of Toronto Press, 1988.

1989

The Global Village: *transformations in world life and media in the 21st century*. With Bruce R. Powers. New York: Oxford University Press, 1989.

1995

Essential McLuhan. Edited by Eric McLuhan and Frank Zingrone. Toronto: Anansi, 1995.

Other Works by Marshall McLuhan

1930

Macaulay: What a Man! *The Manitoban* (University of Manitoba student newspaper), 28 October 1930.

1933

Canada and Internationalism. *The Manitoban*, 1 December 1933.

George Meredith. *The Manitoban*, 21 November 1933.

German Character. *The Manitoban*, 7 November 1933.

Germany and Internationalism. *The Manitoban*, 27 October 1933.

Germany's Development. *The Manitoban*, 3 November 1933.

Public School Education. *The Manitoban*, 17 October 1933.

1934

Adult Education. *The Manitoban*, 16 February 1934.

De Valera. *The Manitoban*, 9 January 1934.

George Meredith as a Poet and Dramatic Novelist. M.A. thesis, University of Manitoba, 1934.

The Groupers. *The Manitoban,* 23 January 1934.

Morticians and Cosmeticians. *The Manitoban*, 2 March 1934.

Not Spiritualism but Spiritism. *The Manitoban*, 19 January 1934.

Tomorrow and Tomorrow. *The Manitoban*, 16 May 1934.

1936

G. K. Chesterton: A Practical Mystic. *The Dalhousie Review* 15 (1936), 455–64.

1937

The Cambridge English School. *Fleur de Lis* (Saint Louis University student literary magazine), 1937, 21–25.

1938

Peter or Peter Pan. *Fleur de Lis*, May 1938, 7–9.

Review of *The Culture of Cities* by Lewis Mumford. *Fleur de Lis*, December 1938, 38–39.

1940

Apes and Angles. *Fleur de Lis*, December 1940, 7–9.

Review of *Art and Prudence* by Mortimer J. Adler. *Fleur de Lis*, October 1940.

1941

Review of *American Renaissance* by F. O. Matthiessen. *Fleur de Lis*, October 1941.

Review of *Poetry and the Modern World* by David Daiches. *Fleur de Lis*, March 1941.

1943

Aesthetic Pattern in Keats' Odes. *University of Toronto Quarterly* 12, 2 (1943), 167–79. Reprinted in Eugene McNamara, ed., *The Literary Criticism of Marshall McLuhan 1943–1962*, 99–113.

Education of Free Men in Democracy: The Liberal Arts. *St. Louis Studies in Honor of St. Thomas Aquinas*, 1943, 47–50.

Herbert's *Virtue. The Explicator* 2, 1 (1943), 4. Reprinted in L. G. Locke, W M.Gibson, and G. Arms, eds., *Readings for Liberal Education* (New York: Rinehart, 1948).

The Place of Thomas Nashe in the Learning of His Time. Ph.D. dissertation, Cambridge University, April 1943.

1944

Dagwood's America. *Columbia*, January 1944, 3, 22.

Edgar Poe's Tradition. *Sewanee Review* 52, 1 (1944), 24–33. Reprinted in Eugene McNamara, ed., *The Literary Criticism of Marshall McLuhan 1943–1962*, 211–21.

Eliot's The Hippopotamus. The Explicator 2, 7 (1944), 50.

Henley's *Invictus. The Explicator* 3, 3 (1944), 22.

Kipling and Forster. *Sewanee Review* 52, 3 (1944), 332–43.

Poetic vs. Rhetorical Exegesis. The Case for Leavis against Richards and Empson. *Sewanee Review* 52, 2 (1944), 266–76.

Wyndham Lewis: Lemuel in Lilliput. *Saint Louis Studies in Honor of St. Thomas Aquinas*, 1944, 58–72.

1945

The Analogical Mirrors. In *Gerard Manley Hopkins: The Kenyon Critics Edition*. Norfolk, CT: New Directions Books, 1945, 15–27.

Reprinted in Eugene McNamara, ed., *The Literary Criticism of Marshall McLuhan 1943–1962*, 63–73.

Another Aesthetic Peep-Show. Review of *The Aesthetic Adventure* by William Gaunt. *Sewanee Review* 53 (1945), 674–77.

The New York Wits. *Kenyon Review* 7 (1945), 12–28.

1946

An Ancient Quarrel in Modern America (Sophists vs. Grammarians). *The Classical Journal*, January 1946, 156–62. Reprinted in Eugene McNamara, ed., *The Literary Criticism of Marshall McLuhan 1943–1962*, 223–34.

Footprints in the Sands of Crime. *Sewanee Review* 54 (1946), 617–34.

Out of the Castle into the Counting-House. *Politics*, September 1946, 277–79.

Review of *William Ernest Henley* by Jerome Hamilton Buckley. *Modern Language Quarterly* 7 (1946), 368–70.

1947

American Advertising. *Horizon* 93, 4 (October 1947), 132–41. Reprinted in Eric McLuhan and Frank Zingrone, eds., *Essential McLuhan* (Toronto: Anansi, 1995), 13–20.

Inside Blake and Hollywood. *Sewanee Review* 55 (1947), 710–15.

Introduction to *Paradox in Chesterton* by Hugh Kenner. New York: Sheed and Ward, 1947.

Mr. Connolly and Mr. Hook. Review of *The Condemned Playground. Essays 1927–1944* by Cyril Connolly and *Education for Modern Man* by Sidney Hook. *Sewanee Review* 55 (1947), 167–72.

The Southern Quality. *Sewanee Review* 55 (1947), 357–83. Reprinted in Eugene McNamara, ed., *The Literary Criticism of Marshall McLuhan 1943–1962*, 185–209.

1948

Henry IV, A Mirror for Magistrates. *University of Toronto Quarterly* 17, 2 (1948), 152–60.

1949

The 'Colour-Bar' of BBC English. *Canadian Forum*, April 1949, 9–10.

Mr. Eliot's Historical Decorum. *Renascence* 2, 1 (1949), 9–15. Reprinted in *Renascence* 25, 4 (1972–73), 183–89.

1950

Pound's Critical Prose. In Peter Russell, ed., *Examination of Ezra Pound: A Collection of Essays* (London: Peter Nevill, 1950), 165–71. Reprinted in Eugene McNamara, ed., *The Literary Criticism of Marshall McLuhan* 1943–1962, 75–81.

T S. Eliot [Review of eleven books about Eliot]. *Renascence* 3, 1 (1950), 43–48.

1951

John Dos Passos: Technique vs. Sensibility. In Charles Gardiner, ed., *Fifty Years of the American Novel: A Christian Appraisal* (New York: Charles Scribner's Sons, 1951), 151–164. Reprinted in Eugene McNamara, ed., *The Literary Criticism of Marshall McLuhan 1943–1962*, 49–62.

Joyce, Aquinas and the Poetic Process. *Renascence* 4, 1 (1951), 3–11.

Review of three books on Ezra Pound. *Renascence* 3, 2 (1951), 200–02.

A Survey of Joyce Criticism. *Renascence* 4, 1 (1951), 12–18.

Tennyson and Picturesque Poetry. *Essays in Criticism* 1, 3 (1951), 262–82. Reprinted in Eugene McNamara, ed., *The Literary Criticism of Marshall McLuhan 1943–1962*, 135–55.

1952

Advertising as a Magical Institution. *Commerce Journal*, January 1952, 25–29.

The Aesthetic Moment in Landscape Poetry. In Alan Downe, ed., *English Institute Essays* (New York: Columbia University Press, 1952), 168–81. Reprinted in Eugene McNamara, ed., *The Literary*

Criticism of Marshall McLuhan 1943–1962, 157–67.

Defrosting Canadian Culture. *American Mercury*, March 1952, 91–97.

Review of *Auden: An Introductory Essay* by Richard Hoggart. *Renascence* 4, 2 (1952), 220–21.

Review of *The Poetry of Ezra Pound* by Hugh Kenner. *Renascence,* 4, 2 (1952), 215–17.

Review of *Word Index to James Joyce's Ulysses* by Miles L. Hanley. *Renascence* 4, 2 (1952), 186–87.

Technology and Political Change. *International Journal* 7, 3 (Summer 1952), 189–95.

1953

The Age of Advertising. *Commonweal*, 11 September 1953, 555–57.

Comics and Culture. *Saturday Night*, February 1953, 1, 19–20.

Culture Without Literacy. *Explorations: Studies in Culture and Communications*, no. 1, December 1953, 117–27. Reprinted in Eric McLuhan and Frank Zingrone, eds., *Essential McLuhan* (Toronto: Anansi, 1995), 302–13.

From Eliot to Seneca. *University of Toronto Quarterly* 22, 2 (1953), 199–202.

James Joyce: Trivial and Quadrivial. *Thought*, Spring 1953, 75–98. Reprinted in Eugene McNamara, ed., *The Literary Criticism of Marshall McLuhan 1943–1962*, 23–47.

The Later Innis. *Queen's Quarterly* 60, 3 (1953), 385–94.

Maritain on Art. *Renascence* 6, 1 (1953), 40–44.

The Poetry of George Herbert and Symbolist Communication. *Thought*, Autumn 1953.

Review of *Light on a Dark Horse: An Autobiography 1901–1935* by Roy Campbell. *Renascence* 5, 2 (1953), 157–59.

Wyndham Lewis: His Theory of Art and Communication. *Shenandoah*, Autumn 1953, 77–88. Reprinted in Eugene McNamara, ed., *The Literary Criticism of Marshall McLuhan 1943–1962*, 83–94.

1954

Catholic Humanism and Modern Letters. In *Christian Humanism in Letters*: *The McAuley Lectures, Series 2*, 1954 (West Hartford, CT: St. Joseph College, 1954), 49–67.

Joyce, Mallarmé, and the Press. *Sewanee Review* 62 (1954), 38–55. Reprinted in Eugene McNamara, ed., *The Literary Criticism of Marshall McLuhan 1943–1962*, 5–21. Reprinted in Eric McLuhan and Frank Zingrone, eds., *Essential McLuhan* (Toronto: Anansi, 1995), 60–71.

Media as Art Forms. *Explorations*, no. 2, April 1954, 6–13.

New Media as Political Forms. *Explorations*, no. 3, August 1954, 120–26.

Poetry and Society. Review of *Dream and Responsibility* by Peter Viereck. *Poetry* 84, 2 (May 1954), 93–95.

Through Emerald Eyes. Review of *Three Great Irishmen: Shaw, Yeats, Joyce* by Aarland Ussher. *Renascence* 6, 2 (1954), 157–58.

1955

Five Sovereign Fingers Taxed the Breath. *Explorations*, no. 4, February 1955. Reprinted in *Shenandoah*, Autumn 1955, 50–52.

Nihilism Exposed. Review of *Wyndham Lewis* by Hugh Kenner. *Renascence* 8, 2 (1955), 97–99.

Paganism on Tip-toe. Review of *The Poetry of T. S. Eliot* by D. E. S. Maxwell. *Renascence* 7, 3 (1955), 158.

Radio and Television vs. The ABCED-Minded. *Explorations*, no. 5, June 1955, 12–18.

Space, Time, and Poetry. *Explorations*, no. 4, February 1955, 56–62.

1956

Educational Effects of Mass Media of Communication. *Teachers College Record*, March 1956, 400–03.

The Media Fit the Battle of Jericho. *Explorations*, no. 6 (July 1956), 15–19. Reprinted in Eric McLuhan and Frank Zingrone, eds., *Essential McLuhan* (Toronto: Anansi, 1995), 298–302.

Music and Silence. Review of two books on Joyce. *Renascence* 8, 3 (1956), 152–53.

'Stylistic.' Review of *Mimesis*: *The Representation of Reality in Western Literature* by Erich Auerbach. *Renascence* 9, 2 (1956), 99–100.

1957

Brain Storming (and other essays). *Explorations*, no. 8, October 1957 (unpaginated).

Characterization in Western Art, 1600–1900. *Explorations*, no. 8, October 1957, unpaginated.

Classical Treatment. Review of *Eliot's Poetry and Plays* by Grover Smith. *Renascence*, 10, 2 (1957), 102–3.

Classrooms Without Walls. *Explorations*, no. 7, March 1957, 22–26.

Coleridge as Artist. In Clarence D. Thorpe, Carlos Baker, and Bennett Weaver, eds., *The Major English Romantic Poets: A Symposium in Reappraisal* (Carbondale: Southern Illinois University Press, 1957), 83–99. Reprinted in Eugene McNamara, ed., *The Literary Criticism of Marshall McLuhan 1943–1962*, 115–33.

Compliment Accepted [Review of six books on James Joyce]. *Renascence* 10, 2 (1957), 106–08.

David Riesman and the Avant-Garde. *Explorations*, no. 7, March 1957, 112–16.

Eternal Ones of the Dream. (with Edmund Carpenter). *Explorations*, no. 7, March 1957, unpaginated.

Jazz and Modern Letters *Explorations*, no. 7, March 1957, 74–76.

Manifestos. *Explorations*, no. 8 (October 1957), unpaginated.

The Organization Man. *Explorations*, no. 8, October 1957, unpaginated.

People of the Word. *Explorations*, no. 8, October 1957, unpaginated.

Sight, Sound, and the Fury. In Bernard Rosenberg and David Manning White, eds., *Mass Culture: The Popular Arts in America* (London: Collier-Macmillan, 1957), 489–95.

Soviet Novels. (with Edmund Carpenter). *Explorations*, no. 7, (March 1957), 123–24.

Third Program in the Human Age. *Explorations*, no. 8, October 1957, 16–18.

1958

One Wheel, All Square [Review of five books on James Joyce]. *Renascence* 10, 4 (1958), 196–200.

Our New Electronic Culture: The Role of Mass Communications in Meeting Today's Problems. *National Association of Educational Broadcasters Journal*, October 1958, 19–20, 24–26.

1959

Joyce or No Joyce. Review of *Joyce among the Jesuits* by Kevin Sullivan. *Renascence* 12, 1 (1959), 53–54.

Myth and Mass Media. *Daedalus*, Spring 1959, 339–48.

Virgil, Yeats, and 13,000 Friends. Review of *On Poetry and Poets* by T. S. Eliot. *Renascence* 11, 2 (1959), 94–95.

Yeats and Zane Grey. Review of *The Letters of William Butler Yeats*, edited by Allan Wade. *Renascence*, 11, 3 (1959), 166–68.

1960

Acoustic Space. (with Edmund Carpenter). In Edmund Carpenter and Marshall McLuhan, eds., *Explorations in Communication: An Anthology* (Boston: Beacon Press, 1960), 65–70.

Another Eliot Party. Review of *T S. Eliot: A Symposium for His Seventieth Birthday*, edited by Neville Braybrooke. *Renascence* 12, 3 (1960), 156–57.

Around the World, Around the Clock. Review of *The Image Industries* by William Lynch. *Renascence* 12, 4 (1960), 204–05.

A Critical Discipline. Review of *Wyndham Lewis: A Portrait of the Artist as the Enemy* by Geoffrey Wagner. *Renascence* 12, 2 (1960), 93–95.

Flirting with Shadows. Review of *The Invisible Poet* by Hugh Kenner. *Renascence* 12, 4 (1960), 212–14.

Joyce as Critic. Review of *The Critical Writings of James Joyce* edited by Ellsworth Mason and Richard Ellmann. *Renascence* 12, 4 (1960), 202–03.

Melodic and Scribal. Review of *Song in the Works of James Joyce* by J. C. Hodgart and Mabel P. Worthington. *Renascence* 13, 1 (1960), 51.

The Personal Approach. Review of *Shakespeare and Company* by Sylvia Beach. *Renascence* 13, 1 (1960), 42–43.

Romanticism Reviewed. Review of *Romantic Image* by Frank Kermode. *Renascence* 12, 4 (1960), 207–09.

1961

The Electric Culture. The Books at the Wake. *Renascence* 13,4 (1961), 219–20.

The Humanities in the Electronic Age. *Humanities*, Fall 1961, 3–11.

Inside the Five Sense Sensorium. *Canadian Architect*, June 1961, 49–51.

The New Media and the New Education. *Basilian Teacher*, December 1961, 93–100.

Producers and Consumers. Review of *James Joyce* by Richard Ellmann. *Renascence* 13, 4 (1961), 217–19.

1962

The Chaplin Bloom. Review of *James Joyce: The Poetry of Conscience* by Mary Parr. *Renascence* 14, 4 (1962), 216–17.

A Fresh Perspective on Dialogue. *The Superior Student* 4, 7 (1962), 2–6.

Joyce, Aquinas, and the Poetic Process. In Thomas E. Connolly, ed., *Joyce's Portrait: Criticisms and Critiques* (New York: Appleton-Century-Crofts, 1962), 249–56.

Phase Two. Review of *The Art of James Joyce* by A. Walton Litz. *Renascence* 14, 3 (1962), 166–67.

Prospect of America. *University of Toronto Quarterly* 32, 1 (1962), 107–08.

1963

Empson, Milton, and God. Review of *Milton's God* by William Empson. *Renascence* 15, 2 (1963), 112.

Printing and the Mind. *The Times Literary Supplement*, 19 July 1963.

1964

Introduction to *The Bias of Communication* by Harold A. Innis. Reprint edition. Toronto: University of Toronto Press, 1964. Also appears in *Explorations* 25 (June 1969).

Murder by Television. *Canadian Forum*, January 1964, 222–23.

Notes on Burroughs [Review of *Naked Lunch* and *Nova Express* by William Burroughs]. *Nation*, 28 December 1964, 517–19.

1965

Address at Vision 65. *American Scholar* 35 (1965–66), 196–205. Reprinted in Eric McLuhan and Frank Zingrone, eds., *Essential McLuhan* (Toronto: Anansi, 1995), 219–32.

Art as Anti-Environment. *Art News Annual* 31 (1965).

T S. Eliot. The *Canadian Forum*, February 1965, 243–44.

Wordfowling in Blunderland. *Saturday Night*, August 1965, 23–27.

1966

The All-at-Once World of Marshall McLuhan. *Vogue* 123 (August 1966), 70–73,111.

Cybernation and Culture. In Charles Dechert, ed., *The Social Impact of Cybernetics* (South Bend, IN: University of Notre Dame Press, 1966), 95–108.

Electronics and the Psychic Drop-Out. *This Magazine Is About Schools* 1, 1 (April 1966), 37–42.

The Emperor's Old Clothes. In Gyorgy Kepes, ed., *The Man-Made Object* (New York: G. Braziller, 1966), 90–95.

The Invisible Environment. *Canadian Architect*, May 1966, 71–74.

Questions and Answers with Marshall McLuhan. *Take One*, November/December 1966, 7–10.

Television in a New Light. In Stanley T. Donner, ed., *The Meaning of Commercial Television* (Austin: University of Texas Press, 1966), 87–107.

1967

The Future of Education. (with George B. Leonard). *Look*, 21 February 1967, 23–25.

The Future of Sex. (with George B. Leonard). *Look*, 25 July 1967, 56–63.

Love. *Saturday Night*, February 1967, 25–28.

Marshall McLuhan Massages the Medium. *Nation's Schools*, June 1967, 36–37.

The Relation of Environment to Anti-Environment. In Floyd W. Matson and Ashley Montagu, eds., *The Human Dialogue: Perspectives on Communication* (New York: Free Press, 1967), 39–47.

1968

Adopt a University. *This Magazine Is about Schools* 2, 4 (Autumn 1968), 50–55.

All the Candidates Are Asleep. *Saturday Evening Post*, August 1968, 34–36.

Guaranteed Income in the Electric Age. In Richard Kostelanetz, ed., *Beyond Left and Right: Radical Thought for Our Times* (New York: William Morrow, 1968), 72–83.

The Reversal of the Overheated Image. *Playboy*, December 1968.

Review of *Federalism and the French Canadians* by Pierre Trudeau. *New York Times Book Review*, 17 November 1968.

1969

Playboy Interview: Marshall McLuhan—A Candid Conversation with the High Priest of Popcult and Metaphysician of Media. *Playboy*, March 1969, 53–54, 59–62, 64–66, 68, 70, 72, 74, 158. Reprinted in Eric McLuhan and Frank Zingrone, eds., *Essential McLuhan* (Toronto: Anansi, 1995), 233–69.

Salt and Scandal in the Gospels. (with Joe Keogh). *Explorations* 26 (December 1969), 82–85.

Wyndham Lewis. *Atlantic Monthly,* December 1969, 93–98.

1970

Cicero and the Renaissance Training for Prince and Poet. *Renaissance and Reformation* (Toronto) 6, 3 (1970), 38–42. Reprinted in Eric McLuhan and Frank Zingrone, eds., *Essential McLuhan* (Toronto: Anansi, 1995), 313–18.

The Man Who Came to Listen. (with Barrington Nevitt). In Tony Bonaparte and John Flaherty, eds., *Peter Drucker: Contributions to Business Enterprise* (New York: New York University Press, 1970), 35–55.

1972

Foreword to A. J. Kirshner, *Training That Makes Sense* (San Rafael, CA: Academic Therapy Publications, 1972), 5–7.

The Popular Hero and Anti-Hero. In Ray B. Browne et al., eds., *Heroes of Popular Culture* (Bowling Green, OH: Bowling Green University Popular Press, 1972).

1973

The Argument: Causality in the Electric World. (with Barrington Nevitt). *Technology and Culture* 14, 1 (1973), 1–18.

Do Americans Go to Church to Be Alone? *The Critic*, January/February 1973, 14–23.

The Medium Is the Message. In C. David Mortensen, ed., *Basic Readings in Communication Theory* (New York: Harper and Row, 1973), 139–52.

Mr. Nixon and the Dropout Strategy. *New York Times*, 29 July 1973.

Understanding McLuhan—and Fie on Any Who Don't. *The Globe and Mail*, 10 September 1973.

Watergate as Theatre. *Performing Arts in Canada*, Winter 1973, 14–15.

1974

English Literature as Control Tower in Communication Study. *English Quarterly* (University of Waterloo), Spring 1974, 3–7.

Medium Meaning Message. (with Barrington Nevitt). *Communication* (UK) 1 (1974), 27–33. Reprinted in Barrington Nevitt, The *Communication Ecology: Re-presentation versus Replica* (Toronto: Butterworths, 1982), 140–44.

A Media Approach to Inflation. *New York Times*, 21 September 1974.

Mr. Eliot and the Saint Louis Blues. *The Antigonish Review*, 18 (Summer 1974), 23–27.

There Is Panic in Abortion Thinking: McLuhan. *Toronto Daily Star*, 31 July 1974.

1975

Letter to *The Listener*, 22 October 1975.

McLuhan's Laws of the Media. *Technology and Culture,* January 1975, 74–78.

1976

The Debates. *New York Times*, 23 September 1976.

The Violence of the Media. *Canadian Forum*, September 1976, 9–12.

1977

Alphabet, Mother of Invention. (with R. K. Logan). *Et Cetera*: *A Review of General Semantics*, December 1977, 373–83.

Canada: The Borderline Case. In David Staines, ed., *The Canadian Imagination: Dimensions of a Literary Culture* (Cambridge, MA: Harvard University Press, 1977), 226–48.

Laws of the Media. *Et Cetera*: *A Review of General Semantics*, June 1977, 173–78.

The Rise and Fall of Nature. *Journal of Communication* 27, 4 (1977), 80–81.

1978

The Brain and the Media: The "Western" Hemisphere. *Journal of Communication* 28, 4 (1978), 54–60.

Figures and Grounds in Linguistic Criticism. Review of Mario J. Valdes and Owen J. Miller, *Interpretation of Narrative* (Toronto: University of Toronto Press, 1978). *Et Cetera: A Review of General Semantics* 36, 3 (1979), 289–94.

A Last Look at the Tube. *New York*, 3 April 1978, 45.

1979

The Double Bind of Communication and the World Problematique. (with Robert K. Logan). *Human Futures*, Summer 1979, 1–3.

Pound, Eliot, and the Rhetoric of *The Waste Land. New Literary History* 10, 3 (1979), 557–80.

1980

Foreword to Karl Appel, *Karl Appel: Works on Paper*. (New York: Abbeville Press, 1980).

1981

Electronic Banking and the Death of Privacy. (with Bruce Powers). *Journal of Communication* 31, 1(1981), 164–69.

Subject Index

Name Index

Herbert Marshall McLuhan

(1911-1980)

One of the most controversial and original thinkers of our time, Marshall McLuhan is universally regarded as the father of communications and media studies. A charismatic figure, whose remarkable perceptions propelled him onto the international stage, McLuhan became the prophet of the new information age.

In his own time he drew both accolades and criticism for his intuitive vision. His steady stream of thought-provoking metaphors regarding culture and technology fascinated his listeners, especially those involved in advertising, politics, journalism, the arts, and media. The information superhighway fulfilled his perceptive observation that the world would ultimately become a "global village".

Marshall McLuhan's work continues to influence a number of disciplines, in particular media studies, modern art, and semiotics.

Marshall McLuhan received his M.A. in English literature from Trinity Hall, Cambridge. His doctoral thesis (Cambridge) is a critique of communication through the ages, recounting the rise and fall of poetics, rhetoric and dialectic. Styles of discourse and learning ultimately influence perception and thereby determine the content of dialogue.

Marshall McLuhan

PUBLISHED BY GINGKO PRESS

The Mechanical Bride: Folklore of Industrial Man
with a new Introduction by Philip B. Meggs

The Medium is the Massage
with Quentin Fiore and Jerome Agel

War and Peace in the Global Village
with Quentin Fiore and Jerome Agel

The Book of Probes
with David Carson
Edited by Eric McLuhan, William Kuhns, and Mo Cohen

Marshall McLuhan Unbound: Volume One
Essays selected by Eric McLuhan
Edited by Eric McLuhan and W. Terrence Gordon

Counterblast – Facsimile Edition
with Harley Parker

BOOKS IN PREPARATION 2004

Through the Vanishing Point:
Space in Poetry and Painting
with Harley Parker

The Letters of Marshall McLuhan
Edited by Matie Molinaro and Corinne McLuhan

Marshall McLuhan Unbound: Volume Two
The Gutenberg Era

Thesis: The Trivium
Thomas Nashe and the Learning of His Time
Edited by W. Terrence Gordon

BOOKS IN PREPARATION 2005

For Marshall McLuhan: Exhibition Catalogue
Edited by Dominique Carré

Experience: Understanding Media
Various Artists

The Complete Mechanical Bride
Early Versions and Outtakes
with CD ROM

Culture is Our Business

From Cliché to Archetype
with Wilfred Watson

EXPLORE EXPLAIN 10/14

TECHNOLOGY &G ON T 29

TECHNOLOGY PRISONS 34
3 STAGES OF ALARM 43

NOTES